September, 1987

Introduction to
QUANTITATIVE
GENETICS

Introduction to
QUANTITATIVE GENETICS

Second Edition

D. S. FALCONER
Department of Genetics and
Agricultural Research Council
Unit of Animal Genetics
University of Edinburgh

Longman
London and New York

Longman Group Limited

Longman House, Burnt Mill, Harlow,
Essex CM20 2JE, England

Associated companies throughout the world

*Published in the United States of America
by Longman Inc., New York*

First published by Oliver and Boyd 1960
Fifth reprint 1972
Reprinted by Longman Group Ltd 1975, 1976
Second edition 1981
Reprinted 1982
Reprinted with amendments 1983
Reprinted 1985

British Library Cataloguing in Publication Data

Falconer, Douglas Scott
  Introduction to quantitative genetics. — 2nd ed.
  1. Population genetics
  I. Title
  575.1   GH455   80–40775
  ISBN 0–582–44195–1

Printed in Great Britain
at The Pitman Press, Bath

CONTENTS

Note

In the reprinting of 1983 I have been able to make some amendments to take account of recent conclusions about the effect of mutation on selection responses. An outline of this work is added to Chapter 12 on p. 206, and consequential amendments have been made on pp. 195, 198, 204 and 313.

D. S. Falconer
December 1982

PREFACE

This book was originally intended to provide an introductory textbook of quantitative genetics, with the emphasis on general principles rather on practical applications. I tried to make the book useful to as wide a range of readers as possible, particularly to biologists who, like myself, have no more than ordinary mathematical ability. In preparing this revised edition, my aims have been: (1) to keep the character of the book, and its length, unchanged; (2) to include some account of all the main developments of the last twenty years; and (3) to be less neglectful of plants. I hope that the compromise made between these conflicting aims will be one that will prove useful. My main regret has been the impossibility of mentioning more than a very few of the experimental studies that have illuminated the subject since the book first appeared.

The inclusion of new material means that rather more than before will be beyond the needs of those for whom the subject forms part of a course of general genetics. The section headings should, however, facilitate the selection of what is relevant. The level of mathematics needed is not more than simple algebra: neither calculus nor matrix methods are used. Some knowledge of statistics, however, is needed, particularly of the analysis of variance and of correlation and regression.

Acknowledgements. Dr W. G. Hill read all the draft and offered many comments and suggestions which have been an immense help. Major improvements to many chapters were made as a result of his advice. I am deeply grateful to him. I owe much also to Dr R. C. Roberts, Professor Alan Robertson and Professor N. W. Simmonds, who read substantial parts of the draft, and to many other colleagues who have advised me on particular points. Without the help that I have received, the book would have had many more blemishes. The errors and misconceptions that remain are, of course, entirely my own. I shall be grateful to have these pointed out to me. I am indebted to the authors and publishers for permission to reproduce material from the sources cited. Finally, I would like to thank Mrs D. J. Bogie for her skilful typing of the manuscript.

Institute of Animal Genetics
West Mains Road   **D. S. Falconer**
Edinburgh 9,   *January 1980*
Scotland

INTRODUCTION

Quantitative genetics is concerned with the inheritance of those differences between individuals that are of degree rather than of kind, quantitative rather than qualitative. These are the individual differences which, as Darwin wrote, 'afford materials for natural selection to act on and accumulate, in the same manner as man accumulates in any given direction individual differences in his domestic productions'. An understanding of the inheritance of these differences is thus of fundamental significance in the study of evolution and in the application of genetics to animal and plant breeding; and it is from these two fields of enquiry that the subject has received the chief impetus to its growth.

Virtually every organ and function of any species shows individual differences of this nature, the differences of size among ourselves or our domestic animals being an example familiar to all. Individuals form a continuously graded series from one extreme to the other and do not fall naturally into sharply demarcated types. Qualitative differences, in contrast, divide individuals into distinct types with little or no connexion by intermediates. Examples are the differences between blue-eyed and brown-eyed individuals, between the blood groups, or between normally coloured and albino individuals. The familiar Mendelian ratios, which display the mechanism of inheritance, can be seen only when a gene difference at a single locus gives rise to a readily detectable difference in some such property of the organism. Quantitative differences, in so far as they are inherited, depend on genes whose effects are small in relation to the variation arising from other causes. Furthermore, quantitative differences are usually, though not necessarily always, influenced by gene differences at many loci. Consequently the individual genes, whether few or many, cannot be identified by their segregation; the Mendelian ratios are not displayed, and the methods of Mendelian analysis cannot be applied.

It is, nevertheless, a basic premiss of quantitative genetics that the inheritance of quantitative differences depends on genes subject to the same laws of transmission and having the same general properties as the genes whose transmission and properties are displayed by qualitative differences. Quantitative genetics is therefore an extension of Mendelian genetics, resting squarely on Mendelian principles as its foundation.

The methods of study in quantitative genetics differ from those employed in Mendelian genetics in two respects. In the first place, since ratios cannot be observed, single progenies are uninformative, and the unit of study must be extended to 'populations', that is, larger groups of individuals comprising many progenies. And, in the second place, the nature of the quantitative differences to be studied requires the measurement, and not just the classification, of the individuals. The extension of Mendelian genetics into quantitative genetics may thus be made in two stages, the first introducing new concepts connected with the genetic properties of 'populations' and the second introducing concepts connected with the inheritance of measurements. This is how the subject is presented in this book. In the first part, which occupies Chapters 1 to 5, the genetic properties of populations are described by reference to genes causing easily identifiable, and therefore qualitative, differences. Quantitative differences are not discussed until the second part, which starts in Chapter 6. These two parts of the subject are often distinguished by different names, the first being referred to as 'population genetics' and the second as 'quantitative genetics' or 'biometrical genetics'.

The theoretical basis of quantitative genetics was established round about 1920 by the work of Fisher (1918), Haldane (summarized 1932) and Wright (1921). The development of the subject over the succeeding years, by these and many other geneticists and statisticians, has been mainly by elaboration, clarification, and the filling in of details, so that today we have a substantial body of theory accepted by the majority as valid.

The theory consists of the deduction of the consequences of Mendelian inheritance when extended to the properties of populations and to the simultaneous segregation of genes at many loci. The premiss from which the deductions are made is that the inheritance of quantitative differences is by means of genes, and that these genes are subject to the Mendelian laws of transmission and may have any of the properties known from Mendelian genetics. The property of 'variable expression' assumes great importance and might be raised to the status of another premiss: that the expression of the genotype in the phenotype is modifiable by non-genetic causes. Other properties whose consequences are taken into account include dominance, epistasis, pleiotropy, linkage, and mutation. The theory then allows us to deduce what will be the genetic properties of a population if the genes have the properties postulated. It allows us also to predict the consequences of any specified breeding plan, including those of natural selection. It therefore forms the basis for understanding evolutionary change. The main practical use of the theory is in comparing the merits of alternative procedures for animal and plant improvement.

The experimental side of quantitative genetics has three roles, complementary to the theoretical side. First, experimental study of populations allows us to deduce the properties of the genes associated with quantitative variation. Second, experimental breeding allows us to test the validity of the theory. And third, there are some consequences of breeding procedures that cannot be predicted from the theory, and questions about these can be answered only by experiment. There is now a large body of experimental data which substantiates the theory in considerable detail, showing that the genes concerned with quantitative variation do have the properties known from Mendelian genetics, and that the outcome of most breeding procedures can be predicted with some confidence. The aim here is to describe all that is reasonably firmly established and, for the sake of clarity, to simplify as far as is possible without being misleading. Consequently, the emphasis is on the theoretical side. Though conclusions will often be drawn directly from experimental data, the experimental side of the subject is presented chiefly in the form of examples, chosen with the purpose of illustrating the theoretical conclusions. These examples, however, cannot always be taken as substantiating the postulates that underlie the conclusions they illustrate. Too often the results of experiments are open to more than one interpretation. The experimental work mentioned is only a very small, and far from random, sample of what has been done. In particular, a great deal more experimentation has been done with plants and farm animals than would appear from its representation among the work cited.

No attempts has been made to give exhaustive references to published work in any part of the subject; or to indicate the origins, or trace the history of the ideas. To have done this would have required a much longer book, and a considerable sacrifice of clarity. Most of the material in the book is covered more fully in one or other of the sources listed below. These sources are not regularly cited in the text. References are given in the text when any conclusion is stated without full explanation of its derivation. These references are not always to the original papers, but rather to the more recent papers where the reader will find a convenient point of entry to the topic under discussion.

Chief sources

(For full bibliographical details see list of References)

Becker (1975) *Manual of Quantitative Genetics.*
Crow and Kimura (1970) *An Introduction to Population Genetics Theory.*
Jacquard (1974) *The Genetic Structure of Populations.*
Kempthorne (1957) *An Introduction to Genetic Statistics.*
Lewontin (1974) *The Genetic Basis of Evolutionary Change.*
Li (1976) *First Course in Population Genetics.*
Mather and Jinks (1971) *Biometrical Genetics.*
    (1977) *Introduction to Biometrical Genetics.*
Wright (1968–78) *Evolution and the Genetics of Populations,* Vols 1–4.

I GENETIC CONSTITUTION OF A POPULATION

Frequencies of genes and genotypes

To describe the genetic constitution of a group of individuals we should have to specify their genotypes and say how many of each genotype there were. This would be a complete description, provided the nature of the phenotypic differences between the genotypes did not concern us. Suppose for simplicity that we were concerned with a certain autosomal locus, A, and that two different alleles at this locus, A_1 and A_2 were present among the individuals. Then there would be three possible genotypes, A_1A_1, A_1A_2, and A_2A_2. (We are concerned here, as throughout the book, exclusively with diploid organisms.) The genetic constitution of the group would be fully described by the proportion, or percentage, of individuals that belonged to each genotype, or in other words by the frequencies of the three genotypes among the individuals. These proportions or frequencies are called *genotype frequencies*, the frequency of a particular genotype being its proportion or percentage among the individuals. If, for example, we found one-quarter of the individuals in the group to be A_1A_1, the frequency of this genotype would be 0.25, or 25 per cent. Naturally, the frequencies of all the genotypes together must add up to unity, or 100 per cent.

Example 1.1 The M-N blood groups is man are determined by two alleles at a locus, and the three genotypes correspond with the three blood groups, M, MN, and N. The following figures, taken from the tabulation of Mourant (1954), show the blood group frequencies among Eskimos of East Greenland and among Icelanders as follows:

		Blood group			*Number of individuals*
		M	MN	N	
Frequency,%	Greenland	83.5	15.6	0.9	569
	Iceland	31.2	51.5	17.3	747

Clearly the two populations differ in these genotype frequencies, the N blood group being rare in Greenland and relatively common in Iceland. Not only is this locus a source of variation within each of the two populations, but it is also a source of genetic difference between the populations.

A population, in the genetic sense, is not just a group of individuals, but a breeding group; and the genetics of a population is concerned not only with the genetic constitution of the individuals but also with the transmission of the genes from one generation to the next. In the transmission the genotypes of the parents are broken down and a new set of genotypes is constituted in the progeny, from the genes transmitted in the gametes. The genes carried by the population thus have continuity from generation to generation, but the genotypes in which they appear do not. The genetic constitution of a population, referring to the genes it carries, is described by the array of *gene frequencies*; that is, by specification of the alleles present at every locus and the numbers or proportions of the different alleles at each locus. If, for example, A_1 is an allele at the A locus, then the frequency of A_1 genes, or the gene frequency of A_1, is the proportion or percentage of all genes at this locus that are the A_1 allele. The frequencies of all the alleles at any one locus must add up to unity, or 100 per cent.

The gene frequencies at a particular locus among a group of individuals can be determined from a knowledge of the genotype frequencies. To take a hypothetical example, suppose there are two alleles, A_1 and A_2, and we classify 100 individuals and count the numbers in each genotypes as follows:

		A_1A_1	A_1A_2	A_2A_2	Total	
Number of individuals		30	60	10	100	
	$\{A_1$	60	60	0	120 $\}$	200
Number of genes	$\{A_2$	0	60	20	80 $\}$	

Each individual contains two genes, so we have counted 200 representatives of the genes at this locus. Each A_1A_1 individual contains two A_1 genes and each A_1A_2 contains one A_1 gene. So there are $120\,A_1$ genes in the sample, and $80\,A_2$ genes. The frequency of A_1 is therefore 60 per cent or 0.6, and the frequency of A_2 is 40 per cent or 0.4. To express the relationship in a more general form, let the frequencies of genes and of genotypes be as follows:

	Genes		Genotypes		
	A_1	A_2	A_1A_1	A_1A_2	A_2A_2
Frequencies	p	q	P	H	Q

so that $p + q = 1$ and $P + H + Q = 1$. Since each individual contains two genes, the frequency of A_1 genes is $\frac{1}{2}(2P + H)$, and the relationship between gene frequency and genotype frequency among the individuals counted is as follows:

$$\left. \begin{aligned} p &= P + \tfrac{1}{2}H \\ q &= Q + \tfrac{1}{2}H \end{aligned} \right\} \dots [1.1]$$

Example 1.2 To illustrate the calculation of gene frequencies from genotype frequencies we may take the M-N blood group frequencies given in Example 1.1. The M and N blood groups represent the two homozygous genotypes and the MN group the heterozygote. The frequency of the M gene in Greenland is, from equation [1.1]

$0.835 + \frac{1}{2}(0.156) = 0.913$, and the frequency of the N gene is $0.009 + \frac{1}{2}(0.156) = 0.087$, the sum of the frequencies being 1.000 as it should be. Doing the same for the Iceland sample, we find the following gene frequencies in the two populations, expressed now as percentages:

	Gene	
	M	N
Greenland	91.3	8.7
Iceland	57.0	43.0

Thus the two populations differ in gene frequency as well as in genotype frequencies.

Causes of change

The genetic properties of a population are influenced in the process of transmission of genes from one generation to the next by a number of agencies. These form the chief subject-matter of the next four chapters, but we may briefly review them here in order to have some idea of what factors are being left out of consideration in this chapter. The agencies through which the genetic properties of a population may be changed are these:

Population size. The genes passed from one generation to the next are a sample of the genes in the parent generation. Therefore the gene frequencies are subject to sampling variation between successive generations, and the smaller the number of parents the greater is the sampling variation. The effects of sampling variation will be considered in Chapters 3–5, and meantime we shall exclude it from the discussion by supposing always that we are dealing with a 'large population', which means simply one in which sampling variation is so small as to be negligible. For practical purposes a 'large population' is one in which the number of adult individuals is in the hundreds rather than in the tens.

Differences of fertility and viability. Though we are not at present concerned with the phenotypic effects of the genes under discussion, we cannot ignore their effects on fertility and viability, because these influence the genetic constitution of the succeeding generation. The different genotypes among the parents may have different fertilities, and if they do they will contribute unequally to the gametes out of which the next generation is formed. In this way the gene frequency may be changed in the transmission. Further, the genotypes among the newly formed zygotes may have different survival rates, and so the gene frequencies in the new generation may be changed by the time the individuals are adult and themselves become parents. These processes are called selection, and will be described in Chapter 2. Meanwhile we shall suppose they are not operating. Human blood-group genes may be taken for the purpose of illustration since the selective forces acting on them are probably not strong. Genes that produce a mutant phenotype which is abnormal in comparison with the wild-type are, in contrast, usually subject to much more severe selection.

Migration and mutation. The gene frequencies in the population may also be changed by immigration of individuals from another population, and by gene mutation. These processes will be described in Chapter 2, and at this stage will also be supposed not to operate.

Mating system. The genotypes in the progeny are determined by the union of the gametes in pairs to form zygotes, and the union of gametes is influenced by the mating of the parents. So the genotype frequencies in the offspring generation are influenced by the genotypes of the pairs that mate in the parent generation. We shall at first suppose that mating is at random with respect to the genotypes under discussion. *Random mating*, or *panmixia*, means that any individual has an equal chance of mating with any other individual in the population. The important points are that there should be no special tendency for mated individuals to be alike in genotype, or to be related to each other by ancestry. If a population covers a large geographic area, individuals inhabiting the same locality are more likely to mate than in individuals inhabiting different localities, and so the mated pairs tend to be related by ancestry. A widely spread population is therefore likely to be subdivided into local groups and mating is random only within the groups. The properties of subdivided populations depend on the size of the local groups and will be described under the effects of population size in Chapters 3–5.

Hardy–Weinberg equilibrium

The Hardy–Weinberg law
In a large random-mating population with no selection, mutation, or migration, the gene frequencies and the genotype frequencies are constant from generation to generation; and, furthermore, there is a simple relationship between the gene frequencies and the genotype frequencies. These properties of a population are derived from a theorem, or principle, known as the *Hardy–Weinberg* law after Hardy and Weinberg, who independently demonstrated them in 1908. A population with constant gene and genotype frequencies is said to be in *Hardy–Weinberg equilibrium.* The relationship between gene frequencies and genotype frequencies is of the greatest importance because many of the deductions about population genetics and quantitative genetics rest on it. The relationship is this: if the gene frequencies of two alleles among the parents are p and q, then the genotype frequencies among the progeny are p^2, $2pq$, and q^2, thus:

	Genes in parents		Genotypes in progeny			
	A_1	A_2	A_1A_1	A_1A_2	A_2A_2	
Frequencies	p	q	p^2	$2pq$	q^2	... [1.2]

(The relationship above refers to autosomal genes; sex-linked genes are not quite so simple and will be explained later.) The conditions of random mating and no selection, required for the Hardy–Weinberg law to hold, refer only to the genotypes under consideration. There may be preferential mating with respect to other attributes, and genotypes of other loci may be subject to selection, without affecting the issue. Two additional conditions are that the genes segregate normally in gametogenesis and that the gene frequencies are the same in males and females. The reasons for these requirements will be seen in the proof.

The proof of the Hardy–Weinberg law involves four steps, which are summarized in Table 1.1, with the conditions that must hold for the deduction at each step to be valid. The details of the four steps are as follows.

1. *From gene frequency in parents to gene frequency in gametes.* Let the parent generation have gene and genotype frequencies as follows:

	Genes		Genotypes		
	A_1	A_2	A_1A_1	A_1A_2	A_2A_2
Frequencies	p	q	P	H	Q

Two types of gamete are produced, those bearing A_1 and those bearing A_2. A_1A_1 individuals produce only A_1 gametes. A_1A_2 individuals, provided segregation is normal, produce equal numbers of A_1 and A_2 gametes. Then, provided all genotypes are equally fertile, the frequency of A_1 among all the gametes produced by the whole population is $P + \frac{1}{2}H$, which by equation [1.1] is the gene frequency of A_1 in the parents producing the gametes. Thus the gene frequency in the whole gametic output is the same as in the parents. This is step 1*a* in Table 1.1. Only some of the gametes form zygotes that will become individuals in the next generation. The gene frequency in the zygotes is unchanged provided the gametes carrying different alleles do not differ in their fertilizing capacity, and provided the zygotes formed represent a large sample of the parental gametes. This is step 1*b*.

2. *From gene frequency in gametes to genotype frequencies in zygotes.* Random mating between individuals is equivalent to random union among

Table 1.1 Steps of deduction in the proof of the Hardy–Weinberg law, and the conditions that must hold.

Step	Deduction from: to	Conditions
1*a*	Gene frequency in parents	(1) Normal gene segregation
		(2) Equal fertility of parents
	Gene frequency in all gametes	
1*b*		(3) Equal fertilizing capacity of gametes
		(4) Large population
	Gene frequency in gametes forming zygotes	
2		(5) Random mating
		(6) Equal gene frequencies in male and female parents
	Genotype frequencies in zygotes	
3		(7) Equal viability
	Genotype frequencies in progeny	
4	Gene frequency in progeny	

their gametes. The genotype frequencies among the zygotes (fertilized eggs) are then the products of the frequencies of the gametic types that unite to produce them. The genotype frequencies among the progeny produced by random mating can therefore be determined simply by multiplying the frequencies of the gametic types produced by each sex of parents. Provided the gametic frequencies are the same in each sex, the zygotes produced are as shown in Table 1.2. The union of A_1 eggs with A_2 sperms need not be distinguished from that of A_2 eggs with A_1 sperms; so the genotype frequencies of the zygotes are:

	Genotype		
	A_1A_1	A_1A_2	A_2A_2
Frequency	p^2	$2pq$	q^2

3. *From zygotes to adults.*   The genotype frequencies in the zygotes deduced above are the Hardy–Weinberg frequencies, as stated in [1.2]. This is, however, not quite the end of the proof because the frequencies will not be observable unless the zygotes survive equally well, at least until they can be classified for genotype. This step may seem trivial, but it must be recognized if the effects of differential viability are to be understood.

4. *From genotype frequencies to gene frequency in progeny.*   This final step proves that the gene frequency has not changed. Provided the different genotypes in the progeny survive equally well to adulthood when they can become parents, their frequencies will be as above. The gene frequency in the adult progeny can then be found by equation [1.1]. The frequency of A_1 is $p^2 + \frac{1}{2}(2pq)$ $= p(p+q) = p$, which is the same as in the parent generation. This proves the constancy of the gene frequency from one generation to the next.

Two further aspects of the Hardy–Weinberg law can now be stated. First, since the gene frequencies are the same in parents and progeny, the relationship between gene frequencies and genotype frequencies in [1.2] applies to a single generation. Second, the genotype frequencies in the progeny depend only on the gene frequencies in the parents and not on the genotype frequencies. This can be

Table 1.2

		Female gametes and their frequencies	
		A_1 p	A_2 q
Male gametes and their frequencies	A_1 p	A_1A_1 p^2	A_1A_2 pq
	A_2 q	A_1A_2 pq	A_2A_2 q^2

seen from step 1 above, where the frequencies of the gametic types were shown to be equal to the parental gene frequencies, no matter what the genotype frequencies are. Consequently, parents with any genotype frequencies, provided they mate at random and provided the gene frequency is the same in males and females, produce progeny in Hardy–Weinberg proportions. If the gene frequency is known not to be the same in males and females, it is easy to deduce the genotype frequencies in the progeny by putting the appropriate gametic frequencies in Table 1.2 at step 2. The gene frequency of an autosomal gene becomes equal in the two sexes of the progeny, and a second generation of random mating produces Hardy–Weinberg genotype frequencies corresponding to the mean of the original gene frequencies.

The relationship between gene frequencies and genotype frequencies in a population in Hardy–Weinberg equilibrium is shown in Fig. 1.1. The graphs of genotype frequencies show two important features. First the frequency of the heterozygotes cannot be greater than 50 per cent, and this maximum occurs when the gene frequencies are $p = q = 0.5$. Second, when the gene frequency of an allele is low, the rare allele occurs predominantly in heterozygotes and there are very few homozygotes. This has important consequences for the effectiveness of selection, as will be seen in the next chapter.

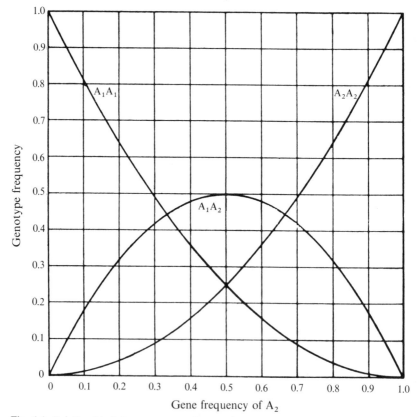

Fig. 1.1. Relationship between genotype frequencies and gene frequency for two alleles in a population in Hardy–Weinberg equilibrium.

Applications of the Hardy–Weinberg law

There are three ways in which the Hardy–Weinberg law is particularly useful, which will now be illustrated.

Gene frequency of recessive allele.   At the beginning of the chapter we saw, in equation [1.1], how the gene frequencies among a group of individuals can be determined from their genotype frequencies; but for this it was necessary to know the frequencies of all three genotypes. Consequently, the relationship in equation [1.1] cannot be applied to the case of a recessive allele, when the heterozygote is indistinguishable from the dominant homozygote. If the genotypes are in Hardy–Weinberg proportions, however, we do not need to know the frequencies of all three genotypes. Let *a*, for example, be a recessive gene with a frequency of q; then the frequency of *aa* homozygotes is q^2, and the gene frequency is the square-root of the homozygote frequency. Example 1.3 illustrates the calculation. For this way of estimating the gene frequency to be a valid one, it is obviously essential that there should be no selective elimination of homozygotes before they are counted. It should be noted also that the estimation of gene frequency in this way is rather sensitive to the effects of non-random mating.

Frequency of 'carriers'.   It is often of interest to know the frequency of heterozygotes, or 'carriers', of recessive abnormalities, and this can be calculated if the gene frequency is known. If Hardy–Weinberg equilibrium can be assumed, the frequency of heterozygotes among all individuals, including homozygotes, is given by $2q(1 - q)$. It is, however, often more relevant to know the frequency among normal individuals, though this will not be very different if homozygotes are rare. The frequency of heterozygotes among normal individuals, denoted by H', is the ratio of genotype frequencies $Aa/(AA + Aa)$, where *a* is the recessive allele. So, when q is the frequency of *a*,

$$H' = \frac{2q(1 - q)}{(1 - q)^2 + 2q(1 - q)} = \frac{2q}{1 + q} \qquad \ldots [1.3]$$

Example 1.3 Phenylketonuria (PKU) is a human metabolic disease due to a single recessive gene. Homozygotes can be detected a few days after birth, and selective elimination before then will be assumed to be negligible. Tests of babies born in Birmingham, UK, over a 3-year period detected 5 cases in 55,715 babies (Raine *et al.*, 1972). The frequency of homozygotes in the sample is 90×10^{-6} or about 1/11,000. The Hardy–Weinberg frequency of homozygotes is q^2, so the gene frequency is $q = \sqrt{(90 \times 10^{-6})} = 9.5 \times 10^{-3} = 0.0095$.

The frequency of heterozygotes in the whole population is $2q(1 - q)$, and among normal individuals is $2q/(1 + q)$. Both work out to be 0.019, approximately. Thus about 2 per cent of normal people, or 1 in 50, are carriers of PKU. It comes as a surprise to most people to discover how common heterozygotes of a rare recessive abnormality are. The point has already been noted as a conclusion drawn from Fig. 1.1.

Test of Hardy–Weinberg equilibrium.   If data are available for a locus where all the genotypes are recognizable, the observed frequencies of the genotypes can be tested for agreement with a population in Hardy–Weinberg equilibrium. According to the Hardy–Weinberg law, the genotype frequencies of progeny are determined by the gene frequency in their parents. If the population is in equilibrium, the gene frequency is the same in parents and progeny, so the gene frequency observed in the progeny can be used as if it were the parental gene frequency to calculate the genotype frequencies expected by the Hardy–Weinberg law. The procedure is illustrated in Example 1.4.

Example 1.4 The M-N blood group frequencies in Iceland were given in Example 1.1. The observed numbers in the sample were as in the following table. The gene frequencies in the sample are first calculated from the observed numbers by equation [1.1]. Then the Hardy–Weinberg genotype frequencies, p^2, $2pq$ and q^2 are calculated from the gene frequencies by equation [1.2], and each is multiplied by the total number to get the numbers expected. For example, the expectation for MM is $(0.5696)^2 \times 747$. Comparing the observed with expected numbers shows a deficiency of both homozygotes and an excess of heterozygotes. The χ^2 tests how well, or how badly, the observed numbers agree with the expected. The discrepancy is not significant and could easily have arisen by chance in the sampling. Note that this χ^2 has only 1 degree of freedom because the gene frequency has been estimated from the data, so that the observed and expected numbers must agree in their gene frequencies as well as in their totals.

	Genotypes				Gene frequencies	
	MM	MN	NN	Total	M	N
Numbers observed	233	385	129	747	0.5696	0.4304
Numbers expected	242.36	366.26	138.38	747		
$\chi_1^2 = 1.96$	$P \sim 0.2$					

The test for agreement with an equilibrium population is a test of whether the conditions for the production of Hardy–Weinberg genotype frequencies have been fulfilled. The conclusions that can be drawn from the test, however, are limited. When good agreement is found, the test gives no reason to doubt the fulfilment of all the conditions. Tests made with blood-group genes nearly always show very good agreement, as in Example 1.4. But there is one condition whose non-fulfilment will not lead to a discrepancy, and that is equal fertility among the parents. The reason for this will be explained in a moment. If the test reveals a discrepancy between the observed and expected frequencies, we can conclude that one or more of the conditions has not been fulfilled. But the nature of the discrepancy does not allow us to identify its source, or decide which condition has not been met. The reason for this is that the same discrepancy can arise from different causes. For example, an excess of heterozygotes can result from selective elimination of homozygotes, or from the gene frequency being different in males and females of the parental generation. The test is not as simple as it seems, and we must look more closely at what it does.

The Hardy–Weinberg law relates genes in parents to genotypes in progeny. Therefore, to test it fully, we need to know the gene frequency in the parents and to calculate the expected genotype frequencies in the progeny from the parental gene frequency. But for the test described we have only the progeny. We find the gene frequency in them by counting. We then say: if this was the gene frequency among the gametes that produced these progeny, the genotypes should be in the Hardy–Weinberg proportions as calculated from the observed gene frequency. If the gene frequency was not the same in the parents as in the progeny, we have used the wrong gene frequency to calculate the expectations. Reference to Table 1.1 will show that the conditions tested are random mating, equal gene frequencies in the two sexes of parents, and equal viability among the progeny; but equal fertility among the parents is not tested. Selection could therefore be acting through fertility and not be detected by this test. Selection acting through the viability of the progeny will lead to disagreement between the observed and expected frequencies. It is not possible, however, to identify the genotype or genotypes that have reduced viability. The reason for this will be explained in the next chapter, after the effects of selection have been dealt with. For fuller discussions of the limitations of the test see Wallace (1958, 1968), Prout (1965); and for fuller consideration of its statistical aspects see Smith (1970).

Mating frequencies and another proof of the Hardy–Weinberg law

Let us now look more closely into the breeding structure of a random-mating population, distinguishing the types of mating according to the genotypes of the pairs, and seeing what are the genotype frequencies among the progenies of the different types of mating. This provides a general method for relating genotype frequencies in successive generations, which will be used in a later chapter. It also provides another proof of the Hardy–Weinberg law; a proof more cumbersome than that already given but showing more clearly how the Hardy–Weinberg frequencies arise from the Mendelian laws of segregation. The procedure is to obtain first the frequencies of all possible mating types according to the frequencies of the genotypes among the parents, and then to obtain the frequencies of genotypes among the progeny of each type of mating according to the Mendelian ratios.

Consider a locus with two alleles, and let the frequencies of genes and genotypes in the parents be, as before.

	Genes		Genotypes		
	A_1	A_2	A_1A_1	A_1A_2	A_2A_2
Frequencies	p	q	P	H	Q

There are altogether nine types of mating, and their frequencies when mating is random are found by multiplying together the marginal frequencies as shown in Table 1.3. Since the sex of the parent is irrelevant in this context, some of the types of mating are equivalent, and the number of different types reduces to six. By summation of the frequencies of equivalent types, we obtain the frequencies of mating types in the first two columns of Table 1.4. Now we have to consider the genotypes of offspring produced by each type of mating, and find the frequency of each genotype in the total progeny, assuming, of course, that all types of mating are equally fertile and all genotypes equally viable. This is done in the right-hand side of Table 1.4. Thus, for example, matings of the type $A_1A_1 \times A_1A_1$ produce only A_1A_1 offspring. So, of the total progeny, a proportion P^2 are A_1A_1

Table 1.3

			Genotype and frequency of female parent		
			A_1A_1	A_1A_2	A_2A_2
			P	H	Q
Genotype and frequency of male parent	A_1A_1	P	P^2	PH	PQ
	A_1A_2	H	PH	H^2	HQ
	A_2A_2	Q	PQ	HQ	Q^2

Table 1.4

Mating		Genotype and frequency of progeny		
Type	Frequency	A_1A_1	A_1A_2	A_2A_2
$A_1A_1 \times A_1A_1$	P^2	P^2	—	—
$A_1A_1 \times A_1A_2$	$2PH$	PH	PH	—
$A_1A_1 \times A_2A_2$	$2PQ$	—	$2PQ$	—
$A_1A_2 \times A_1A_2$	H^2	$\frac{1}{4}H^2$	$\frac{1}{2}H^2$	$\frac{1}{4}H^2$
$A_1A_2 \times A_2A_2$	$2HQ$	—	HQ	HQ
$A_2A_2 \times A_2A_2$	Q^2	—	—	Q^2
	Sums	$(P+\frac{1}{2}H)^2$	$2(P+\frac{1}{2}H)(Q+\frac{1}{2}H)$	$(Q+\frac{1}{2}H)^2$
	$=$	p^2	$2pq$	q^2

genotypes derived from this type of mating. Similarly, one-quarter of the offspring of $A_1A_2 \times A_1A_2$ matings are A_1A_1. So this type of mating, which has a frequency of H^2, contributes a proportion $\frac{1}{4}H^2$ of the total A_1A_1 progeny. To find the frequency of each genotype in the total progeny we add the frequencies contributed by each type of mating. The sums, after simplification, are given at the foot of Table 1.4, and from the identity given in equation [1.1] they are seen to be equal to p^2, $2pq$, and q^2. These are the Hardy–Weinberg equilibrium frequencies, and we have shown that they are attained by one generation of random mating, irrespective of the genotype frequencies among the parents.

Multiple alleles

Restriction of the treatment to two alleles at a locus suffices for many purposes. If we are interested in one particular allele, as often happens, then all the other alleles at the locus can be treated as one. Formulation of the situation in terms of two alleles is therefore often possible even if there are in fact more than two. If we are interested in more than one allele we can still, if we like, treat the situation as a two-allele system by considering each allele in turn and lumping the others together. But the treatment can be easily extended to cover more than two alleles, and no new principle is introduced. In general, if q_1 and q_2 are the frequencies of any two alleles, A_1 and A_2, of a multiple series, then the genotype frequencies under Hardy–Weinberg equilibrium are as follows:

Genotype		
A_1A_1	A_1A_2	A_2A_2
Frequency: q_1^2	$2q_1q_2$	q_2^2

These frequencies are also attained by one generation of random mating. This can readily be seen by reducing the situation to a two-allele system, and considering each allele in turn. Or it can be proved, though somewhat more laboriously, by the method explained above for the two-allele system.

Example 1.5 The ABO blood groups in man are determined by a series of allelic genes. For the purpose of illustration we shall recognize three alleles, A, B, and O, and show how the gene frequencies can be estimated from the blood-group frequencies. Since O is recessive to both A and B, Hardy–Weinberg frequencies have to be assumed for estimating the gene frequencies. Let the frequencies of the A, B, and O genes be p, q, and r respectively, so that $p + q + r = 1$. The following table shows: (1) the blood groups (i.e. phenotypes); (2) the genotypes represented in each group; (3) the expected frequencies of the blood groups in terms of p, q, and r, on the assumption of Hardy–Weinberg equilibrium; (4) observed frequencies of blood groups in a sample of 190,177 U.K. airmen, quoted by Race and Sanger (1954).

(1) Blood group	(2) Genotype	Frequency	
		(3) expected	(4) observed %
A	AA + AO	$p^2 + 2pr$	41.716
B	BB + BO	$q^2 + 2qr$	8.560
O	OO	r^2	46.684
AB	AB	$2pq$	3.040

Calculation of the gene frequencies is rather more complicated than in the case of two alleles. The following is the simplest method: other methods, giving maximum-likelihood estimates, are described by Yasuda and Kimura (1968) and Elandt-Johnson (1971). First, the frequency of the O gene is simply the square root of the frequency of the O group. Next it will be seen that the sum of the frequencies of the B and O groups is $q^2 + 2qr + r^2 = (q + r)^2 = (1 - p)^2$. So $p = 1 - \sqrt{(B + O)}$, where B and O are the frequencies of the blood groups B and O. In the same way $q = 1 - \sqrt{(A + O)}$, and we have seen that $r = \sqrt{O}$. This method gives the following gene frequencies in the sample:

A gene:	$p = 0.2567$
B gene:	$q = 0.0598$
O gene:	$r = 0.6833$
Total	0.9998

The reason why the estimates of the gene frequencies do not add up exactly to unity is that the genotypes in the sample are not in the exact Hardy–Weinberg proportions. The AB group has not been used in arriving at the estimates, so some information has been lost. If the unused phenotype had been at a higher frequency, the loss of information might have been more serious and a more exact method might have been needed.

Sex-linked genes

With sex-linked genes the situation is rather more complex than with autosomal genes. The relationship between gene frequency and genotype frequency in the homogametic sex is the same as with an autosomal gene, but the heterogametic sex has only two genotypes and each individual carries only one gene instead of two. For this reason two-thirds of the sex-linked genes in the population are carried by the homogametic sex and one-third by the heterogametic. For the sake of brevity the heterogametic sex will be referred to as male. Consider two alleles, A_1 and A_2, with frequencies p and q, and let the genotypic frequencies be as follows:

	Females			Males	
	A_1A_1	A_1A_2	A_2A_2	A_1	A_2
Frequency:	P	H	Q	R	S

The frequency of A_1 among the females is then $p_f = P + \frac{1}{2}H$, and the frequency among the males is $p_m = R$. The frequency of A_1 in the whole population is

$$\left.\begin{aligned}
\bar{p} &= \tfrac{2}{3}p_f + \tfrac{1}{3}p_m \\
&= \tfrac{1}{3}(2p_f + p_m) \\
&= \tfrac{1}{3}(2P + H + R)
\end{aligned}\right\} \quad \dots [1.4]$$

Now, if the gene frequencies among males and among females are different, the population is not in equilibrium. The gene frequency in the population as a whole does not change, but its distribution between the two sexes oscillates as the population approaches equilibrium. The reason for this can be seen from the following consideration. Males get their sex-linked genes only from their mothers; therefore p_m is equal to p_f in the previous generation. Females get their sex-linked genes equally from both parents; therefore p_f is equal to the mean of p_m and p_f in the previous generation. Using primes to indicate the progeny generation, we have

$$p'_m = p_f$$
$$p'_f = \tfrac{1}{2}(p_m + p_f)$$

The difference between the frequencies in the two sexes is

$$p'_f - p'_m = \tfrac{1}{2}(p_m + p_f) - p_f$$
$$= -\tfrac{1}{2}(p_f - p_m)$$

i.e. half the difference in the previous generation, but in the other direction. Therefore the distribution of the genes between the two sexes oscillates, but the difference is halved in successive generations and the population rapidly approaches an equilibrium in which the frequencies in the two sexes are equal. Figure 1.2 illustrates the approach to equilibrium with a gene frequency of 2/3,

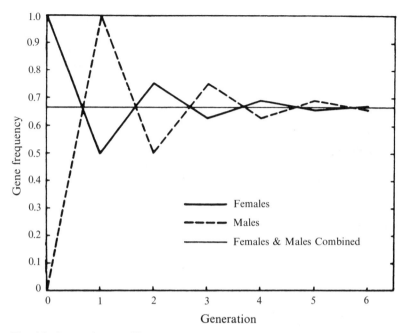

Fig. 1.2. Approach to equilibrium under random mating for a sex-linked gene, showing the gene frequency among females, among males, and in the two sexes combined. The population starts with females all of one sort ($q_f = 1$), and males all of the other sort ($q_m = 0$).

when the population is started by mixing females of one sort (all A_1A_1) with males of another sort (all A_2) and letting them breed at random.

Example 1.6 Searle (1949) gives the frequencies of a number of genes in a sample of cats in London. The animals examined were sent to clinics for destruction; they were therefore not necessarily a random sample. Among the genes studied was the sex-linked gene formerly known as 'yellow' but now called 'orange' (O). All three genotypes in females are recognizable, the heterozygote being 'tortoiseshell' or 'calico'. The data were tested against the Hardy–Weinberg expectations, to see particularly if there was any evidence of non-random mating. The first test is to see whether the gene frequency is the same in the two sexes. Then the genotypes in females

	Numbers of individuals						
	Females				Males		
	+ +	+ O	OO	Total	+	O	Total
Observed	277	54	7	338	311	42	353
Expected	273.4	61.2	3.4	338			
	$\chi^2_{[1]} = 4.6$; $P = 0.04$						

	Numbers of genes			frequencies of O-gene
	+	O	Total	q
in females	608	68	676	0.101
in males	311	42	353	0.119

are tested against the Hardy–Weinberg law in the same way as was done in Example 1.4. The numbers in each phenotypic class are shown in the table, with the gene frequencies calculated from them. The gene frequency is a little higher in males, but not significantly so. There is therefore no reason so far to think the population was not in equilibrium. The appropriate gene frequency for calculating the expected genotype frequencies in females is taken, for simplicity, to be the gene frequency observed in females. The expectations, calculated in the same way as in Example 1.4. are given in the table. The numbers observed do not agree very well with expectation, but with the small expected numbers the discrepancy is only doubtfully significant. The discrepancy, if real, might possibly have been due to non-random mating, but it might also have been due to human preferences for the colours having biased the sample and made it unrepresentative of the breeding population. For a more extensive analysis and discussion of cat populations, see Metcalfe and Turner (1971).

More than one locus

The attainment of the equilibrium in genotype frequencies after one generation of random mating is true of all autosomal loci considered separately. But it is not true of the genotypes with respect to two or more loci considered jointly. To illustrate the point, suppose there were two populations, one consisting entirely of $A_1A_1 B_1B_1$ genotypes and the other entirely of $A_2A_2 B_2B_2$ genotypes. Suppose that these two populations were mixed, with equal numbers of each sex, and allowed to mate at random. With two alleles at each of two loci there are nine possible genotypes, but only three of these would appear in the first-generation progeny, the two original double homozygotes and the double heterozygote. There would be complete association between the traits determined by the two loci, and the two traits would appear to be determined by a single gene difference. With continued random mating the missing genotypes would appear in subsequent generations, but not immediately at their equilibrium frequencies, and the initial association between the traits would be progressively reduced. If the two loci were linked, the attainment of equilibrium frequencies would take longer because the appearance of the missing genotypes depends on recombination between the two loci. Disequilibrium with respect to two or more loci is called *gametic phase disequilibrium*, or *linkage disequilibrium*, irrespective of whether the loci are linked or not. Disequilibrium can arise from intermixture of populations with different gene frequencies, or from chance in small populations. Disequilibrium can also be produced, and maintained, by selection favouring one combination of alleles over another. The rate at which a random breeding population approaches equilibrium can be deduced as follows.

We first need a measure of the amount of disequilibrium. This is best expressed in terms of the frequencies of gametic types, rather than of zygotic genotypes. Consider two loci, each with two alleles, and gene frequencies as shown in Table 1.5. There are then four types of gamete. The population is in equilibrium if the gametes contain random combinations of the genes. The gametic frequencies at equilibrium therefore depend only on the gene frequencies, and are as shown in the table. Let the actual, non-equilibrium, frequencies be r, s, t, and u, as shown. Each of these differs from the equilibrium frequency by an amount D, two gametic types having a positive, and two a negative, deviation. The value of D for each gametic type is necessarily the same, except for the sign. The amount of disequilibrium is measured by D. The disequilibrium can be

Table 1.5

Genes	A_1	A_2	B_1	B_2
Gene frequencies	p_A	q_A	p_B	q_B
Gametic types	$A_1 B_1$	$A_1 B_2$	$A_2 B_1$	$A_2 B_2$
Frequencies, equilibrium	$p_A p_B$	$p_A q_B$	$q_A p_B$	$q_A q_B$
Frequencies, actual	r	s	t	u
Difference from equilibrium	$+D$	$-D$	$-D$	$+D$

expressed by reference to genotypes by comparing the frequencies of coupling and of repulsion double heterozygotes. The genotype $A_1 B_1 / A_2 B_2$ can be called a coupling heterozygote, whether the two loci are linked or not. Its frequency is $2ru$. The repulsion heterozygote is $A_1 B_2 / A_2 B_1$ and its frequency is $2st$. If the population is in equilibrium, these two genotypes have equal frequencies. The relationship with D is

$$D = ru - st$$

Thus D is equal to half the difference in frequency between coupling and repulsion heterozygotes.

When a population in linkage disequilibrium mates at random, the amount of disequilibrium is progressively reduced with each succeeding generation. The rate at which this happens depends on the frequency of gametic types in two successive generations. This is perhaps easiest to visualize if the two loci are thought of as being linked on the same chromosome. The disequilibrium D in the progeny generation can be obtained from the frequency of any of the four gametic types, so let us consider only the $A_1 B_1$ type. This can appear in the progeny gametes in two ways. First, it can be produced as a non-recombinant from the genotype $A_1 B_1 / A_x B_x$, the subscript x meaning that either of the two alleles can be present. The frequency with which $A_1 B_1$ is produced in this way is $r(1 - c)$, r being the frequency of $A_1 B_1$ in the parental gametes and c the recombination frequency. Or, second, it can be produced as a recombinant from the genotype $A_1 B_x / A_x B_1$. The frequency of the $A_1 B_x$ chromosome is p_A and that of the $A_x B_1$ chromosome is p_B. So the frequency with which $A_1 B_1$ arises in this way is $p_A p_B c$. Therefore the frequency of $A_1 B_1$ in the progeny gametes is

$$r' = r(1 - c) + p_A p_B c$$

and the disequilibrium in the progeny generation is

$$\begin{aligned} D' &= r' - p_A p_B \\ &= r(1 - c) - p_A p_B(1 - c) \\ &= (r - p_A p_B)(1 - c) \\ &= D(1 - c) \end{aligned}$$

If we take the process one generation further we get

$$D'' = D'(1 - c) = D(1 - c)^2$$

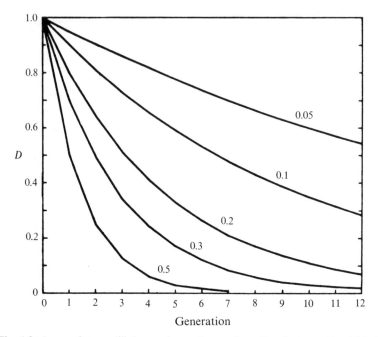

Fig. 1.3. Approach to equilibrium under random mating of two loci, considered jointly. The graphs show the amount of disequilibrium, D, relative to the disequilibrium in generation 0. The five graphs refer to different degrees of linkage between the two loci, as indicated by the recombination frequency shown alongside each graph. The graph marked 0.5 refers to unlinked loci.

Thus, after any number t of generations, the disequilibrium is given by

$$D_t = D_0(1 - c)^t \qquad \qquad \dots [1.5]$$

The loci do not have to be linked to be in disequilibrium. With unlinked loci $c = \frac{1}{2}$ and the amount of disequilibrium is halved by each generation of random mating. With linked loci the disequilibrium disappears more slowly. Figure 1.3 shows how the disequilibrium is reduced over 12 generations, with different degrees of linkage.

The approach to equilibrium given by the above equation applies equally to the disequilibrium of any number of loci considered jointly, provided $(1 - c)$ is defined as the probability of a gamete passing through a generation without recombination between any of the loci. The probability of no recombination becomes smaller the larger the number of loci, particularly if the loci are unlinked. With three unlinked loci the probability of no recombination is 1/4, with four it is 1/8, and so on. The approach to equilibrium is thus faster, the more loci there are under consideration. A practical consequence of this is that when a large number of loci are available for study, disequilibrium is more likely to be found with pairs of loci than with larger numbers considered jointly. For methods of estimating the amount of disequilibrium in populations, see Weir and Cockerham (1979).

Non-random mating

There are two distinct forms of non-random mating. The first is when mated individuals are related to each other by ancestral descent. This tends to increase the frequencies of homozygotes at all loci. Its effect will be described in Chapter 3. The second is when individuals tend to mate preferentially with respect to their genotypes at any particular locus under consideration. This form of non-random mating is dealt with briefly here.

Assortative mating

If mated pairs are of the same phenotype more often than would occur by chance, this is called *assortative mating*, and if less often it is called *disassortative mating*. To the extent that the phenotype reflects the genotype, assortative or disassortative mating affects the genotype frequencies. The effects are described by Crow and Kimura (1970) and will be only briefly outlined here. Assortative mating is of some importance in human populations, where it occurs with respect to stature, intelligence, and other characters. These, however, are not single gene differences such as can be discussed in the present context. Disassortative mating is widespread in the self-sterility systems of plants.

The consequences of assortative mating with a single locus can be deduced from Table 1.4 by appropriate modification of the frequencies of the types of mating to allow for the increased frequency of matings between like phenotypes. The effect on the genotype frequencies among the progeny is to increase the frequencies of homozygotes and reduce that of heterozygotes. In effect the population becomes partially subdivided into two groups, mating taking place more frequently within than between the groups. If assortative mating is continued in successive generations, the population approaches an equilibrium at which the genotype frequencies remain constant.

Disassortative mating has consequences that are, in general, opposite to those of assortative mating: it leads to an increase of heterozygotes and a reduction of homozygotes. Disassortative mating, however, usually has the additional consequence of changing the gene frequency. If mating is predominantly between unlike phenotypes, then the rarer phenotype has a better chance of success in mating than has the commoner phenotype. Consequently the rarer alleles are favoured, and the gene frequency changes toward intermediate values at which the phenotypes are equal in frequency. A familiar example of disassortative mating is the bisexual mode of reproduction which leads immediately to a gene, or chromosome, frequency of 0.5. Self-sterility mechanisms of plants are based on multiple alleles, and the favouring of the rarer alleles results in the coexistence of a large number of alleles, all at more or less equal frequencies.

2 CHANGES OF GENE FREQUENCY

We have seen that a large random-mating population is stable with respect to gene frequencies and genotype frequencies, in the absence of agencies tending to change its genetic properties. We can now proceed to a study of the agencies through which changes of gene frequency, and consequently of genotype frequencies, are brought about. There are two sorts of process: *systematic processes*, which tend to change the gene frequency in a manner predictable both in amount and in direction; and the *dispersive process*, which arises in small populations from the effects of sampling, and is predictable in amount but not in direction. In this chapter we are concerned only with the systematic processes, and we shall consider only large random-mating populations in order to exclude the dispersive process from the picture. There are three systematic processes: *migration, mutation,* and *selection.* We shall study these separately at first, assuming that only one process is operating at a time, and then we shall see how the different processes interact.

Migration

The effect of migration is very simply dealt with and need not concern us much here, though we shall have more to say about it later, in connection with small populations. Let us suppose that a large population consists of a proportion m of new immigrants in each generation, the remainder, $1 - m$, being natives. Let the frequency of a certain gene be q_m among the immigrants and q_0 among the natives. Then the frequency of the gene in the mixed population, q_1, will be

$$q_1 = mq_m + (1 - m)q_0$$
$$= m(q_m - q_0) + q_0 \qquad \ldots [2.1]$$

The change of gene frequency, Δq, brought about by one generation of immigration is the difference between the frequency before immigration and the frequency after immigration. Therefore

$$\Delta q = q_1 - q_0$$
$$= m(q_m - q_0) \qquad \ldots [2.2]$$

Thus the rate of change of gene frequency in a population subject to immigration depends, as must be obvious, on the immigration rate and on the difference of gene frequency between immigrants and natives.

Mutation

The effect of mutation on the genetic properties of the population differs according to whether we are concerned with a mutational event so rare as to be virtually unique, or with a mutational step that recurs repeatedly. The first produces no permanent change in a large population, whereas the second does.

Non-recurrent mutation

Consider first a mutational event that gives rise to just one representative of the mutated gene or chromosome in the whole population. This sort of mutation is of very little importance as a cause of change of gene frequency, because the product of a unique mutation has only a very small chance of surviving in a large population. The original mutated gene is present in a heterozygote and its chance of being lost in the next generation is one-half. If it survives, it may be represented by one or more copies, but each copy has only a one-half chance of surviving to the third generation. The loss is permanent, so the chance of indefinite survival is very small indeed, and is zero in an infinitely large population. Because real populations are not infinitely large, unique mutations must be expected very occasionally to survive indefinitely and lead to a change of gene frequency. More will be said about this in Chapter 4.

Recurrent mutation

It is with the second type of mutation – recurrent mutation – that we are chiefly concerned as an agent for causing change of gene frequency. Each mutational event recurs regularly with characteristic frequency, and in a large population the frequency of a mutant gene is never so low that complete loss can occur from sampling. We have, then, to find out what is the effect of this 'pressure' of mutation on the gene frequency in the population.

Suppose gene A_1 mutates to A_2 with a frequency u per generation. (u is the proportion of all A_1 genes that mutate to A_2 between one generation and the next.) If the frequency of A_1 in one generation is p_0, the frequency of newly mutated A_2 genes in the next generation is up_0. So the new gene frequency of A_1 is $p_0 - up_0$, and the change of gene frequency is $-up_0$. Now consider what happens when the genes mutate in both directions. Suppose for simplicity that there are only two alleles, A_1 and A_2, with initial frequencies p_0 and q_0. A_1 mutates to A_2 at a rate u per generation, and A_2 mutates to A_1 at a rate v. Then after one generation there is a gain of A_2 genes equal to up_0 due to mutation in one direction, and a loss equal to vq_0 due to mutation in the other direction. Stated in symbols, we have the situation:

$$\text{Mutation rate} \qquad A_1 \underset{v}{\overset{u}{\rightleftharpoons}} A_2$$

$$\text{Initial gene frequencies} \quad p_0 \qquad q_0$$

Then the change of gene frequency in one generation is

$$\Delta q = up_0 - vq_0 \qquad\qquad \dots [2.3]$$

It is easy to see that this situation leads to an equilibrium in gene frequency at which no further change takes place, because if the frequency of one allele increases fewer of the other are left to mutate in that direction and more are available to mutate in the other direction. The point of equilibrium can be found by equating the change of frequency, Δq, to zero. Thus at equilibrium

or

and

$$pu = qv$$

$$\frac{p}{q} = \frac{v}{u}$$

$$q = \frac{u}{u + v}$$

$$\dots [2.4]$$

Two conclusions can be drawn from the effect of mutation on gene frequency. Mutation rates are generally very low – about 10^{-5} or 10^{-6} per generation for most loci in most organisms. This means that between about 1 in 100,000 and 1 in 1,000,000 gametes carry a newly mutated allele at any particular locus. With normal mutation rates, therefore, mutation alone can produce only very slow changes of gene frequency; on an evolutionary time-scale they might be important, but they could scarcely be detected by experiment except with microorganisms. The second conclusion concerns the equilibrium between mutation in the two directions. Studies of reverse mutation (from mutant to wild type) indicate that it is usually less frequent than forward mutation (from wild type to mutant), on the whole about one-tenth as frequent (Muller and Oster, 1957; Schlager and Dickie, 1967). The equilibrium gene frequencies for such loci, resulting from mutation alone, would therefore be about 0.1 of the wild-type allele and 0.9 of the mutant; in other words the 'mutant' would be the common form and the 'wild type' the rare form. Since this is not the situation found in natural populations, it is clear that the frequencies of such genes are not the product of mutation alone. We shall see in the next section that the rarity of mutant alleles is attributable to selection.

Selection

Hitherto we have supposed that all individuals in the population contribute equally to the next generation. Now we must take account of the fact that individuals differ in viability and fertility, and that they therefore contribute different numbers of offspring to the next generation. The contribution of offspring to the next generation is called the *fitness* of the individual, or sometimes the *adaptive value*, or *selective value*. If the differences of fitness are in any way associated with the presence or absence of a particular gene in the individual's genotype, then *selection* operates on that gene. When a gene is subject to selection its frequency in the offspring is not the same as in the parents, since parents of different genotypes pass on their genes unequally to the next generation. In this way selection causes a change of gene frequency, and consequently also of genotype frequency. The change of gene frequency resulting from selection is more complicated to describe than that resulting from mutation, because the differences of fitness that give rise to the selection are an aspect of the phenotype. We therefore have to take account of the degree of dominance shown by the genes in question. Dominance, in this connection, means dominance with respect to fitness, and this is not necessarily the same as the dominance with respect to the main visible effects of the gene. Most mutant genes, for example, are completely recessive to the wild type in their visible effects, but this does not necessarily mean that the heterozygote has a fitness equal to that of the wild-type homozygote. The meaning of the different degrees of dominance with which we shall deal is illustrated in Fig. 2.1.

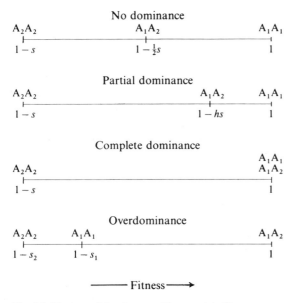

Fig. 2.1. Degrees of dominance with respect to fitness.

It is most convenient to think of selection acting against the gene in question, in the form of selective elimination of one or other of the genotypes that carry it. This may operate either through reduced viability or through reduced fertility in its widest sense, including mating ability, or through both. In the life-cycle of individuals, selection acts first through viability, then through fertility. We therefore have to deduce the change of gene frequency from the zygote stage of one generation to the zygote stage of the progeny generation. Gene frequencies cannot be observed in zygotes, so there are practical difficulties in deducing the selective forces from observed changes of gene frequency, but we shall return to these later. The strength of the selection is expressed as the *coefficient of selection*, s, which is the proportionate reduction in the gametic contribution of a particular genotype compared with a standard genotype, usually the most favoured. The contribution of the favoured genotype is taken to be 1, and the contribution of the genotype selected against is then $1 - s$. This expresses the fitness of one genotype relative to the other. Suppose, for example, that the coefficient of selection is $s = 0.1$; the fitness is then 0.9, which means that for every 100 zygotes produced by the favoured genotype, only 90 are produced by the genotype selected against. Fitness, defined in this way as the proportionate contribution of offspring, should strictly speaking be called *relative fitness*, but it will be referred to as fitness throughout what follows.

The fitness of a genotype with respect to any particular locus is not necessarily the same in all individuals. It depends on the environmental circumstances in which the individual lives, and also on the genotype with respect to genes at other loci. When we assign a certain fitness to a genotype, this refers to the average fitness of this genotype in the whole population. Though differences of fitness between individuals result in selection being applied to many, perhaps to all, loci simultaneously, we shall limit our attention here to the effects of selection on the genes at a single locus, supposing that the average fitness of the different genotypes remains constant despite the changes resulting from selection applied simultaneously to other loci. The conclusions we shall reach apply equally to natural selection occurring under natural conditions without the intervention of man, and to artificial selection imposed by the breeder or experimenter through his choice of individuals as parents and through the number of offspring he chooses to rear from each parent.

Change of gene frequency under selection

We have first to derive the basic formulae for the change of gene frequency brought about by one generation of selection. Then we can consider what they tell us about the effectiveness of selection. The different conditions of dominance have to be taken account of, but the method is the same for all, and it will be illustrated by reference to the case of complete dominance with selection acting against the recessive homozygote. Table 2.1 shows the genotypes with their Hardy–Weinberg frequencies before selection. A_2A_2 is the recessive homozygote with a

Table 2.1 Selection against a recessive gene.

	Genotypes			
	A_1A_1	A_1A_2	A_2A_2	Total
Initial frequencies	p^2	$2pq$	q^2	1
Coefficient of selection	0	0	s	
Fitness	1	1	$1-s$	
Gametic contribution	p^2	$2pq$	$q^2(1-s)$	$1-sq^2$

coefficient of selection s acting against it. The next line gives the fitness of each genotype. Multiplying the initial frequency by the fitness gives the frequency of each genotype after selection. This is entered as the 'gametic contribution' in order to allow for selection to operate over the whole life-cycle. Note that after selection the total frequency is no longer unity, because there has been a proportionate loss of sq^2 due to the selection. To find the frequency of A_2 gametes produced – and so the frequency of A_2 genes in the progeny – we take the gametic contribution of A_2A_2 individuals plus half that of A_1A_2 individuals and divide by the new total, i.e., we apply equation [1.1]. Thus the new gene frequency is

$$q_1 = \frac{q^2(1-s)+pq}{1-sq^2}$$

This can be simplified by substituting $p = (1 - q)$. Rearrangement then gives

$$q_1 = \frac{q - sq^2}{1 - sq^2} \qquad \dots [2.5]$$

The change of gene frequency, Δq, resulting from one generation of selection is

$$\Delta q = q_1 - q$$

Substituting for q_1 from equation [2.5], and after some rearrangement, this becomes

$$\Delta q = -\frac{sq^2(1 - q)}{1 - sq^2} \qquad \dots [2.6]$$

From this we see that the effect of selection on gene frequency depends not only on the intensity of selection s, but also on the initial gene frequency. But both relationships are somewhat complex, and the examination of their significance will be postponed till after the other situations have been dealt with.

Expressions for the new gene frequency and for the change of gene frequency, with different conditions of dominance, are given in Table 2.2. The general expression (5) in the table allows Δq to be worked out for any degree of dominance with respect to fitness. Two expressions (3 and 4) are given for a completely dominant gene, according to the direction of selection. The first,

Table 2.2 Change of gene frequency by one generation of selection, with different conditions of dominance for fitness, as specified below. The initial gene frequency of A_2 is q.

Initial frequencies and fitness of genotypes			New gene frequency	Change of gene frequency
A_1A_1	A_1A_2	A_2A_2	q_1	$\Delta q = q_1 - q$
p^2	$2pq$	q^2		
(1) $\quad 1$	$1 - \frac{1}{2}s$	$1 - s$	$\dfrac{q - \frac{1}{2}sq - \frac{1}{2}sq^2}{1 - sq}$	$-\dfrac{\frac{1}{2}sq(1 - q)}{1 - sq}$
(2) $\quad 1$	$1 - hs$	$1 - s$	$\dfrac{q - hspq - sq^2}{1 - 2hspq - sq^2}$	$-\dfrac{spq[q + h(p - q)]}{1 - 2hspq - sq^2}$
(3) $\quad 1$	1	$1 - s$	$\dfrac{q - sq^2}{1 - sq^2}$	$-\dfrac{sq^2(1 - q)}{1 - sq^2}$
(4) $\quad 1 - s$	$1 - s$	1	$\dfrac{q - sq + sq^2}{1 - s(1 - q^2)}$	$+\dfrac{sq^2(1 - q)}{1 - s(1 - q^2)}$
(5) $\quad 1 - s_1$	1	$1 - s_2$	$\dfrac{q - s_2q^2}{1 - s_1p^2 - s_2q^2}$	$+\dfrac{pq(s_1p - s_2q)}{1 - s_1p^2 - s_2q^2}$

(1) No dominance; selection against A_2.
(2) Partial dominance of A_1; selection against A_2.
(3) Complete dominance of A_1; selection against A_2.
(4) Complete dominance of A_1; selection against A_1.
(5) Overdominance; selection against A_1A_1 and A_2A_2. (Applicable also to any degree of dominance with fitnesses expressed relative to A_1A_2.)

which was derived above, is for selection against the recessive homozygote. If, in contrast, selection is against the dominant phenotype, Δq is not quite the same. The difference may best be appreciated by considering the effects of total elimination, when $s = 1$. The formula for selection against the dominant phenotype then reduces to $q_1 = 1$, which expresses the fact that if only the recessive homozygotes breed the gene frequency goes to 1 immediately. Total elimination of the recessive homozygote, on the other hand, will leave all of the recessive genes that are present in heterozygotes. The difference between the effects of selection in opposite directions becomes less marked as the value of s decreases. All the forms of selection mentioned so far tend in the end to eliminate one or other allele from the population. Overdominance for fitness, where heterozygotes are superior in fitness, in contrast, tends to maintain both alleles in the population. This form of selection will be given more detailed attention later.

The expressions for Δq in Table 2.2 are rather cumbersome and it is often useful to simplify them by an approximation that is good enough for many purposes. If either the coefficient of selection s, or the gene frequency q, is small, then the denominators of the equations in Table 2.2 become very nearly unity, and we can use the numerators alone as expressions for Δq. Then for selection in either direction we have, with no dominance:

$$\Delta q = \pm \tfrac{1}{2} sq(1 - q) \qquad \text{(approx.)} \qquad \qquad \ldots [2.7]$$

and with complete dominance:

$$\Delta q = \pm sq^2(1 - q) \qquad \text{(approx.)} \qquad \qquad \ldots [2.8]$$

Effectiveness of selection

We see from the formulae that the effectiveness of selection, i.e., the magnitude of Δq, depends on the initial gene frequency q. The nature of this relationship is best appreciated from graphs showing Δq at different values of q. Figure 2.2 shows these graphs for the cases of no dominance and complete dominance. They also distinguish between selection in the two directions. A value of $s = 0.2$ was chosen for the coefficient of selection because, for reasons given in Chapter 12, this seems to be the right order of magnitude for the coefficient of selection operating on genes concerned with metric characters in laboratory selection experiments. First we may note that with this value of s there is never a great difference in Δq according to the direction of selection. The two important points about the effectiveness of selection that these graphs demonstrate are: (1) selection is most effective at intermediate gene frequencies and becomes least effective when q is either large or small; (2) selection for or against a recessive gene is extremely ineffective when the recessive allele is rare. This is the consequence of the fact, noted earlier, that when a gene is rare it is represented almost entirely in heterozygotes.

Another way of looking at the effect of the initial gene frequency on the effectiveness of selection is to plot a graph showing the course of selection over a number of generations, starting from one or other extreme. Such graphs are shown in Fig. 2.3. They were constructed directly from those of Fig. 2.2, and refer again to a coefficient of selection $s = 0.2$. They show that the change due to selection is at first very slow, whether one starts from a high or a low initial gene

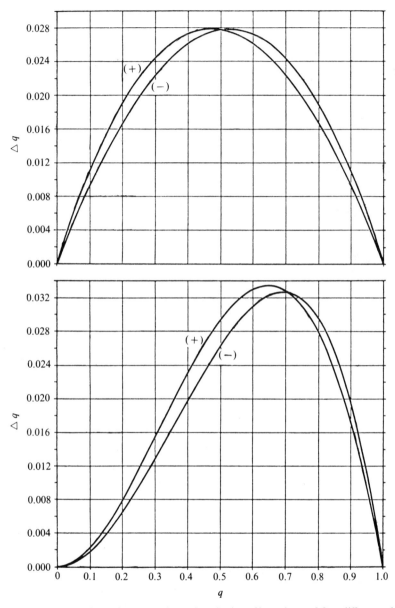

Fig. 2.2. Change of gene frequency, Δq, under selection of intensity $s = 0.2$, at different values of initial gene frequency, q. Upper figure: a gene with no dominance. Lower figure: a gene with complete dominance. The graphs marked $(-)$ refer to selection against the gene whose frequency is q, so that Δq is negative. The graphs marked $(+)$ refer to selection in favour of the gene, so that Δq is positive. (*After Falconer, 1954.*)

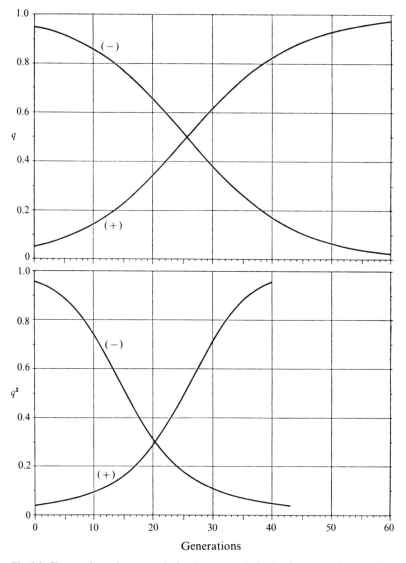

Fig. 2.3. Change of gene frequency during the course of selection from one extreme to the other. Intensity of selection, $s = 0.2$. Upper figure: a gene with no dominance. Lower figure: a gene with complete dominance, q being the frequency of the recessive allele and q^2 that of the recessive homozygote. The graphs marked $(-)$ refer to selection against the gene whose frequency is q, so that q or q^2 decreases. The graphs marked $(+)$ refer to selection in favour of the gene, so that q or q^2 increases. (*After Falconer, 1954.*)

frequency; it becomes more rapid at intermediate frequencies and falls off again at the end. In the case of a fully dominant gene one is chiefly interested in the frequency of the homozygous recessive genotype, i.e., q^2. For this reason the graph shows the effect of selection on q^2 instead of on q.

Example 2.1 Figure 2.4 shows the change of gene frequency of an autosomal recessive lethal in *Drosophila melanogaster* described by Wallace (1963). The

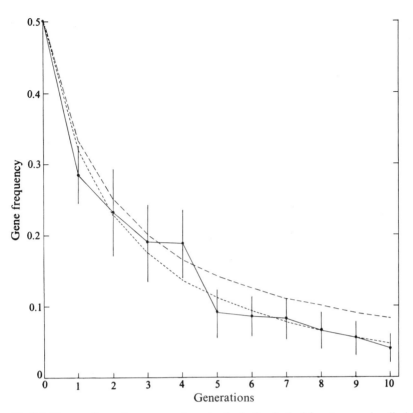

Fig. 2.4. Change of gene frequency under natural selection in the laboratory, as described in Example 2.1. (*Adapted from Wallace, 1963.*)

population was started from flies that were all heterozygotes, and the gene frequency in generation 0 was consequently $q = 0.5$. The parents of each subsequent generation were a random sample of the surviving progeny of the previous generation. Only heterozygotes and normal homozygotes survived. Heterozygotes were identified by test matings. About 100 to 200 flies were tested in each generation, giving a count of about 200 to 400 genes from which to estimate the gene frequency. The observed gene frequency ± two standard errors is plotted for each of 10 successive generations.

Expected gene frequencies were calculated for each generation by the formulae for q_1 in Table 2.2. Two expectations were calculated. The first, shown by the broken line, assumes the lethal to be completely recessive. With $s = 1$, the formula in line (3) of Table 2.2 reduces to $q_1 = q/(1 + q)$. The observed results suggest that the gene frequency was reduced a little faster than would be expected for a completely recessive lethal gene. The second expectation, shown by the dotted line, assumes the fitness of heterozygotes was reduced by 10 per cent. With $s = 1$, the formula in line (2) of Table 2.2 reduces to $q_1 = pq(1 - h)/[p^2 + 2pq(1 - h)]$, and this was evaluated with $h = 0.1$. The results agree well with this expectation.

Number of generations required

How many generations of selection would be needed to effect a specified change of gene frequency? An answer to this question may be required in connection with breeding programmes or proposed eugenic measures. We shall consider only the case of selection against a recessive when elimination of the unwanted homozygote is complete, i.e., $s = 1$. This would apply to natural selection against a recessive lethal, and to artificial selection against an unwanted recessive in a breeding programme. We shall also, for the moment, suppose that there is no mutation. The expression for the new gene frequency after one generation was given in equation [2.5] (and in line (3) of Table 2.2).

Substituting $s = 1$ in this equation and writing $q_0, q_1, q_2, \ldots, q_t$ for the gene frequency after $0, 1, 2, \ldots, t$ generations of selection, we have

$$q_1 = \frac{q_0}{1 + q_0}$$

and

$$q_2 = \frac{q_1}{1 + q_1}$$

$$= \frac{q_0}{1 + 2q_0}$$

by substituting for q_1 and simplifying. So in general

$$q_t = \frac{q_0}{1 + tq_0} \qquad \ldots [2.9]$$

and the number of generations, t, required to change the gene frequency from q_0 to q_t is

$$t = \frac{q_0 - q_t}{q_0 q_t}$$

$$= \frac{1}{q_t} - \frac{1}{q_0} \qquad \ldots [2.10]$$

The example below illustrates the point, already made, that when the frequency of a recessive gene is low, selection is very slow to change it.

Example 2.2 It is sometimes suggested, as a eugenic measure, that those suffering from serious inherited defects should be prevented from reproducing, since in this way the frequency of such defects would be reduced in future generations. Before deciding whether the proposal is a good one, we ought to know what it would be expected to achieve. We cannot properly discuss this problem without taking mutation into account, as we shall do later; the answer we get, ignoring mutation, shows what is the best that could be hoped for. Albinism will serve as an example, though it is not a very serious defect. Supposing albinism to be due to a single recessive gene, how long would it take to reduce the frequency of albino individuals to half its present value? The present frequency among European people is about 1/20,000. This is q_0^2, and it gives $q_0 = 1/141$. The objective is $q_t^2 = 1/40,000$, $q_t = 1/200$. So, from equation [2.10], $t = 200 - 141 = 59$ generations. With 25 years to a generation it would take nearly 1,500 years to achieve this modest objective. Albinism is not, in fact, a single genetic entity, but can be caused by at least two different recessive genes, each at a frequency lower than 1/141. So in reality the elimination would be even slower.

In domesticated species, of course, the elimination of deleterious genes can be greatly speeded up by progeny testing. Test-matings are made to known heterozygotes, and this identifies heterozygotes among those tested. The gene can then be eliminated very quickly. The gene persists only in heterozygotes that have been misclassified as normal through an inadequate number of progeny.

Average fitness and load

When the gene frequency is changed by selection, some individuals must suffer 'genetic death' by their failure to survive or to reproduce, and the average fitness of the population is thereby reduced. The proportion of the population that suffer genetic death is called the *load* borne by the population as a consequence of the presence of the deleterious gene in it. If L is the load, the average fitness of the population is $1 - L$. The average fitness and the load were deduced in Table 2.1 without being specifically pointed out. The average fitness is the total of the genotype frequencies after selection and is the denominator of all the expressions for q_1 or Δq given in Table 2.2. For a recessive gene, for example, the average fitness is $1 - sq^2$ and the load is sq^2. The average fitness is again relative fitness, relative to a population that does not have the deleterious gene in it. The load is not necessarily a real detriment to the population, because most species produce more offspring than the resources of its environment can support, and the death of some individuals from genetic causes leaves room for others that would otherwise have died from lack of food or some other cause. There is a species of *Drosophila*, for example (*D. tropicalis*, from Central America), in which 50 per cent of individuals in a certain locality suffer genetic death, and yet the population flourishes (Dobzhansky and Pavlovsky, 1955).

Equilibria

Balance between mutation and selection

Having described the effects of mutation and selection separately, we must now compare them and consider them jointly. Which is the more effective process in causing change of gene frequency? Is it reasonable to attribute the low frequency of deleterious genes that we find in natural populations to the balance between mutation tending to increase the frequency and selection tending to decrease it? The expressions already obtained for the change of gene frequency under mutation or selection alone show that both depend on the initial gene frequency, but in different ways. Mutation to a particular gene is most effective in increasing its frequency when the mutant gene is rare (because there are more of the unmutated genes to mutate); but selection is least effective when the gene is rare. The relative effectiveness of the two processes depends therefore on the gene frequency, and if both processes operate for long enough a state of equilibrium will eventually be reached. So we must find what the gene frequency will be when equilibrium is reached. This is done by equating the two expressions for the change of gene frequency, because at equilibrium the change due to mutation will be equal and opposite to the change due to selection.

Let us consider first a fully recessive gene with frequency q, mutation rate to it u, and from it v, and selection coefficient against it s. Then from equations

[2.3] and [2.6], we have at equilibrium

$$u(1 - q) - vq = \frac{sq^2(1 - q)}{1 - sq^2} \qquad \ldots [2.11]$$

This equation is too complicated to give a clear answer to our question. But we can make two simplifications with only a trivial sacrifice of accuracy. We are specifically interested in genes at low equilibrium frequencies. If q is small, the term vq representing back mutation is relatively unimportant and can be neglected; and we can use the approximate expression (equation [2.8]) for the selection effect. Making these simplifications we have the equilibrium condition for selection against a recessive gene:

$$u(1 - q) = sq^2(1 - q) \quad \text{(approx.)}$$
$$u = sq^2 \qquad \qquad \text{(approx.)} \qquad \ldots [2.12]$$
$$q = \sqrt{\frac{u}{s}} \qquad \quad \text{(approx.)} \qquad \ldots [2.13]$$

For a gene with no dominance, similar reasoning from line (1) in Table 2.2 gives the equilibrium condition

$$q = \frac{2u}{s} \qquad \text{(approx.)} \qquad \ldots [2.14]$$

Finally, consider selection against a completely dominant gene, the frequency of the dominant gene being $1 - q$, and the mutation rate to it being v. In this case $1 - q$ is very small and the term $u(1 - q)$ in equation [2.11] is negligible. We have therefore at equilibrium

$$vq = sq^2(1 - q) \qquad \text{(approx.)}$$
$$q(1 - q) = \frac{v}{s} \qquad \text{(approx.)}$$

or $$H = \frac{2v}{s} \qquad \text{(approx.)} \qquad \ldots [2.15]$$

where H is the frequency of heterozygotes. If the mutant gene is rare, H is very nearly the frequency of the mutant phenotype in the population.

> **Example 2.3**   If the equilibrium state is accepted as applicable, we can use it to get an estimate of the mutation rate of dominant abnormalities for which the coefficient of selection is known. Among some human examples described by Haldane (1949) is the case of dominant dwarfism (chondrodystrophy) studied in Denmark. The frequency of dwarfs was estimated at 10.7×10^{-5}, and their fitness $(1 - s)$ at 0.196. The estimate of fitness was made from the number of children produced by dwarfs compared with their normal sibs. The mutation rate, by equation [2.15], comes out at 4.3×10^{-5}. Though there is a possibility of serious error in the estimate of frequency owing to prenatal mortality of dwarfs, the mutation rate is almost certainly estimated within the right order of magnitude. The mutation rate to recessives cannot be reliably estimated in this way because the estimate is very sensitive to small departures from equilibrium. (For more about mutation rates in man, see Stern, 1973).

These expressions for the equilibrium gene frequency under the joint action of mutation and selection show that the gene frequency can have any value at equilibrium, depending on the relative magnitude of the mutation rate and the coefficient of selection. But if mutation rates are of the order of magnitude commonly accepted, i.e., 10^{-5} or thereabouts, then only a mild selection against the mutant gene will be needed to hold it at a very low equilibrium frequency. For example, if a gene mutates at the rate of 10^{-5}, a selective disadvantage of 10 per cent is enough to hold the frequency of the recessive homozygote at 1 in 10,000; and a 50 per cent disadvantage will hold it at 1 in 50,000. It is quite clear therefore that the low frequency of deleterious mutants in natural populations is in accord with what would be expected from the joint action of mutation and selection.

Let us now consider the load, or proportion of genetic deaths, when a population is in equilibrium. The load from a recessive gene is sq^2, as explained earlier. The equilibrium equation [2.12] therefore shows that the load at equilibrium is

$$L = u \qquad\qquad \dots [2.16]$$

Thus the load depends only on the mutation rate and not at all on how seriously deleterious the gene is. The reason for this surprising conclusion is that a more deleterious gene comes to equilibrium at a lower gene frequency; there are therefore fewer homozygotes, though more of them die. With a less deleterious gene there are more homozygotes, but fewer of them die. Since the load from each locus does not depend on the selection coefficient, the total load from recessive alleles at all loci is simply the sum of the mutation rates, $\Sigma\, u$. With deleterious dominant genes, the homozygotes are so rare that they can be neglected. The load therefore comes from the death of heterozygotes and is therefore $L = sH$, where H is the frequency of heterozygotes. Substituting the equilibrium frequency of heterozygotes from equation [2.15] gives the load at equilibrium as

$$L = 2v \qquad\qquad \dots [2.17]$$

where v is the mutation rate to the dominant allele. Again the load is not affected by the harmfulness of the gene. Comparison of equations [2.16] and [2.17] raises another question. Why should the load from a dominant gene be twice that from a recessive with the same mutation rate? The reason is that the death of a mutant homozygote removes two genes from the population whereas the death of a heterozygote removes only one mutant gene. Equation [2.17] seems to suggest that the loss of one gene by the death of a heterozygote balances the introduction of two genes by mutation. This is not so because the loss by death is expressed per individual, whereas the gain by mutation is expressed per gamete: the mutation rate per individual is $2v$. The load from partially dominant alleles is between u and $2u$, and the total load from all loci is between $\Sigma\, u$ and $2\Sigma\, u$, where u is the mutation rate to alleles with any degree of dominance. If mutation rates are about 10^{-5}, an organism with 10,000 loci capable of mutation to deleterious alleles would have a total load of between 10 and 20 per cent; that is to say, about 1 or 2 zygotes in 10 would die as a result of mutation.

The fact that recessive genes at low frequencies respond only very slowly to selection makes it very unlikely that rare recessives are at their equilibrium frequencies in real populations. Unless the environmental conditions remain exceptionally constant over a long period, selection coefficients are likely to change faster than selection can adjust the frequency to each new equilibrium value. This, of course, would not apply to genes that are lethal under all conditions. There is also another reason, which applies to lethals too, for thinking that present-day human populations are not in equilibrium. Modern civilization has reduced the subdivision into local, partially inbreeding, groups, and this has reduced the frequency of homozygotes as will be explained in the next chapter. In consequence, both the gene frequencies and the homozygote frequencies are below their equilibrium values, and must be presumed to be at present increasing slowly toward new equilibria at higher values. Deductions about rare recessives based on the supposition of equilibrium therefore cannot be made, particularly for human populations.

Changes of equilibrium

Mutation rates can be increased by artificially produced radiation or environmental chemicals; selection coefficients can be reduced by medical treatment or by domestication, or they can be increased by eugenic measures. What effects would be expected from these changes? Whatever the change, there will be a new equilibrium gene frequency toward which the population will start to move. The effect on the frequency of affected individuals, when the new equilibrium is reached, can readily be seen from equations [2.13] and [2.15]: for example, doubling the mutation rate, or halving the selection coefficient, would eventually double the frequency of affected individuals. The immediate affect of increasing the mutation rate depends on the coefficient of selection against the gene – the lower the selection coefficient, the slower the approach to equilibrium. The consequences of an increased mutation rate would unquestionably be harmful. (For an assessment of the consequences see Crow, 1957). The consequences of changing the selection coefficient one way or the other, however, need some comment.

Intensification of selection has sometimes been advocated as a eugenic measure for human populations. Example 2.2 showed how extremely slow such measures applied to a recessive gene would be to make a worthwhile reduction of its frequency. When mutation is taken into consideration the prospects are seen to be even worse. Not only is mutation hindering the selection, but it puts a limit – the equilibrium for $s = 1$ – below which the frequency cannot be reduced. Serious defects, moreover, have already a fairly strong natural selection working on them, and the addition of artificial selection can do no more than make the coefficient of selection s equal to 1. This would probably seldom do more than double the present coefficient of selection, and the incidence of defects would be reduced to not less than half their present values.

Perhaps the reduced intensity of natural selection under modern conditions should give us more concern. Minor genetic defects, such as colour-blindness, must presumably have had some selective disadvantage in the past but now have very little, if any, effect on fitness. Moreover, medical treatment removes, or reduces, the selection pressure against susceptibility to a variety of diseases that have at least some degree of genetic causation. This relaxation of natural selection suggests that the frequencies of the genes concerned will increase toward new equilibria at higher values. If this is true we must expect the incidence of minor genetic defects to increase in the future, and also the proportion of people who need medical treatment for a variety of diseases. By applying humanitarian principles for our own good now we are perhaps laying up a store of inconvenience for our descendants in the distant future.

Selection favouring heterozygotes

We have considered the effects of selection operating on genes that are partially or fully dominant with respect to fitness; but, though the appropriate formula was given in Table 2.2, we have not yet discussed the consequences of overdominance with respect to fitness; that is, when the heterozygote has a higher fitness than either homozygote. At first sight it may seem rather improbable that selection should favour the heterozygote of two alleles rather than one or other of the homozygotes. There is good evidence, however, that it does occur, though opinion is divided on how common a situation it is. Let us first examine the consequences of this form of selection, and then consider how it might operate.

Selection operating on a gene with partial or complete dominance tends toward the total elimination of one or other allele, the final gene frequency, in the absence of mutation, being 0 or 1. When selection favours the heterozygote, however, the gene frequency tends toward an equilibrium at an intermediate value, both alleles remaining in the population, even without mutation. The reason is as follows. The change of gene frequency after one generation was given in Table 2.2 as being

$$\Delta q = \frac{pq(s_1 p - s_2 q)}{1 - s_1 p^2 - s_2 q^2}$$

The condition for equilibrium is that $\Delta q = 0$, and this is fulfilled when $s_1 p = s_2 q$. The gene frequencies at this point of equilibrium are therefore

$$\frac{p}{q} = \frac{s_2}{s_1} \qquad \qquad \text{...[2.18]}$$

or

$$q = \frac{s_1}{s_1 + s_2} \qquad \qquad \text{...[2.19]}$$

Now, if q is greater than its equilibrium value (but not 1), and p therefore less, $s_1 p$ will be less than $s_2 q$, and Δq will be negative; that is to say q will decrease. Similarly, if q is less than its equilibrium value (but not 0) it will increase. Therefore when the gene frequency has any value, except 0 or 1, selection changes it toward the intermediate point of equilibrium given in equation [2.19], and both alleles remain permanently in the population. Three or more alleles at a locus are maintained in the same way, provided the heterozygote of any pair is superior in fitness to both homozygotes of that pair (Kimura, 1956). A feature of the equilibrium worthy of note is that the gene frequency depends not on the degree of superiority of the heterozygote but on the relative disadvantage of one homozygote compared with that of the other. Therefore there is a point of equilibrium at some more or less intermediate gene frequency whenever a heterozygote is superior to both the homozygotes, no matter by how little.

The load resulting from overdominance for fitness is $s_1p^2 + s_2q^2$, from the denominator of the expression for Δq. Substituting the equilibrium value for q from equation [2.19], and the analogous value for p, leads to the following expression for the load at equilibrium:

$$L = \frac{s_1 s_2}{s_1 + s_2} \qquad \ldots [2.20]$$

Substitution of $s_2 = s_1 p/q$ and separately of $s_1 = s_2 q/p$ from equation [2.18] leads to

$$L = s_1 p = s_2 q \qquad \ldots [2.21]$$

Thus the load depends on the selection coefficients, unlike the load due to recurrent mutation, and the total load from all overdominant loci cannot be obtained in any simple way by summation.

An example of heterozygote advantage is described below, and the possible causes of overdominance for fitness are then discussed. First, however, we must examine the effects of selection on observed genotype frequencies.

Selection and the Hardy–Weinberg test.   In the previous chapter a test was described which, by comparing observed with expected genotype frequencies, tests for the fulfilment of the conditions required for generating Hardy–Weinberg frequencies. Since the absence of selection is one condition, disagreement between observed and expected frequencies can provide evidence that selection has operated. But the conclusions that can be drawn are limited, and we must now look more closely into what the test can and cannot reveal. The test requires a locus at which heterozygotes can be distinguished, so that the gene frequency can be determined by counting. Usually the genotypes observed are those of adults in only one generation. They have been subject to selection through viability differences, but not yet to selection through fertility differences. The test does not reveal selection on fertility. If the parents differ in fertility, random mating produces Hardy–Weinberg frequencies in the zygotes that will become the adults observed. The test therefore can only reveal selection acting through viability. It is tempting to believe that the relative viabilities of the genotypes can be deduced from their deviations from expectation, particularly to see an excess of heterozygotes as evidence of heterozygote advantage. This, however, is not a valid deduction, for the following reason. The expectations are calculated from the observed gene frequency, and they are expectations based on the supposition that the gene frequency in the zygotes was the same as that observed in the adults. This supposition may not be correct, in which case the expectations will have been wrongly calculated. The consequence of calculating wrong expectations is that an apparent excess of heterozygotes can result from selection acting against only one homozygote, i.e., from a gene that is completely recessive with respect to fitness. The numerical example in Table 2.3 will make clear what happens. The zygotes are in Hardy–Weinberg frequencies corresponding to a gene frequency of 0.4 for the A_2 allele. Of the 16 $A_2 A_2$ zygotes, 6 survive to be counted as adults, and the gene frequency is reduced to 1/3. With this observed gene frequency the Hardy–Weinberg expectations are calculated as shown. The result is that the

Table 2.3. Spurious heterozygote advantage from the Hardy–Weinberg test.

	Genotypes				Gene frequency of A_2
	A_1A_1	A_1A_2	A_2A_2	Total	
Number of zygotes	36	48	16	100	0.4
Number of adults	36	48	6	90	
Frequency of adults (%)	40.0	53.3	6.7	100	0.33
H–W expectation (%)	44.4	44.5	11.1	100	

observed frequency of heterozygotes is above expectation and those of both homozygotes are below their expectations. The test appears to indicate heterozygote superiority, but the selection in fact was against one homozygote only. The only situation in which heterozygote advantage can be inferred from an excess of heterozygotes is when the population is in equilibrium and the gene frequency is not changing and, furthermore, when there are no differences of fertility among the parents. Only then is the gene frequency the same in the adults as it was in the zygotes. It must be remembered, however, that an excess of heterozygotes results also from unequal gene frequencies in the male and female parents. A difference between the sexes can occur by chance if the sample of progeny used for the test is a small one. (This is explained in the next chapter.)

Consider now what the observed genotype frequencies can tell us about the relative viabilities. Let us assign viabilities relative to that of the heterozygote. With Hardy–Weinberg frequencies in the zygotes, the genotype frequencies in the adults will be as follows, when P, H and Q are the observed frequencies.

	Genotype			
	A_1A_1	A_1A_2	A_2A_2	Total
Frequency in adults	$\begin{cases} p^2(1-s_1) \\ P \end{cases}$ $\begin{matrix} 2pq \\ H \end{matrix}$		$\begin{matrix} q^2(1-s_2) \\ Q \end{matrix}$	$\begin{matrix} 1-s_1p^2-s_2q^2 \\ 1 \end{matrix}$

The discrepancy can be expressed as $PQ/(\frac{1}{2}H)^2$, which with Hardy–Weinberg frequencies is equal to 1. When selection is taken into account it can readily be shown that

$$\frac{PQ}{(\frac{1}{2}H)^2} = (1-s_1)(1-s_2) \qquad \qquad \ldots [2.22]$$

This measure of the discrepancy is therefore an estimate of the product of the viabilities of the two homozygotes relative to the heterozygote. If one allele is known to be recessive, then the other homozygote has unreduced viability; e.g., if A_2 is recessive, $(1-s_1) = 1$. Then the viability, $(1-s_2)$, of the recessive homozygote can be estimated.

It will be seen from the brief account given that the estimation of relative fitnesses is not straightforward. For a fuller treatment of the problems, see Prout (1965).

Example 2.4 Sickle-cell anaemia in man is a well-known example of heterozygote advantage. It is particularly useful as an example because the data allow a test of observation with theory. The disease is caused by the abnormal haemoglobin-S. Homozygotes suffer from a severe anaemia from which many die, yet the gene is present among Africans and their descendants in America at frequencies much too high to be accounted for by mutation counterbalancing the selection against homozygotes. The explanation of the high frequencies is that heterozygotes have an advantage over normal homozygotes through an increased resistance to malaria (Allison, 1954). The selective forces can be calculated from data given by Allison (1956) and one can then see how well these can account for the observed gene frequency. Allison classified 287 infants and 654 adults, from a district of Tanzania, for genotype. (Homozygotes are recognized by the presence of red blood cells with a characteristic 'sickle' shape; heterozygotes are recognized by the sickling of their cells when the blood sample is deoxygenated.) The observed numbers and frequencies are shown in the table, with the gene frequencies calculated from them by equation [1.1]. (AA denotes the normal homozygote, AS the heterozygote, and SS the anaemic homozygote.) Most of the differential selection is thought to take place before adulthood, i.e., the surviving genotypes do not differ much in fertility. The infants therefore represent the genotype frequencies before selection and the adults after selection, and so we can calculate the selection coefficients from the observed frequencies. First, however, note that if the gene frequency is in equilibrium it will be the same after selection as it was before, and the data agree well with this expectation. Dividing the frequency of each genotype after selection by its frequency before selection gives the relative fitness of that genotype, as shown in the table. The fitnesses of the homozygotes can then be expressed relative to the heterozygote by dividing each by the heterozygote fitness. The homozygote fitnesses are $1 - s_1$ and $1 - s_2$, from which the selection coefficients work out to be 0.24 against AA and 0.80 against SS, both relative to AS. The equilibrium gene frequency expected to result from this selection against both homozygotes, by equation [2.19], is $q_s = 0.23$, which is reasonably close to the observed value. Thus the selective forces observed in the differential viability do satisfactorily account for the frequency of the sickle-cell gene in this population.

	Genotype			Frequency of S-gene
	AA	AS	SS	
Numbers of infants	189	89	9	
adults	400	249	5	
Frequency in infants	0.6585	0.3101	0.0314	0.1864
adults	0.6116	0.3807	0.0076	0.1980
Relatives fitness	0.9288	1.2277	0.2420	
Fitness relative to AS	0.7565	1	0.1971	
Selection coefficient	$s_1 = 0.2435$		$s_2 = 0.8029$	

$$\text{Expected } q_S = \frac{s_1}{s_1 + s_2} = 0.2327$$

The selective values may be more interesting if expressed relative to the normal homozygote. The fitness of AS is then $1/(1 - s_1) = 1.32$, and that of SS is $(1 - s_2)/(1 - s_1) = 0.26$. Thus the resistance to malaria confers a 32 per cent advantage on the heterozygote, and this balances a 74 per cent disadvantage in the anaemic homozygote when the gene frequency is about 0.2.

Possible causes of overdominance for fitness. Let us now consider some of the ways in which selection might operate so as to favour heterozygotes. One way is through pleiotropy, i.e., the gene having more than one phenotypic effect. To produce overdominance for fitness, the alleles must affect two components of fitness in opposite directions. The heterozygote advantage of sickle-cell anaemia arises in this way; one homozygote reduces fitness through one component, the anaemia, while the other homozygote reduces fitness through another component, susceptibility to malaria. There are a few other genes in man where heterozygote advantage for similar reasons is proved or suspected. Another example is the resistance of wild rats to the anticoagulant poison warfarin (Greaves *et al.*, 1977). The gene conferring resistance is dominant, so that heterozygotes and homozygotes are resistant. Homozygotes, however, have a much increased requirement for vitamin-K, which is not met by the normal diet. So in areas where the poison is being used, one homozygote is selected against by the poison and the other by the vitamin-K deficiency, leading to an equilibrium frequency of the resistance gene, which was about 0.34 in the area studied.

There are many ways in which a locus can affect different components of fitness. For example, the components can be different stages of the life-cycle, different environments encountered by the same individual at different times, or by different individuals in different places, the two sexes, different combinations of genes at other loci which modify the locus in question. The conditions that produce overdominance for fitness are that the alleles affect the components in opposite directions and that there is some degree of dominance on the scale in which the components combine to give fitness. The meaning of the last condition is this: if the components are multiplied together to give fitness, then there must be some degree of dominance on the geometric scale but not necessarily on the arithmetic scale. To take a simple example, a hypothetical locus with two alleles in mice might affect the number born per litter and the number of litters as follows:

	Genotype		
	A_1A_1	A_1A_2	A_2A_2
Number per litter	6	7	8
Number of litters	8	7	6
Total number = fitness	48	49	48

Fitness is the product of the two components and there is overdominance for fitness. In their effects on the components separately, the alleles have no dominance on the arithmetic scale, but a small degree of dominance on the geometric scale, the geometric mean of the homozygous values being 6.9. Overdominance generated in this way is known as *marginal overdominance* (Wallace, 1968), meaning that the overdominance appears only in the margin of the table.

Gametic phase disequilibrium of linked loci can generate pseudo-overdominance in a similar way. If two loci are closely linked so that they appear to be one, and if the favourable alleles are dominant and linked in repulsion, then the heterozygote may be superior to either homozygote. The possibility of pseudo-overdominance being caused by linkage makes it extremely difficult to establish real overdominance at single loci from observations on populations derived from crosses between different strains, because it is formally impossible to exclude the presence of a closely linked but unrecognized locus. Wild populations of many *Drosophila* species have chromosomes with different gene arrangements carried in inverted segments. Inversion heterozygotes are generally superior in fitness to homozygotes (see Wallace, 1968), and this heterozygote superiority is probably due to the linkage of the genes in the inversions.

Finally, overdominance can arise at the molecular level. If a locus codes for an enzyme, the products of the two alleles (allozymes) are likely to have different properties, such as enzymatic activity, heat-stability, or optima for environmental factors such as temperature or pH. The mixture of allozymes may therefore make the heterozygote more versatile than either of the homozygotes with single allozymes, i.e., less susceptible to the impairment of enzyme function by environmental circumstances. Or, if the allozymes differ in activity, the intermediate activity of the heterozygote may be more favourable than the higher or lower activities of the homozygotes. For further details and discussion of the evidence for overdominance at the molecular level, see Berger (1976).

We have seen that there are many ways by which overdominance for fitness could arise. It must be admitted, however, that the cases where it has been proved to occur are very few indeed.

Polymorphism

We saw earlier that the balance between mutation and selection satisfactorily accounts for the presence of deleterious genes at low frequencies, causing the appearance of rare abnormal, or mutant, individuals. Genes of this sort, however, are only a minor part of the genetic variation found in natural populations. There are many genes causing variants that are neither rare nor in any way abnormal, and the presence of these genes cannot easily be accounted for by the simple balance of selection against mutation. The blood-group genes used as examples in the first chapter are of this sort. More striking examples are the colour varieties found in many species, particularly among insects, snails, and fish. The existence of these visible differences caused by genes at intermediate frequencies is called *polymorphism*, and the term is extended to cover all such variants, whether readily discernible or not. The term is also used to describe loci at which there are variant alleles at intermediate frequencies. Electrophoresis and other methods for detecting differences in the amino-acid composition of proteins have shown that very many loci coding for proteins are polymorphic, at least one-third of loci and perhaps much more in most organisms. It is clear therefore that many loci, perhaps the majority, carry allelic differences causing genetic variation between normal individuals. The 'intermediate' gene frequencies by which polymorphic loci are defined are arbitrary, but are usually taken to be in the range of 0.01 to 0.99 or, more strictly, the frequency of the commonest allele is taken to be not more than 0.99. The essential point is that the rarer alleles are at frequencies too high to be regarded as equilibrium frequencies for mutation balanced by selection, unless the selection is extremely weak. We therefore need another explanation for the widespread existence of polymorphic variation. This is a subject of major interest and lively controversy in population genetics. Here we can only give a very brief summary of the mechanisms that may be responsible for polymorphism; for a comprehensive survey of the problem and the evidence, see Lewontin (1974), and for a shorter account, Maynard Smith (1975). The main argument is between those, the 'selectionists', who believe the polymorphisms to be *balanced*, i.e., stable equilibria maintained by selective forces, and others, the 'neutralists', who believe that many polymorphisms have no functional significance, but result from chance events of mutation and finite population size. The various mechanisms that may account for polymorphisms are as follows.

1. *Heterozygote advantage.*   We saw in the previous section that overdominance for fitness maintains an equilibrium gene frequency at intermediate levels, and that there are many ways in which overdominance for fitness can arise. Heterozygote advantage is therefore an attractive explanation for polymorphism. The paucity of proven cases need not be a difficulty because an advantage of the heterozygote so small as to be quite undetectable in practice would be enough to maintain a polymorphism. There are, however, difficulties connected with the effects of inbreeding, and with the number of loci that could be kept in a polymorphic state by heterozygote advantage. For further discussion of the problem, see Lewontin (1974), Berger (1976), and Wills (1978).

2. *Frequency-dependent selection.*   Having a phenotype that is rare may itself be an advantage, irrespective of what the phenotype is. The direction of selection is then dependent on the gene frequency: an allele at low frequency produces the rare phenotype and is favoured, but the same allele at a high frequency is selected against. This leads to a stable equilibrium gene frequency and so to a balanced polymorphism. Many examples of frequency-dependent selection are known. Pollen grains bearing a rare self-sterility allele have a better chance of fertilizing an ovule because the same allele is seldom present in the stigmata of other plants. Birds and fish have been shown to take disproportionately more of the more common type of food when they are offered a choice (see, e.g., Allen, 1975), and this is thought to exert frequency-dependent selection on polymorphic prey, such as snails, giving an advantage to individuals with a rare pattern of colouration (see Clarke, 1969). When *Drosophila* males with different genotypes that affect their courtship behaviour are mixed, the rarer genotype has more success in inseminating females (see Petit and Nouaud, 1976). In general, the development of special methods of attack and of defence in the relationships between predator and prey and between pathogen and host, seem likely to result in frequency-dependent selection. Frequency-dependent selection is reviewed by Ayala and Campbell (1974) and by Clarke (1979).

3. *Heterogeneous environment.*   The environment experienced by individuals of a population is not constant; it differs from place to place and varies with time. If one allele is advantageous in one environment and another in a different environment, stable polymorphism can result without heterozygotes necessarily being on average superior. Selection can be thought of as tending to adapt different individuals to different environments. The situation is complex because the outcome depends on many factors such as the dominance relations, whether individuals choose to breed in the environments to which they are adapted, whether mating is preferentially between individuals from the same environment or

random, what proportions of the whole population inhabit each of the different environments, whether individuals each encounter only one environment ('coarse-grained' environment) or more than one ('fine-grained'), and whether the heterogeneity of the environment is spatial or temporal. A relatively simple form of selection in a heterogeneous environment results in a *cline*. This is a gradient of gene frequency between one locality and another, one allele being at a high frequency at one end of the cline and at a low frequency at the other end. Clines are thought, or in many cases known, to be maintained by selection favouring one allele in one locality and another allele in another locality, with a limited amount of migration which allows mating only between individuals from neighbouring parts of the cline. The selection in opposite directions at the two ends and the 'gene-flow' up and down the cline, maintain the polymorphism. The role of heterogeneous environments in maintaining polymorphism is reviewed by Felsenstein (1976). Evidence that it has an important role was obtained from a survey of the polymorphism and ecology of 243 species (Nevo, 1978).

4. *Transition.* Polymorphisms seen at present might possibly be transitional stages in the evolutionary replacement of one allele by another, which has become more advantageous through some environmental change in the past. This, however, is unlikely to be the explanation of more than a very small proportion of polymorphisms.

5. *Neutral mutation.* All the above mechanisms involve selection as the force responsible for the polymorphism. It is possible, however, that the selection coefficients may be very small indeed, so small that mutation rates become a significant factor. The mutations that might give rise to the protein polymorphisms must be regarded as unique events, because the variant forms of the protein differ at only a few amino-acid sites, and the probability of getting the same amino-acid substitution by recurrent mutation is very small indeed. When each mutation is treated as a unique event, taking the population size into account shows that mutation and chance can give rise to polymorphisms. When the population is not infinite in size a unique mutation has a small, but not zero, chance of not only surviving but of eventually spreading through the whole population. The smaller the population, the greater is the mutant's chance of survival. The vast majority of new mutants will be lost, but a few will survive and replace the original allele. (The chance of survival will be considered further in Chapter 4.) Those few mutants that do survive take a very long time to spread through the population, and during this time they contribute to the polymorphism. Since the changes of gene frequency are dependent on chance, the spread of a mutant through the population is erratic, its frequency sometimes increasing and sometimes decreasing. Furthermore, some mutants will survive and increase in frequency at first, only to disappear later. If a new mutant has a selective advantage its chance of survival is, of course, increased. If it has a selective disadvantage its chance is decreased, but it may nevertheless survive if the disadvantage is small and the population size is small. According to the neutral-mutation theory some of the polymorphisms present at any time represent genes, nearly neutral with respect to fitness, that have mutated in the distant past and are still present in the population.

Opinion is divided about how many polymorphisms can be attributed to neutral mutation. The issue rests primarily on the strength of the selection acting on them, and on the size of the population. The 'neutralists' think that many allelic differences are effectively neutral, while the 'selectionists' take the opposite view. There are many sorts of observation that can be made on populations that bear on the question, but much of the evidence obtained has proved to be consistent with either vewpoint. In many cases selection has been clearly proved to operate, but in many other cases it has not been possible to discriminate between the two hypotheses. Crow (1972), Lewontin (1974) and Ohta (1974) have reviewed the problem.

3 SMALL POPULATIONS: I. Changes of Gene Frequency under Simplified Conditions

We have now to consider the last of the agencies through which gene frequencies can be changed. This is the dispersive process, which differs from the systematic processes in being random in direction, and predictable only in amount. In order to exclude this process from the previous discussions we have postulated always a 'large' population, and we have seen that in a large population the gene frequencies are inherently stable. That is to say, in the absence of migration, mutation, or selection, the gene and genotype frequencies remain unaltered from generation to generation. This property of stability does not hold in a small population, and the gene frequencies are subject to random fluctuations arising from the sampling of gametes. The gametes that transmit genes to the next generation carry a sample of the genes in the parent generation, and if the sample is not large the gene frequencies are liable to change between one generation and the next. This random change of gene frequency is the dispersive process.

In this chapter and the next we shall be concerned with the effects of the dispersive process on gene frequencies. If the deductions to be made about gene frequencies seem to be rather remote from reality, it should be remembered that the properties of a population with respect to any genetically determined character depend on gene frequencies. The conclusions are therefore fully relevant to quantitative characters to be dealt with in later chapters.

There are, broadly speaking, four consequences of the dispersive process, which are to be explained and quantified in this chapter. These are not really different consequences, but rather different ways in which the consequences may be seen. They are:

1. *Random drift.* The random changes of gene frequency are called random drift. If the gene frequency in any one small population is followed, it may be seen to change in an erratic manner from generation to generation, with no tendency to revert to its original value.

2. *Differentiation between sub-populations.* Random drift occurring independently in different sub-populations leads to genetic differentiation between the sub-populations. The inhabitants of a large area seldom in nature constitute a single large population, because mating takes place more often between inhabitants of the same region. Natural populations are therefore more or less

subdivided into local groups or sub-populations, and these come to differ in gene frequencies if the number of individuals in the group is small. Domesticated or laboratory populations, in the same way, are often subdivided – for example, into herds or strains – and in them the subdivision and genetic differentiation are often more marked.

3. *Uniformity within sub-populations.*   Genetic variation within each sub-population becomes progressively reduced, and the individuals become more and more alike in genotype. This genetic uniformity is the reason for the widespread use of inbred strains of laboratory animals in many areas of biological research. (An inbred strain is one maintained as a small population over many generations.)

4. *Increased homozygosity.*   Homozygotes increase in frequency at the expense of heterozygotes. This, coupled with the general tendency for deleterious alleles to be recessive, is the genetic basis for the loss of fertility and viability that almost always results from inbreeding.

There are two different ways of looking at the dispersive process and of deducing its consequences. One is to regard it as a sampling process and to describe it in terms of sampling variance. The other is to regard it as an inbreeding process and describe it in terms of the genotypic changes resulting from matings between related individuals. Of these, the first is probably the simpler for a description of how the process works, but the second provides a more convenient means of quantifying its consequences. The plan to be followed here is first to describe the general nature of the dispersive process from the point of view of sampling. This will show how the four consequences come about. Then we shall approach the process afresh from the point of view of inbreeding, and show how the two viewpoints connect with each other. In all this we shall confine our attention to the simplest possible situation, with migration, mutation, and selection excluded. Thus we shall see what happens in small populations in the absence of other factors influencing gene frequency. In the next chapter we shall extend the conclusions to more realistic situations by removing the restrictive simplifications, and in Chapter 5 we shall consider the special cases of pedigreed populations and very small populations maintained by regular systems of close inbreeding.

The idealized population

In order to reduce the dispersive process to its simplest form we imagine an idealized population as follows. We suppose there to be initially one large population in which mating is random, and this population becomes subdivided into a large number of sub-populations. The subdivision might arise from geographical or ecological causes under natural conditions, or from controlled breeding in domesticated or laboratory populations. The initial random-mating population will be referred to as the *base population*, and the sub-populations will be referred to as *lines*. All the lines together constitute the whole population, and each line is a 'small population' in which gene frequencies are subject to the dispersive process. When a single locus is under discussion we cannot properly understand what goes on in one line except by considering it as one of a large number of lines. But what happens to the genes at one locus in a number of lines happens equally to those at a number of loci in one line, provided they all start at the same gene frequency. So the consequences of the process apply equally to a single line provided we consider many loci in it.

The simplifying conditions specified for the idealized population are the following:

1. Mating is restricted to members of the same line. The lines are thus isolated in the sense that no genes can pass from one line to another. In other words migration is excluded.

2. The generations are distinct and do not overlap.

3. The number of breeding individuals in each line is the same for all lines and in all generations. Breeding individuals are those that transmit genes to the next generation.

4. Within each line mating is random, including self-fertilization in random amount.

5. There is no selection at any stage.

6. Mutation is disregarded.

The situation implied by these conditions is represented diagrammatically in Fig. 3.1, and may be described thus: All breeding individuals contribute equally to a pool of gametes from which zygotes will be formed. Union of gametes is strictly random. Out of a potentially large number of zygotes, only a limited number survive to become breeding individuals in the next generation, and this is the stage at which the sampling of the genes transmitted by the gametes takes place. Survival of zygotes is random, and consequently the contribution of the parents to the next generation is not uniform, but varies according to the chances of survival of their progeny. Since the population size is constant from generation to generation, the average number of progeny that reach breeding age is one per individual parent or two per mated pair of parents. In this scheme the sampling is seen as a single event, the reduction of a large number of gametes to a small number of breeding progeny. The reduction of numbers may take place in several stages. This makes no difference to the theoretical consequences deduced from the final number of breeding progeny, provided the sampling at each stage is

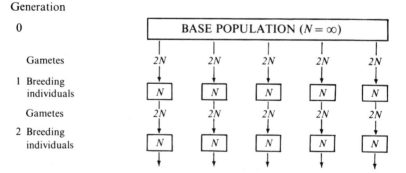

Fig. 3.1. Diagrammatic representation of the subdivision of a single large population – the base population – into a number of sub-populations, or lines.

random. The observed consequences, however, would be affected if a population were enumerated at stages before the reduction of numbers was complete.

The following symbols will be used in connection with the idealized population.

$N =$ the number of breeding individuals in each line and generation. This is the *population size*.

$t =$ time, in generations, starting from the base population at t_0.

$q =$ frequency of a particular allele at a locus.

$p = 1 - q =$ frequency of all other alleles at that locus. q and p refer to the frequencies in any one line; $\bar{q}$ and $\bar{p}$ refer to the frequencies in the whole population and are the means of q and p; q_0 and p_0 are the frequencies in the base population.

Since all systematic processes tending to change the gene frequency have been excluded, the mean gene frequency among all lines at any stage must be the same as the initial frequency. Thus $\bar{q} = q_0$, and the two can be used interchangeably in this chapter.

It is obvious that the conditions specified for the idealized population do not hold in real populations. The conclusions to be drawn in this chapter, however, can be made applicable to real populations by the simple device of replacing the population size N by the 'effective' population size N_e, a concept to be introduced in the next chapter.

Sampling

Variance of gene frequency

The change of gene frequency resulting from sampling is random in the sense that its direction is unpredictable. But its magnitude can be predicted in terms of the variance of the change. Consider the formation of the lines from the base population. Each line is formed from a sample of N individuals drawn from the base population. Since each individual carries two genes at a locus, the subdivision of the population represents a series of samples each of $2N$ genes, drawn at random from the base population. The gene frequencies in these samples will have an average value equal to that in the base population, i.e. q_0, and will be distributed about this mean with a variance $p_0 q_0/2N$, which is the binomial variance of sample means, the sample size being in this case $2N$. This variance is the variance of q_1, the gene frequency in the different lines after one generation. Since the initial gene frequency q_0 is the same for all lines, it is also the variance of $(q_1 - q_0)$, which is the change of gene frequency. Thus the change of gene frequency, Δq, resulting from sampling in one generation, can be stated in terms of its variance as

$$\sigma^2_{\Delta q} = \frac{p_0 q_0}{2N} \qquad \dots [3.1]$$

This variance of Δq expresses the magnitude of the change of gene frequency resulting from the dispersive process. It expresses the expected change in any one line, or the variance of gene frequencies that would be found among many lines after one generation. Its effect is a dispersion of gene frequencies among the lines;

in other words, the lines come to differ in gene frequency, though the mean in the population as a whole remains unchanged.

In the next generation the sampling process is repeated, but each line now starts from a different gene frequency and so the second sampling leads to a further dispersion. The variance of the change now differs among the lines, since it depends on the gene frequency q_1 in the first generation of each line separately. The effect of continued sampling through successive generations is that each line fluctuates irregularly in gene frequency, and the lines spread apart progressively, thus becoming differentiated. These are the first two consequences of the process, and they are exemplified in Fig. 3.2. If there were only one small population or line, one would see only the random drift in the erratic changes of gene frequency from generation to generation. Having several lines, as in Fig. 3.2, one sees random drift in each line and also the progressive differentiation between them as they drift apart. The differentiation between lines is more clearly seen in Fig. 3.3, from a different experiment, showing the distributions of gene frequency in successive generations.

Increasing differentiation among the lines is equivalent to increasing variance of the gene frequency among them. The variance of the gene frequency,

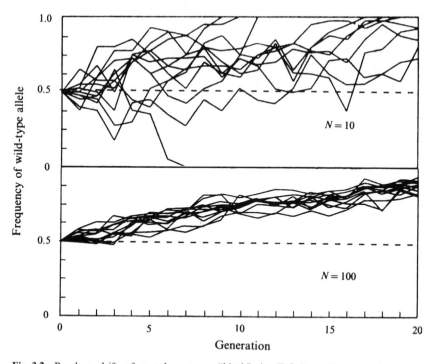

Fig. 3.2. Random drift of a colour gene ('black') in *Tribolium*. Heterozygotes were recognizable, so the gene frequencies were estimated exactly by counting. The figure shows the results with two population sizes, $N = 10$ and $N = 100$. There were 12 lines with each population size. Natural selection favoured the wild-type allele and led to an overall increase in its frequency, random drift causing variation of the lines around the mean, more marked in the smaller than in the larger populations. (*After Rich, Bell, and Wilson, 1979.*)

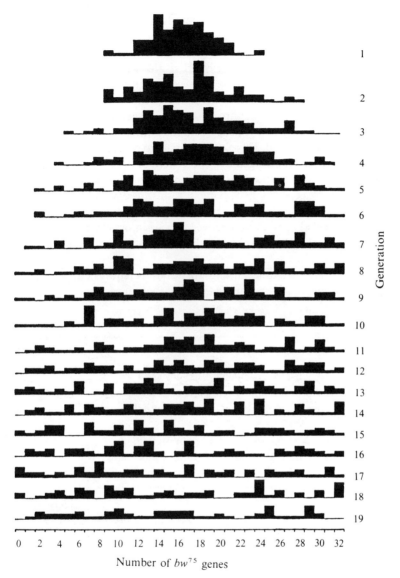

Fig. 3.3. Distributions of gene frequencies in 19 consecutive generations among 105 lines of *Drosophila melanogaster*, each of 16 individuals. The gene frequencies refer to two alleles at the 'brown' locus (bw^{75} and bw), with initial frequencies of 0.5. The height of each black column shows the number of lines having the gene frequency shown on the scale below, previously fixed lines being excluded. (*After Buri, 1956.*)

σ_q^2, among the lines, at any generation t, is given by

$$\sigma_q^2 = p_0 q_0 \left[1 - \left(1 - \frac{1}{2N} \right)^t \right] \qquad \dots [3.2]$$

(The derivation of this expression will be explained later because it is more easily understood by consideration of the inbreeding aspect of the process.) Figure 3.4

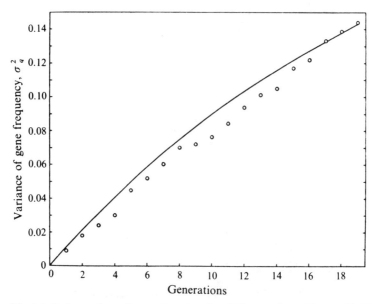

Fig. 3.4. Variance of gene frequencies among lines in the experiment illustrated in Fig. 3.3. The circles are the observed values, and the smooth curve shows the expected variance as given by equation [3.2]. The value taken for N is 11.5, which is the 'effective number', N_e, as explained in the next chapter. (*Data from Buri, 1956.*)

shows the variance of gene frequencies observed in the experiment of Fig. 3.3, with the expected variance calculated from equation [3.2]. We may note here a fact that will be needed later, and is obvious from equation [3.2], namely that $\sigma_p^2 = \sigma_q^2$.

Examination of the distributions of gene frequencies in Fig. 3.3 shows that the distributions change in shape, becoming eventually quite flat, with all frequencies equally probable. (This is not true of the limiting frequencies of 0 and 1, which are discussed in the next section.) Theoretical considerations show that there are two phases in the dispersion. During the initial phase the gene frequencies are spreading out from the initial value. This phase is followed by a steady phase, when the gene frequencies are evenly spread out over the range and all gene frequencies except the two limits are equally probable. This uniform distribution of the steady phase is attained even if the initial gene frequency is not 0.5, though it takes longer to reach it. The duration, in generations, of the initial phase is a small multiple of the population size, depending on the initial gene frequency. With $q_0 = 0.5$ it lasts about $2N$ generations, and with $q_0 = 0.1$ it lasts about $4N$ generations (Kimura, 1955). The theoretical distributions of gene frequency during the initial phase, with $q_0 = 0.5$ and $q_0 = 0.1$, are shown in Fig. 3.5. The observed distributions in Fig. 3.3 agree well with the theoretical distributions for $q_0 = 0.5$.

Fixation

There are limits to the spreading apart of the lines that can be brought about by the dispersive process. The gene frequency cannot change beyond the limits of 0 or 1, and sooner or later each line must reach one or other of these limits.

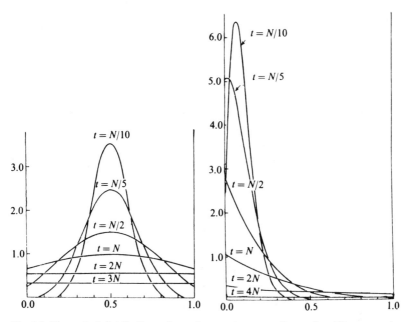

Fig. 3.5. Theoretical distributions of gene frequencies among lines, after different numbers of generations, t, expressed in terms of the population size of the lines, N. In the left-hand figure $q_0 = 0.5$; on the right $q_0 = 0.1$. Previously fixed lines (see next section) are excluded. The horizontal scale is the gene frequency, q, in any line. The vertical axis is the probability, scaled to make the area under each curve equal to the proportion of unfixed lines. (*After Kimura, 1955.*)

Moreover, the limits are 'traps' or points of no return, because once the gene frequency has reached 0 or 1 it cannot change any more in that line. When a particular allele has reached a frequency of 1 it is said to be *fixed* in that line, and when it reaches a frequency of 0 it is *lost*. When an allele reaches fixation, no other allele can be present in that line, and the line may then be said to be fixed. When a line is fixed, all individuals in it are of identical genotype with respect to that locus. Eventually all lines, and all loci in a line, become fixed. The individuals of a line are then genetically identical, and this is the basis of the genetic uniformity of highly inbred strains.

When the process has gone to completion and all lines are fixed, the mean gene frequency is still unchanged and equal to the initial gene frequency. Therefore the proportion of the lines in which different alleles at a locus are fixed is equal to the initial frequencies of the alleles. If the base population contains two alleles A_1 and A_2 at frequencies p_0 and q_0 respectively, then A_1 will be fixed in the proportion p_0 of the lines, and A_2 in the remaining proportion, q_0. The variance of the gene frequency among the lines is then $p_0 q_0$, as many be seen from equation [3.2] by putting t equal to infinity. (In Fig. 3.3 the lines in which fixation or loss has just occurred are shown, but not those in which it occurred earlier.)

When concerned with the attainment of genetic uniformity one wants to know how soon fixation takes place; what is the probability of a particular locus being fixed, or what proportion of all loci in a line will be fixed, after a certain number of generations. Consideration of the progressive nature of the dispersion, as illustrated in Fig. 3.3, will show that fixation does not start immediately; the dispersion of gene frequencies must proceed some way before any line is likely to reach fixation. To deduce the probability of fixation is mathematically complicated and only an outline of the conclusions can be given here. By the time the uniform distribution of the steady phase has been reached, a locus with an initial gene frequency of $q_0 = 0.5$ will have been fixed in about 50 per cent of lines, while a locus starting at $q_0 = 0.1$ will have been fixed in over 90 per cent of lines. After the steady phase has been reached fixation proceeds at a constant rate: a proportion $1/2N$ of the lines previously unfixed become fixed in each generation. After the earliest stages of fixation, the proportion of lines in which a gene with initial frequency q_0 is expected to be fixed, lost, or to be still segregating is approximately as follows (Wright, 1952):

$$
\left.
\begin{array}{ll}
\text{fixed:} & q_0 - 3p_0q_0\,P \\
\text{lost:} & p_0 - 3p_0q_0\,P \\
\text{neither:} & 6p_0q_0\,P
\end{array}
\right\}
\quad \text{where } P = \left(1 - \frac{1}{2N}\right)^t
\qquad \ldots [3.3]
$$

Figure 3.6 shows the progress of fixation and loss in an experiment with *Drosophila*. The expectation calculated from equation [3.3] fits the data very well after about generation 8, when 5 per cent of lines had been fixed.

Genotype frequencies

Change of gene frequency leads to change of genotype frequencies; so the genotype frequencies in small populations follow the changes of gene frequency

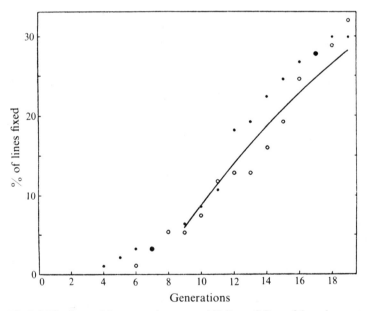

Fig. 3.6. Fixation and loss occurring among 107 lines of *Drosophila melanogaster*, during 19 generations. This is not the same experiment as that illustrated in Figs. 3.3 and 3.4, but was similar in nature. There were 16 parents per generation in each line, and the effective number (see Ch. 4) was 9. The closed circles show the percentage of lines in which the bw^{75} allele has become fixed; the open circles show the percentage in which it has been lost and the bw allele fixed. The smooth curve is the expected amount of fixation of one or other allele, computed from the effective number by equation [3.3]. (*Data from Buri, 1956.*)

resulting from the dispersive process. In the idealized population, which we are still considering, mating is random within each of the lines. Consequently the genotype frequencies in any one line are the Hardy–Weinberg frequencies appropriate to the gene frequency in the previous generation of that line. (There is, in fact, a small deviation from Hardy–Weinberg frequencies within lines, which will be explained at the end of this chapter, but it can be ignored for the moment.) As the lines drift apart in gene frequency they become differentiated also in genotype frequencies. But differentiation is not the only aspect of the change: the general direction of the change is toward an increase of homozygous, and a decrease of heterozygous, genotypes. The reason for this is the dispersion of gene frequencies from intermediate values toward the extremes. Heterozygotes are most frequent at intermediate gene frequencies (see Fig. 1.1), so the drift of gene frequencies toward the extremes leads, on the average, to a decline in the frequency of heterozygotes. It also leads to a higher proportion of the individuals in a line having the same genotype, and so to an increased genetic uniformity within lines.

The genotype frequencies in the population as a whole can be deduced from a knowledge of the variance of gene frequencies in the following way. If an allele has a frequency q in one particular line, homozygotes of that allele will have a frequency of q^2 in that line. The frequency of these homozygotes in the population as a whole will therefore be the mean value of q^2 over all lines. We shall write this mean frequency of homozygotes as $(\overline{q^2})$. The value of $(\overline{q^2})$ can be found from a knowledge of the variance of gene frequencies among the lines, by noting that the variance of a set of observations is found by deducting the square of the mean from the mean of the squared observations. Thus

$$\sigma_q^2 = (\overline{q^2}) - \bar{q}^2$$

and
$$(\overline{q^2}) = \bar{q}^2 + \sigma_q^2 \qquad \qquad \ldots [3.4]$$

where σ_q^2 is the variance of gene frequencies among the lines, as given in equation [3.2], and $\bar{q}^2$ is the square of the mean gene frequency. Since the mean gene frequency $\bar{q}$ is equal to the original q_0, it follows that $\bar{q}^2$ or q_0^2 is the original frequency of homozygotes in the base population. Thus in the population as a whole the frequency of homozygotes of a particular allele increases, and is always in excess of the original frequency by an amount equal to the variance of the gene frequency among the lines. In a two-allele system the same applies to the other allele, and the frequency of heterozygotes is reduced correspondingly. Noting from equation [3.2] $\sigma_p^2 = \sigma_q^2$, we therefore find the genotypic frequencies for a locus with two alleles as follows:

Genotype	Frequency in whole population	
A_1A_1	$p_0^2 + \sigma_q^2$	
A_1A_2	$2p_0q_0 - 2\sigma_q^2$	$\ldots [3.5]$
A_2A_2	$q_0^2 + \sigma_q^2$	

These genotype frequencies are no longer the Hardy–Weinberg frequencies appropriate to the original or mean gene frequency. The Hardy–Weinberg

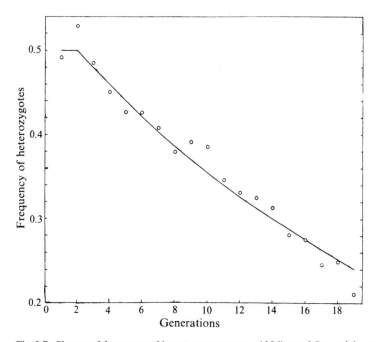

Fig. 3.7. Change of frequency of heterozygotes among 105 lines of *Drosophila melanogaster*, each with 16 parents. The same experiment as is illustrated in Figs. 3.3. and 3.4. The frequency of heterozygotes refers to the population as a whole, all lines taken together. The smooth curve is the expected frequency of heterozygotes. (*Data from Buri, 1956.*)

relationships between gene frequency and genotype frequencies, though they hold good within each line separately, do not hold if the lines are taken together and regarded as a single population. This fact causes some difficulty in relating gene and genotype frequencies in natural populations, because they are often more or less subdivided and the degree of subdivision is seldom known. An example of the decrease of heterozygotes resulting from the dispersion of gene frequencies is shown in Fig. 3.7.

The foregoing account of genotype frequencies describes the situation in terms of one locus in many lines. It can be regarded equally as referring to many loci in one line; then the change in any one line or small population is an increase in the number of loci at which individuals are homozygous and a corresponding decrease in the number at which they are heterozygous – in short, an increase of homozygotes at the expense of heterozygotes. This change of genotype frequencies resulting from the dispersive process is the genetic basis of the phenomenon of inbreeding depression, of which a full explanation will be found in Chapter 14.

We have now surveyed the general nature of the dispersive process and its four major consequences – random drift, differentiation of sub-populations, genetic uniformity within sub-populations, and overall increase in the frequency of homozygous genotypes. Let us now look at the process from another viewpoint, as an inbreeding process. Instead of regarding the increase of homozygotes as a consequence of the dispersion of gene frequencies, we shall now look directly at the manner in which the additional homozygotes arise.

Inbreeding

Inbreeding means the mating together of individuals that are related to each other by ancestry. That the degree of relationship between the individuals in a population depends on the size of the population will be clear by consideration of the numbers of possible ancestors. In a population of bisexual organisms every individual has two parents, four grand-parents, eight great-grandparents, etc., and t generations back it has 2^t ancestors. Not very many generations back, the number of individuals required to provide separate ancestors for all the present individuals becomes larger than any real population could contain. Any pair of individuals must therefore be related to each other through one or more common ancestors in the more or less remote past; and the smaller the size of the population in previous generations the less remote are the common ancestors, or the greater their number. Thus pairs mating at random are more closely related to each other in a small population than in a large one. This is why the properties of small populations can be treated as the consequences of inbreeding.

The essential consequence of two individuals having a common ancestor is that they may both carry replicates of one of the genes present in the ancestor; and if they mate they may pass on these replicates to their offspring. Thus inbred individuals – that is to say, offspring produced by inbreeding – may carry two genes at a locus that are replicates of one and the same gene in a previous generation. Consideration of this consequence of inbreeding shows that there are two sorts of identity among allelic genes, and two sorts of homozygote. The sort of identity we have hitherto considered is a functional identity. If two genes cannot be distinguished by their phenotypic effects, or by any other functional criterion, they are regarded as being the same allele. An individual carrying a pair of such genes is a homozygote in the ordinary sense. The new sort of identity is one of origin by replication. Two genes that have originated from the replication of one single gene in a previous generation may be called *identical by descent*, or simply *identical*. Two genes that are not identical are *independent* in descent. Homozygotes of identical genes may be called *identical homozygotes*. Other terms in use are *autozygous* to describe identical homozygotes and *allozygous* to describe homozygotes that are not known to be autozygous. It is the production of identical homozygotes that gives rise to the increase of homozygotes as a consequence of inbreeding.

Identity by descent provides the basis for a measure of the dispersive process, through the degree of relationship between the mating pairs. The measure is the *coefficient of inbreeding*, which is the probability that the two genes at any locus in an individual are identical by descent. It refers to an individual and expresses the degree of relationship between the individual's parents. If the parents of any generation have mated at random then the coefficient of inbreeding of the progeny is the probability that two gametes taken at random from the parent generation carry identical genes at a locus. This is the average coefficient of inbreeding of all the progeny. Individuals of different families will have different inbreeding coefficients because with random mating some pairs of parents will be more closely related than other pairs. It is, however, with the average coefficient of inbreeding that we are concerned as a measure of the dispersive process. The coefficient of inbreeding is generally symbolized by F.

The degree of relationship expressed in the inbreeding coefficient is essentially a comparison between the population in question and some specified or implied base population. Without this point of reference it is meaningless, as the following consideration will show. On account of the limitation in the number of independent ancestors in any population not infinitely large, all genes now present at a locus in the population would be found to be identical by descent if traced far enough back into the remote past. Therefore the inbreeding coefficient only becomes meaningful if we specify some time in the past beyond which ancestries will not be pursued, and at which all genes present in the population are to be regarded as independent – that is, not identical by descent. This point is the *base population* and by its definition it has an inbreeding coefficient of zero. The inbreeding coefficient of a subsequent generation expresses the amount of the dispersive process that has taken place since the base population, and compares the degree of relationship between the individuals now, with that between individuals in the base population. Reference to the base population is not always explicitly stated, but is always implied. For example, we can speak of the inbreeding coefficient of a population subdivided into lines. The comparison of relationship is between the individuals of a line and individuals taken at random from the whole population. The base population implied is a hypothetical population from which all the lines were derived.

Inbreeding in the idealized population

Let us now return to the idealized population and deduce the coefficient of inbreeding in successive generations, starting with the base population and its progeny constituting generation 1. The situation may be visualized by thinking of a hermaphrodite marine organism, capable of self-fertilization, shedding eggs and sperm into the sea. There are N individuals, each shedding equal numbers of gametes which unite at random. All the genes at a locus in the base population have to be regarded as being non-identical; so, considering only one locus, among the gametes shed by the base population there are $2N$ different sorts, in equal numbers, bearing the genes A_1, A_2, A_3, etc., at the A locus. The gametes of any one sort carry identical genes; those of different sort carry genes of independent origin. What is the probability that a pair of gametes taken at random carry identical genes? This is the inbreeding coefficient of generation 1. Any gamete has a $(1/2N)$th chance of uniting with another of the same sort, so $1/2N$ is the probability that uniting gametes carry identical genes, and is thus the coefficient of inbreeding of the progeny. Now consider the second generation. There are now two ways in which identical homozygotes can arise, one from the new replication of genes and the other from the previous replication. The probability of newly replicated genes coming together in a zygote is again $1/2N$. The remaining proportion, $1 - 1/2N$, of zygotes carry genes that are independent in their origin from generation 1, but may have been identical in their origin from generation 0. The probability of their identical origin in generation 0 is what we have already deduced as the inbreeding coefficient of generation 1. Thus the total probability of identical homozygotes in generation 2 is

$$F_2 = \frac{1}{2N} + \left(1 - \frac{1}{2N}\right)F_1$$

where F_1 and F_2 stand for the inbreeding coefficients of generations 1 and 2 respectively. The same argument applies to subsequent generations, so that in general the inbreeding coefficient of individuals in generation t is

$$F_t = \frac{1}{2N} + \left(1 - \frac{1}{2N}\right)F_{t-1} \qquad \ldots [3.6]$$

Thus the inbreeding coefficient is made up of two parts: an 'increment', $1/2N$, attributable to the new inbreeding, and a 'remainder', attributable to the previous inbreeding and having the inbreeding coefficient of the previous generation. In the idealized population the 'new inbreeding' arises from self-fertilization, which brings together genes replicated in the immediately preceding generation. Exclusion of self-fertilization simply shifts the replication one generation further back, so that the 'new inbreeding' brings together genes replicated in the grandparental generation; the coefficient of inbreeding is affected, but not very much, as we shall see later. The distinction between 'new' and 'old' inbreeding brings clearly to light a point which we note here in passing because it will be needed later and is often important in practice: if there is no 'new inbreeding', as would happen if the population size were suddenly increased, the previous inbreeding is not undone, but remains where it was before the increase of population size.

Let us call the 'increment' or 'new inbreeding' ΔF, so that

$$\Delta F = \frac{1}{2N} \qquad \ldots [3.7]$$

Equation [3.6] may then be rewritten in the form

$$F_t = \Delta F + (1 - \Delta F)F_{t-1} \qquad \ldots [3.8]$$

Further rearrangement makes clearer the precise meaning of the 'increment' ΔF.

$$\Delta F = \frac{F_t - F_{t-1}}{1 - F_{t-1}} \qquad \ldots [3.9]$$

From the equation written thus we see that the 'increment' ΔF measures the *rate of inbreeding* in the form of a proportionate increase. It is the increase of the inbreeding coefficient in one generation, relative to the distance that was still to go to reach complete inbreeding. This measure of the rate of inbreeding provides a convenient way of going beyond the restrictive simplifications of the idealized population, and it thus provides a means of comparing the inbreeding effects of different breeding systems. When the inbreeding coefficient is expressed in terms of ΔF, equation [3.8] is valid for any breeding system and is not restricted to the idealized population, though only in the idealized population is ΔF equal to $1/2N$.

So far we have done no more than relate the inbreeding coefficient in one generation to that of the previous generation. It remains to extend equation [3.8] back to the base population and so express the inbreeding coefficient in terms of the number of generations. This is made easier by the use of a symbol P for the complement of the inbreeding coefficient $1 - F$, which is known as the *panmictic index.* Substitution of $(1 - P)$ for F in equation [3.9], and rearrangement, leads to

$$\frac{P_t}{P_{t-1}} = 1 - \Delta F \qquad \qquad \ldots [3.10]$$

Thus the panmictic index is reduced by a constant proportion in each generation. Extension back to generation $t - 2$ gives

$$\frac{P_t}{P_{t-2}} = (1 - \Delta F)^2$$

and extension back to the base population gives

$$P_t = (1 - \Delta F)^t P_0 \qquad \qquad \ldots [3.11]$$

where P_0 is the panmictic index of the base population. The base population is defined as having an inbreeding coefficient of 0, and therefore a panmictic index of 1. The inbreeding coefficient in any generation t, referred to the base population, is therefore

$$F_t = 1 - (1 - \Delta F)^t \qquad \qquad \ldots [3.12]$$

The consequences of the dispersive process were described earlier from the viewpoint of sampling variance. Let us now look again at them, applying the rate of inbreeding and the inbreeding coefficient as measures of the process. Strictly speaking we should refer still to the idealized population, but the equating of the two viewpoints is generally valid, unless the population size is different in parent and offspring generations or there is non-random mating within lines.

Variance of gene frequency
First, the variance of the change of gene frequency in one generation, taken from equation [3.1] and expressed in terms of the rate of inbreeding, becomes

$$\sigma^2_{\Delta q} = \frac{p_0 q_0}{2N} = p_0 q_0 \Delta F \qquad \qquad \ldots [3.13]$$

An equivalent way of writing equation [3.13] is in terms of the inbreeding coefficient and the variance of gene frequencies after one generation. It follows that the relationship is the same after any number of generations, so that after t generations

$$\sigma^2_q = p_0 q_0 F_t \qquad \qquad \ldots [3.14]$$

Equation [3.14] can be shown to be equivalent to equation [3.2], which was given without explanation. Replacing F_t in equation [3.14] by $[1 - (1/2N)^t]$ from equations [3.12] and [3.7] gives equation [3.2].

Measures of the dispersive process based on inbreeding are more useful than those based on the variance of gene frequencies because they apply equally to any mean, or initial, gene frequency. Thus ΔF expresses the rate of dispersion, and F the cumulated effect of random drift.

Genotype frequencies
Let us consider next the genotype frequencies in the population as a whole. The genotype frequencies expressed in terms of the variance of gene frequency in equation [3.5] can be rewritten in terms of the coefficient of inbreeding from

equation [3.14]. The frequency of A_2A_2, for example, is

$$(\overline{q^2}) = q_0^2 + \sigma_q^2 = q_0^2 + p_0q_0F$$

The genotype frequencies expressed in this way are entered in the left-hand side of Table 3.1. As was explained before, this way of writing the genotype frequencies shows how the homozygotes increase at the expense of the heterozygotes. Recognition of identity by descent to which the inbreeding viewpoint led us means that we can now distinguish the two sorts of homozygote, identical and independent, among the A_1A_1 and A_2A_2 genotypes. The frequency of identical homozygotes among both genotypes together is by definition the inbreeding coefficient, F; and the division between the two genotypes is in proportion to the initial gene frequencies. So p_0F is the frequency of A_1A_1 identical homozygotes, and q_0F that of A_2A_2 identical homozygotes. The remaining genotypes, both homozygotes and heterozygotes, carry genes that are independent in origin and are therefore the equivalent of pairs of gametes taken at random from the population as a whole. Their frequencies are therefore the Hardy–Weinberg frequencies. Thus, from the inbreeding viewpoint, we arrive at the genotype frequencies shown in the right-hand columns of Table 3.1. This way of writing the genotype frequencies shows how homozygotes are divided between those of independent and those of identical origin. The equivalence of the two ways of expressing the genotype frequencies can be verified from their algebraic identity. Both ways show equally clearly how the heterozygotes are reduced in frequency in proportion to $1 - F$.

The panmictic index, which was defined earlier as $P = 1 - F$, expresses the frequency of heterozygotes in a subdivided population relative to the Hardy–Weinberg frequency expected if the population as a whole mated at random. This can be seen by consideration of the frequency of heterozygotes given in Table 3.1. Let H_t and H_0 be the frequencies of heterozygotes in a subdivided and random-mating population respectively. Then $H_0 = 2p_0q_0$ and $H_t = 2p_0q_0(1 - F) = H_0(1 - F)$. The panmictic index at generation t is therefore

$$P_t = 1 - F_t = \frac{H_t}{H_0} \qquad \dots [3.15]$$

When a real population is sampled, a deficiency of heterozygotes may be the only indication that it is a subdivided population. The observed frequency of heterozygotes, H, relative to the Hardy–Weinberg frequency, $2\bar{p}\bar{q}$, then gives the

Table 3.1 Genotype frequencies for a locus with two alleles, expressed in terms of the inbreeding coefficient F.

	Original frequencies		Change due to inbreeding		Origin:		
					Independent		Identical
A_1A_1	p_0^2	$+$	p_0q_0F	$=$	$p_0^2(1 - F)$	$+$	p_0F
A_1A_2	$2p_0q_0$	$-$	$2p_0q_0F$	$=$	$2p_0q_0(1 - F)$		
A_2A_2	q_0^2	$+$	p_0q_0F	$=$	$q_0^2(1 - F)$	$+$	q_0F

panmictic index as $P = H/2\bar{p}\bar{q}$, $\bar{p}$ and $\bar{q}$ being the observed gene frequencies in the sample as a whole. Caution is needed, however, in regarding the value of P so calculated as anything more than a description of the sample. It is unlikely that all the sub-populations that really existed would be equally represented in the sample and, unless they were, P could not validly be used to estimate, for example, the variance of the gene frequency among the sub-populations by equation [3.14].

Genotype frequencies within lines. Throughout this chapter it has been assumed that mating is at random within lines, and that consequently the genotype frequencies within any line are the Hardy–Weinberg frequencies appropriate to the gene frequency in that line. It was pointed out, however, that the genotype frequencies actually deviate slightly from the Hardy–Weinberg expectations. The reason for this deviation is that the sample of genes passed to the next generation consists, in fact, of two independent samples, one in male parents and the other in female parents, with N genes in each sample. The male and female parents therefore differ, on average, in their gene frequencies. A difference in gene frequency between male and female parents leads to an excess of heterozygotes in the progeny; or, in other words, an expectation calculated from the mean gene frequency is too low. By putting appropriate gametic frequencies in Table 1.2 it can be shown that the expected frequency of heterozygotes within any line is $H = 2pq + \frac{1}{2}\overline{D^2}$, where p and q are the mean gene frequencies in the line, D is the difference in gene frequency between male and female parents, and $\overline{D^2}$ is the mean squared difference. Since $\bar{D} = 0$ it follows, by analogy with equation [3.4], that $\overline{D^2} = \sigma_D^2$. The variance σ_D^2 is the variance of the difference between two binomial samples of size N, which is $2pq/N$. Thus the expected frequency of heterozygotes within lines is

$$\left.\begin{aligned}H &= 2pq + pq/N \\[1mm] &= 2pq\left(1 + \frac{1}{2N}\right)\end{aligned}\right\} \qquad \dots [3.16]$$

(For further details see Robertson, 1965). The excess of heterozygotes is trivial unless N, the number of parents of the sample, is very small. But it can have an appreciable effect if the frequencies observed in a small sample of a single population are tested for agreement with Hardy–Weinberg expectations.

The overall frequency of heterozygotes in the whole of a subdivided population is sometimes used to estimate the amount of inbreeding in the history of the population, rather than as a description of the present state of subdivision. This is done in Example 4.1. If the lines are separately identifiable, and the number of parents sampled in each line is known, correction can be made for the excess of heterozygotes. Substitution of H from equation [3.16] for H_0 in equation [3.15] gives

$$1 - F = \frac{H}{2pq(1 + 1/2N)} \qquad \dots [3.17]$$

where H is the observed frequency of heterozygotes, p and q are the overall observed gene frequencies, and N is the number of parents in each line.

4 SMALL POPULATIONS:
II. Less Simplified Conditions

In order to simplify the description of the dispersive process we confined our attention in the last chapter to an idealized population, and to do this we had to specify a number of restrictive conditions, which could seldom be fulfilled in real populations. The purpose of this chapter is to adapt the conclusions of the last chapter to situations in which the conditions imposed do not hold; in other words, to remove the more serious restrictions and bring the conclusions closer to reality. The restrictive conditions were of two sorts, one sort being concerned with the breeding structure of the population and the other excluding mutation, migration, and selection from consideration. We shall first describe the effects of deviations from the idealized breeding structure, and then consider the outcome of the dispersive process when mutation, migration, or selection are operating at the same time.

Effective population size

If the breeding structure does not conform to that specified for the idealized population, it is still possible to evaluate the dispersive process in terms of either the variance of gene frequencies or the rate of inbreeding. This can be done by the same general methods and no new principles are involved. We shall therefore give the conclusions briefly and without detailed explanation. The most convenient way of dealing with any particular deviation from the idealised breeding structure is to express the situation in terms of the *effective number* of breeding individuals, or the *effective population size*, N_e. This is the number of individuals that would give rise to the calculated sampling variance, or rate of inbreeding, if they bred in the manner of the idealized population. Suppose, for example, that the rate of inbreeding, ΔF, had been calculated for a particular breeding structure from consideration of the probability of identical homozygotes being produced. In the idealized population, ΔF is related to the population size N by equation [3.7] as $\Delta F = 1/2N$. The effective size is related to ΔF in the same way and would therefore be obtained from the calculated ΔF as $N_e = 1/2\Delta F$. Thus all the conclusions drawn in the previous chapter are valid for any breeding structure, and the formulae deduced can be applied, if the effective number N_e is substituted for the actual number N. When the breeding structure is known, the effective number can be derived from the actual number, and the relationships between the two are given below for the most common departures from the idealized breeding structure. The exact expressions are often complicated, but in most circumstances a simple approximation can be used with sufficient accuracy. It is important to note that in these relationships the actual number N is the number of breeding individuals, and it therefore cannot be obtained from a census, unless the different age-groups are distinguished. Knowing the effective population size N_e for any breeding structure, one can then obtain the rate of inbreeding as

$$\Delta F = \frac{1}{2N_e} \qquad \qquad \ldots [4.1]$$

and from ΔF any of the consequences of inbreeding can be calculated by the formulae of the previous chapter.

Exclusion of closely related matings

In bisexual organisms self-fertilization is, of course, impossible. Sib-mating is also excluded in man, and is often deliberately avoided in the maintenance of populations of laboratory and domesticated animals. The exclusion of closely related matings, however, does not make a great deal of difference to the rate of inbreeding, for the following reason. The progeny of a closely related mating have a higher coefficient of inbreeding than those of less closely related matings. Their presence therefore raises the average coefficient of inbreeding of the population at any time. But their higher inbreeding is not permanent: mating at random, they themselves are likely to mate with less closely related individuals, and so their higher-than-average inbreeding is not passed on to their progeny. Thus the exclusion of closely related matings reduces the average coefficient of inbreeding throughout, but it does not much affect the rate at which the inbreeding accumulates. The effect of the exclusions can be quantified approximately in the effective number as follows (Wright, 1969 p. 212).

With self-fertilization excluded,

$$N_e = N + \tfrac{1}{2} \qquad \text{(approx.)} \qquad \qquad \ldots [4.2a]$$

and so, by equation [4.1],

$$\Delta F = 1/(2N + 1) \qquad \text{(approx.)} \qquad \qquad \ldots [4.2b]$$

With sib-mating also excluded,

$$N_e = N + 2 \qquad \text{(approx.)} \qquad \qquad \ldots [4.3a]$$

and $\qquad\qquad\qquad \Delta F = 1/(2N + 4) \qquad \text{(approx.)} \qquad \qquad \ldots [4.3b]$

The approximations introduce very little error in calculating ΔF unless N is very small, as with close inbreeding; but then other methods of deducing ΔF are required, as will be explained in the next chapter.

Different numbers of males and females

In domestic and laboratory animals the sexes are often unequally represented among the breeding individuals, since it is more economical, when possible, to use fewer males than females. The two sexes, however, whatever their relative numbers, contribute equally to the genes in the next generation. Therefore the sampling variance attributable to the two sexes must be reckoned separately. Since the sampling variance is proportional to the reciprocal of the number, the effective number is twice the harmonic mean of the numbers of the two sexes. It is twice the harmonic mean because the population size is $N = N_m + N_f$, where N_m and N_f are the numbers of males and females respectively. The harmonic mean is $1/[\frac{1}{2}(1/N_m + 1/N_f)]$, so

$$\frac{1}{N_e} = \frac{1}{4N_m} + \frac{1}{4N_f} \qquad \text{(approx.)}$$

$$N_e = \frac{4N_m N_f}{N_m + N_f} \qquad \text{(approx.)} \qquad \dots [4.4]$$

The rate of inbreeding is then

$$\Delta F = \frac{1}{8N_m} + \frac{1}{8N_f} \qquad \text{(approx.)} \qquad \dots [4.5]$$

This gives a close enough approximation unless both N_m and N_f are very small, as with close inbreeding. It should be noted that the rate of inbreeding depends chiefly on the numbers of the less numerous sex. For example, if a population were maintained with an indefinitely large number of females but only one male in each generation, the effective number would be only about 4.

Unequal numbers in successive generations

The rate of inbreeding in any one generation is given, as before, by $1/2N$. If the numbers are not constant from generation to generation, then the mean rate of inbreeding is the mean value of $1/2N$ in successive generations. The effective number is the harmonic mean of the numbers in each generation. Over a period of t generations, therefore,

$$\frac{1}{N_e} = \frac{1}{t}\left[\frac{1}{N_1} + \frac{1}{N_2} + \frac{1}{N_3} + \dots + \frac{1}{N_t}\right] \qquad \text{(approx.)} \qquad \dots [4.6]$$

Thus the generations with the smallest numbers have the most effect. The reason for this can be seen by consideration of the new and old inbreeding referred to in connection with equation [3.6]. An expansion in numbers does not affect the previous inbreeding; it merely reduces the amount of new inbreeding. So, in a population with fluctuating numbers, the inbreeding proceeds by steps of varying amount, and the present size of the population indicates only the present rate of inbreeding.

Non-random distribution of family size

This is the most important deviation from the breeding system of the idealized population. Its consequence is usually to render the effective number less than the actual, but in special circumstances it makes it greater. Family size means here the number of progeny of an individual that become breeding individuals in the next generation. It will be remembered that in the idealized population each breeding individual has an equal probability of contributing genes, or progeny, to the next generation. The contribution of progeny is randomly distributed among the parents, and family sizes vary. In real populations the parents seldom have an equal chance of contributing progeny because they differ in fertility and in the survival of their progeny. This variation among parents leads to a greater variation of family size, and this has the consequence that a greater proportion of the next generation come from a smaller number of parents. The effective number is thus reduced. Conversely, the variation of family size may, by special breeding methods, be reduced below the random amount, with a consequent increase of the effective number. The relation of effective number to variation of family size is, briefly, as follows.

Attention will be restricted here to populations of constant size and with males and females in equal numbers. The mean family size $\bar{k}$ of all individuals, whether male or female, must then be 2 because to replace the population each individual must on average have 1 male and 1 female offspring represented among the parents of the next generation. Random variation of family size, as in the idealized population, gives rise to a binomial distribution which, unless N is very small, differs little from a Poisson distribution. A Poisson distribution has a variance equal to the mean so the variance of family size when parents have an equal chance of contributing to the next generation is $V_k = \bar{k} = 2$. When parents do not have an equal chance of contributing to the next generation, through differences of fertility or other reasons, the variance of family size is greater than 2. The way in which the variance of family size influences the effective number can be deduced by consideration of the probability of a zygote being an identical homozygote, in a manner similar to that by which the inbreeding increment was deduced in the last chapter. The effective number is then obtained from the rate of inbreeding. The relationship to which this leads is approximately

$$N_e = \frac{4N}{V_k + 2} \qquad \text{(approx.)} \qquad \qquad \dots [4.7]$$

This reduces to $N_e = N$ for the idealized population in which $V_k = 2$. The relationship in equation [4.7] refers to monogamous mating, when V_k is the same for both sexes. If males can mate with more than one female, V_k is likely to be different for males and females. The effective number is then given by

$$N_e = \frac{8N}{V_{km} + V_{kf} + 4} \qquad \text{(approx.)} \qquad \qquad \dots [4.8]$$

where V_{km} and V_{kf} are the variances of family sizes of males and females respectively (Hill, 1979).

Variation in family size above the random amount, due to differences of reproductive success, is the most important cause of N_e being less than N, having a much larger effect than the other departures from the idealized population. There is some information on how much the effective number is reduced by this cause. The variance of family size V_k has been estimated in a few species (see Crow and Kimura, 1970), from which the ratio N_e/N can be calculated, N being the number of breeding individuals, not the total census number. From the family sizes of women, N_e/N ranged from 0.69 to 0.94 in four sets of data; in the snail *Lymnea* it was 0.75; in *Drosophila* females it was 0.71 and in males 0.48.

The reduction of N_e from all causes can be estimated from the observed effect of inbreeding on the variance of gene frequency among lines. The ratio of N_e/N has been estimated in this way for *Drosophila melanogaster* with the sexes equal in actual numbers. The estimates for the sexes jointly from five experiments ranged from 0.56 to 0.83 (Kerr and Wright, 1954a, b; Wright and Kerr, 1954; Buri, 1956).

Minimal inbreeding

It is often desirable to keep stocks of laboratory animals with the least possible inbreeding. Increasing the number of breeding individuals N as much as possible is not the only thing that can be done. By choice of the individuals to be used as parents, the variance of family size, V_k, can be reduced below its random amount, and the effective number consequently increased. If the individuals are chosen equally from all families, then there is no variation in family size, and $V_k = 0$. Substitution into equation [4.7] shows that the effective number then becomes $N_e = 2N$, approximately. The exact relationship is

$$N_e = 2N - 1 \qquad \qquad \ldots [4.9]$$

which is very nearly twice what it would be in an idealized population of the same size. Equation [4.9] refers to a population bred from equal numbers of males and females. Equalization of family size then means choosing two individuals from the progeny of each pair of parents.

If the sexes are unequal in numbers, the variance of family size can be made zero by choosing as parents one male from each sire's progeny and one female from each dam's progeny. The rate of inbreeding is then given by the following formula (Gowe, Robertson, and Latter, 1959):

$$\Delta F = \frac{3}{32N_m} + \frac{1}{32N_f} \qquad \ldots [4.10]$$

where N_m and N_f are the actual numbers of male and female parents respectively, and females are more numerous than males.

The avoidance of matings between close relatives, such as sibs or cousins, seems at first sight to be an easy way of reducing the rate of inbreeding. This delays the first increment of inbreeding, but very little reduction of the subsequent rate of inbreeding is achieved. The reasons for this were explained earlier and equation [4.3a] gives the effective population size with self-fertilization and sib-mating excluded. If family size is deliberately equalized then the avoidance of closely related matings achieves no further reduction in the rate of inbreeding (Robinson and Bray, 1965). The chief advantages of avoiding matings between close relatives are to make the rate of inbreeding more constant from generation to generation, and to make the inbreeding coefficients of individuals more uniform within generations.

Overlapping generations

In most natural populations, and in domesticated animals, the generations are not discrete but are overlapping. This means that the individuals present at any time are of different ages and at different stages of their life-cycles. Furthermore, individuals differ in length of life and consequently in their opportunities for reproduction. Differences of lifetime therefore add to differences of fertility in increasing the variance of family size, the longer-lived individuals having a greater chance of contributing offspring to the next generation than the shorter-lived. The effect on N_e is dealt with by equations [4.7] or [4.8]. There is, however, a problem in finding what is the total number per generation, i.e., N in equation [4.7]. Provided the population has a stable age-structure the total number per generation can be found as follows. We need to know the number of individuals born within a specified time-interval, which might be one year or any convenient period. This number is the size of the cohort defined by the time-interval. The cohort size N_c is related to the total number alive at any time, N_T, i.e., the census count, by $N_c = N_T/E$, where E is the expectation of life, or the mean age at death, expressed in units of the specified time-interval that defines the cohort (see Emigh and Pollak, 1979). We need to know also the generation length L in units of the specified time-interval, the generation length being the average age of parents at the birth of their offspring. Then the total number per generation is $N = N_c L$, and the effective number per generation is given approximately by

$$N_e = \frac{4N_c L}{V_k + 2} \quad \text{(approx.)} \qquad \qquad \dots [4.11]$$

(Hill, 1979), where V_k is the variance of family size from all causes. If males and females differ in numbers or in generation length, as is often the case with farm animals, equation [4.11] has to be modified in a manner explained by Hill (1979).

The effective number in the human population of the USA has been estimated as $N_e = 0.41 N_T$, but the ratio is probably somewhat lower than this because the estimate did not take account of all the possible sources of variation of fertility (Emigh and Pollak, 1979).

Example 4.1 Data from a mouse experiment (Garnett and Falconer, 1975, and unpublished) will serve to illustrate the use of several of the formulae deduced in this and the previous chapter. Furthermore, by calculating the effective population size independently from the variance and the inbreeding approaches, we can check on the validity of the theory. The population consisted of 18 lines, all originating from the same random-bred base and all maintained by minimal inbreeding with 8 pairs of parents mated in every generation (Falconer, 1973). The data consist of gene and genotype frequencies at 5 polymorphic enzyme loci in each of the lines. The enzyme loci are listed in the table. There were two alleles present at all the loci; all the heterozygotes were distinguishable and the gene frequencies were obtained by counting (equation [1.1]). At generation 27 all the parents were typed, so the gene frequencies at that time were determined without error. For each locus, the variance of gene frequency among the 18 lines was calculated. The table gives, for each locus, the mean gene frequency, $\bar{q}$, the variance of gene frequency, σ_q^2, and the overall frequency of heterozygotes in the population as a whole, H. There is no reason to think that the gene frequencies had changed from their values in the base population, so for the calculations it is assumed that $q_0 = \bar{q}$.

The calculations to be made are: (1) the effective population size N_e, expected from the number of parents N, and the breeding structure; (2) the inbreeding coefficient F, from the variance of gene frequencies σ_q^2, and then N_e from F at generation $t = 27$; (3) F at $t = 27$ from the frequency of heterozygotes H, and then N_e from F again.

1. With 8 pairs of parents, $N = 16$. With minimal inbreeding ($V_k = 0$), equation [4.9] gives $N_e = 2N - 1 = 31$. Equation [4.1] gives $\Delta F = 1/2N_e = 0.0161$, and equation [3.12] gives $F = 1 - (1 - 0.0161)^{27} = 0.355$. These expected values will be realized only if $V_k = 0$ is achieved. In practice some pairs will inevitably be sterile, so V_k will not be zero and N_e will be less than 31.

2. F is related to σ_q^2 by equation [3.14]. Taking *Dip-1* as an example, $F = \sigma_q^2/\bar{p}\bar{q} = 0.077/(0.236 \times 0.764) = 0.427$. Each locus gives an independent estimate of F. They are given in the table in the column headed $F(2)$. The mean is $F = 0.378$. From this mean estimate of F, we get the rate of inbreeding ΔF from equation [3.12]. By rearrangement, $(1 - \Delta F)^t = 1 - F_t$, which with $t = 27$ yields $\Delta F = 0.0174$. The effective population size N_e is found from ΔF by equation [4.1]. This gives $N_e = 1/2\Delta F = 28.7$.

3. Equal numbers of individuals were classified in all lines, so F can be estimated from the overall frequency of heterozygotes, H. This could be done by equation [3.15], but the number of parents per line is small enough to make the expected chance differences of gene frequencies in males and females not negligible. Allowance for this is made in equation [3.17], which gives $(1 - F) = H/2pq(1 + 1/2N)$. Taking *Dip-1* again as an example, $(1 - F) = 0.240/(2 \times 0.236 \times 0.764 \times 1.03125) = 0.646$, and $F = 0.354$. Again, each locus gives an independent estimate of F, as given in the column headed $F(3)$. ΔF and N_e are calculated from F in the same way as under calculation 2 above, and the values are entered at the foot of the table.

Locus	$\bar{q}$	σ_q^2	H	$F(2)$	$F(3)$
Dip-1	0.764	0.077	0.240	0.427	0.355
Id-1	0.370	0.102	0.301	0.438	0.374
Gpi-1	0.297	0.072	0.283	0.345	0.343
Gpd-1	0.215	0.042	0.253	0.249	0.273
Got-2	0.141	0.052	0.134	0.429	0.464
Mean				0.378	0.362

Method	F	ΔF	N_e
1. Expected from breeding structure	0.355	0.0161	31
2. Variance of gene frequency, σ_q^2	0.378	0.0174	28.7
3. Frequency of heterozygotes, H	0.362	0.0165	30.3
4. Pedigrees	0.379	0.0175	28.6

The inbreeding coefficient can be calculated in yet another way – from the pedigree records, in a manner to be explained in the next chapter. This is an exact determination because it is based on the probabilities of identical homozygotes arising from the matings actually made. The calculation was made for each line, and the mean value was $F = 0.379$. This gives $\Delta F = 0.0175$ and $N_e = 28.6$. The ratio of N_e (achieved)/N_e (expected) is $28.6/31 = 0.92$, and the ratio of N_e (achieved)/N is $28.6/16 = 1.79$. Comparing the three estimates of N_e shown at the foot of the table, we see that the estimates from the variance of gene frequencies and from the frequency of heterozygotes agree very well with the pedigrees. It may be noted that if the correction for unequal gene frequencies in male and female parents is not made, the estimate of N_e from the frequency of heterozygotes is 32.7 instead of 30.3.

Mutation, migration, and selection

The description of the dispersive process given so far in this chapter and the previous one is conditional on the systematic processes of mutation, migration, and selection being absent, and its relevance to real populations is therefore limited. So let us now consider the effects of the dispersive and systematic processes when acting jointly. The systematic processes, as we have seen in Chapter 2, tend to bring the gene frequencies to stable equilibria at particular values which would be the same for all populations under the same conditions. The dispersive process, in contrast, tends to scatter the gene frequencies away from these equilibrium values, and if not held in check by the systematic processes it would in the end lead to all genes being either fixed or lost in all populations not infinite in size. The tendency of the systematic processes to change the gene frequency toward its equilibrium value becomes stronger as the frequency deviates further from this value. For this reason the opposing tendencies of the dispersive and systematic processes reach a point of balance: a point at which the dispersion of the gene frequencies is held in check by the systematic processes. When this point of balance is reached there will be a certain degree of differentiation between sub-populations, but it will neither increase nor decrease so long as the conditions remain unchanged. The problem is therefore to find the distribution of gene frequencies among the lines of a subdivided population when this steady state has been reached. The solution is complicated mathematically, and we shall give only the main conclusions, explaining their meaning but not their derivation.

Non-recurrent neutral mutation

We shall first consider briefly the fate of unique mutations that are neutral with respect to fitness, which were discussed in Chapter 2 as a possible source of polymorphism. For further details of neutral mutations as a cause of genetic change, see Kimura and Ohta (1971), Nei (1975), and Kimura (1979). What is the chance that an allelic substitution will occur at any particular locus by the process of mutation and random drift? An 'allelic substitution' means that the allele or alleles present now are all replaced by a new one at some time in the future. If there were no mutation and no selection, all the alleles at any locus present in the population would be identical by descent. In other words, if the complete pedigree of the population could be traced back into the remote past, the present population would be found to be completely inbred. Looking forward from the present to the future, one of the alleles present now would, in the absence of mutation and selection, eventually become fixed in the whole population. The number of representatives of each autosomal locus present at any time is $2N$, where N is the actual population size, and one of these will, in the absence of selection, eventually become fixed. Therefore the chance that any particular one becomes fixed is $1/2N$. Let u be the neutral mutation rate at the locus in question; i.e., the probability that a new neutral allele appears by mutation in any one generation. The total number of new mutants at the locus is then $2Nu$, assuming that each new mutant is initially present in only one copy. For each mutant separately, the chance of fixation is $1/2N$. Therefore the probability that one or another of the new mutants becomes fixed is $2Nu \times 1/2N = u$. This is the probability of an allelic substitution at a locus occurring in any particular generation, and it is simply equal to the mutation rate per generation at the locus in question. The reason why it is independent of the population size is that in a larger population the larger number of new mutants is balanced by the smaller individual chance of survival.

Selection, of course, increases or decreases the chance of fixation, according to whether the new mutant is favourable or unfavourable. The great majority of mutants are expected to be deleterious rather than beneficial. What is the chance that a deleterious mutant gives rise to an allelic substitution? By the same reasoning, this is equal to the mutation rate to 'effectively neutral' alleles; and, according to Kimura (1979), an 'effectively neutral' allele is one with a coefficient of selection s against it in the range from $s = 0$ (i.e. strictly neutral) up to $s = 1/2N_e$. Thus effective neutrality depends on the effective population size, and an effectively neutral allele is one for which the product $N_e s$ is less than $\frac{1}{2}$.

The number of allelic substitutions that take place depends on the number of loci that can mutate to effectively neutral alleles. If, for example, there were 10,000 such loci, each with a mutation rate of 10^{-5} per generation, the number of substitutions would be 0.1 per generation, or 1 per 10 generations. The rate at which two sub-populations would differentiate is twice as much because substitutions can occur in both populations. In other words, they would differ by one allelic substitution for every 10 generations that they are separated, counting generations down both lines of descent since their separation.

Recurrent mutation and migration

Recurrent mutation and migration can be dealt with together because they change the gene frequency in the same manner. Consider again a population subdivided into many lines, all with an effective size N_e; and let a proportion m of the breeding individuals of every generation in each line be immigrants coming at random from all other lines. Let u and v be the mutation rates in the two directions between two alleles at a locus. The state of dispersion between the sub-populations, when the balance between dispersion on the one hand and migration and mutation on the other is reached, can be expressed as the inbreeding coefficient as follows:

$$F = \frac{1}{4N_e(u + v + m) + 1} \qquad \text{(approx.)} \qquad \ldots [4.12]$$

If the mean gene frequencies were known, the state of dispersion could be expressed as the variance of gene frequency by putting $\sigma_q^2 = F\bar{p}\bar{q}$, from equation [3.14].

The theoretical distributions of gene frequencies corresponding to four equilibrium values of F are shown in Fig. 4.1. These distributions are similar in general form to the distributions that a population goes through during the process of inbreeding without mutation or migration, shown in Fig. 3.5. The effect of mutation or migration can be thought of as arresting the process at a point corresponding to some value of F or of σ_q^2, the variance of gene frequency among sub-populations. The chief difference here is that if F goes beyond 0.33, when all gene frequencies, including fixation, are equally probable, the distribution becomes U-shaped, with more sub-populations being at the extremes and fewer at intermediate gene frequencies. The reason for this is that, with mutation or migration, fixation in any one line is not permanent.

In order to see what equation [4.12] and the distributions in Fig. 4.1 mean, let us consider these questions: at what value of F will the population stabilize if there is mutation at the known rates, but no migration? and: how much migration, with no mutation, would be needed to produce the distributions shown? For this purpose it will be accurate enough to take equation [4.12]

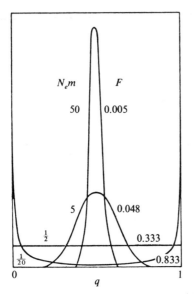

Fig. 4.1. Theoretical distributions of gene frequency among sub-populations, when dispersion is balanced against mutation or migration, and the mean gene frequency is 0.5. The vertical axis is the probability, as in Fig. 3.5. The states of dispersion to which the curves refer are indicated by the values of F in the figure. The values of $N_e m$ are the numbers of immigrants per generation, as explained in the text. (*Based on Wright, 1951.*)

as $N_e = 1/4F(u + v + m)$. Substitute in this the usual mutation rate of $u + v = 10^{-5}$, with $m = 0$, and $F = 0.005$ corresponding to the least dispersed distribution in Fig. 4.1. This gives $N_e = 5 \times 10^6$, which means that sub-populations of size 5 million would differentiate as far as $F = 0.005$ before being stabilized by mutation. Smaller sub-population would differentiate further. For example, the uniform distribution corresponding to $F = 0.333$ would be reached by sub-populations of size 75,000. For many species, perhaps most, even this is an unrealistically large size for sub-populations mating at random within themselves. The conclusion, therefore, is that recurrent mutation is negligible as a factor slowing down or arresting the differentiation of sub-populations by random drift. Migration, however, is quite a different matter. Rearrangement of equation [4.12], with $u + v = 0$, gives $F = 1/(4N_e m + 1)$. Thus the state of dispersion depends on the number of immigrants per generation, which is $N_e m$, irrespective of the population size. This conclusion, which may at first seem paradoxical, can be understood by noting that a smaller population needs a higher rate of immigration than a larger one to be held at the same state of dispersion. Substitution of the values of F corresponding to the distributions in Fig. 4.1 gives the values of $N_e m$ entered in the figure. For example, one immigrant every alternate generation ($N_e m = \frac{1}{2}$) is sufficient to maintain the flat distribution of gene frequencies corresponding to $F = 0.333$. The conclusion is that quite small numbers of immigrants will prevent much differentiation by random drift. The reason why mutation and migration are so different in their effects is the same as was pointed out in Chapter 2: realistic mutation rates are very much smaller than realistic migration rates.

The situation to which the foregoing consideration of migration refers is known as the 'island model'. It pictures a discontinuous population with immigrants to any sub-population coming from any other sub-population with equal probability. A more realistic model is the 'neighbourhood model' or 'isolation by distance'. The population is pictured as being continuously distributed over the area inhabited, but subdivided into 'neighbourhoods' by the limited distance that individuals travel between birth and reproduction. A neighbourhood is the area within which mating is effectively random, and corresponds to a sub-population. Gene frequencies, however, vary continuously from neighbourhood to neighbourhood across the area. Since immigrants to a neighbourhood come from close by more often than from further away, they differ in gene frequency less than immigrants in the island model do. Therefore migration is less effective in counteracting random drift. The conclusion to which the neighbourhood model leads is that a large amount of local differentiation will take place if the effective number in the neighbourhoods is of the order of 20, a moderate amount if it is of the order of 200, but a negligible amount if it is larger than about 1,000.

Selection

Selection operating on a locus in a large population brings the gene frequency to an equilibrium, at an intermediate value when selection favours heterozygotes and at a low value when selection is balanced against mutation. The dispersive process tends to shift the gene frequency away from its equilibrium value. This reduces the average fitness of the population, because the load is minimal at the equilibrium, and some sub-populations may even become fixed for the deleterious allele. The effect of selection is stronger the further the gene frequency is away from the equilibrium value. So the opposing forces of selection and random drift reach a balance at which there is a stable distribution of gene frequencies among sub-populations. The question then is: how small must the sub-populations be to cause appreciable differentiation with its consequent deviations from the optimal gene frequency? The following illustrative cases will have to suffice for an answer, and for an understanding of the joint effects of mutation, selection, and dispersion the reader must consult other sources.

Consider first selection favouring heterozygotes. The effect depends on the equilibrium gene frequency. When the two homozygotes are at an equal disadvantage and the equilibrium gene frequency is consequently 0.5, the distributions of gene frequencies look roughly like those in Fig. 4.1. The least dispersed one, corresponding to $F = 0.005$, would be attained by a selection coefficient of $s = 0.1$ against both homozygotes in sub-populations of size $N_e = 1,000$. More dispersion would need less selection or smaller populations. The most dispersed distribution, with a substantial amount of fixation, would need very roughly $s = 0.01$ with $N_e = 100$. Thus selection for heterozygotes does not allow much random drift unless the selection is very weak (around 1 per cent) or the population size very small (around 100). If, however, the equilibrium gene frequency is not 0.5, the selection is less effective in preventing the random drift. When the equilibrium gene frequency is above roughly 0.8 or below 0.2 the selection actually accelerates the random drift (Robertson, 1962). The reason for

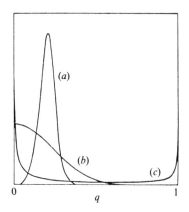

Fig. 4.2. Theoretical distributions of gene frequency among sub-populations when the dispersion is balanced by mutation and selection. The graphs refer to a recessive gene with $u = v = \frac{1}{20}s$, in populations of size: (a) $N_e = 50/s$, (b) $N_e = 5/s$, and (c) $N_e = 0.5/s$. (*Based on Wright, 1942.*)

this is that one homozygote is then much fitter than the other and the selection increases the probability of fixation of the more fit homozygote.

Next consider selection against a recessive allele balanced by recurrent mutation. This is difficult to illustrate because with realistic values of the selection coefficient, the equilibrium gene frequency will be very low, and the distributions are squeezed up against the limit near $q = 0$. Figure 4.2 shows three stable distributions for very weak selection with an equilibrium gene frequency of about $q = 0.2$. Mutation rate is taken to be the same in both directions, and the coefficient of selection s is 20 times the mutation rate. If we assume a mutation rate of 10^{-5}, then $s = 20 \times 10^{-5}$ and the population sizes to which the distributions refer are (a) 250,000, (b) 25,000, and (c) 2,500. The conclusion is, again, that selection does not allow much random drift unless the selection is weak or the population size very small. The amount of random drift depends approximately on the product of the population size and the selection coefficient $N_e s$, and if we are content to be very imprecise we can say that a substantial amount of random drift occurs only if $N_e s$ is less than about 1/4.

Example 4.2. The opposing forces of dispersion and selection are illustrated in Fig. 4.3, from an experiment with *Drosophila melanogaster* (Wright and Kerr, 1954). The frequency of the sex-linked gene 'Bar' was followed for 10 generations in 108 lines each maintained by 4 pairs of parents. (On account of the complication of sex-linkage, which increases the rate of dispersion, the theoretical effective number was 6.765: the effective number as judged from the actual rate of dispersion was $N_e = 4.87$.) The initial gene frequency was 0.5. The circles in the figure show the distribution of the gene frequency among the lines in the fourth to tenth generations, when the distribution had reached its steady form. The smooth curve shows the theoretical distribution based on $N_e = 5$ and a coefficient of selection against Bar of $s = 0.17$. Previously fixed lines are not included in the distributions. Altogether, at the tenth generation, 95 of the 108 lines had become fixed for the wild-type allele and 3 for Bar while 10 remained unfixed. Thus, despite a 17 per cent selective disadvantage, the deleterious allele was fixed in about 3 per cent of the lines.

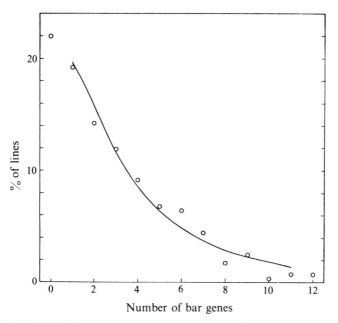

Fig. 4.3. Distribution of gene frequencies under inbreeding and selection, as explained in Example 4.2. (*Data from Wright and Kerr, 1954.*)

Random drift in natural populations

Having described the dispersive process and its theoretical consequences, we may now turn to the more practical question of how far these consequences are actually seen in natural populations. The answering of this question is beset with difficulties, and the following comments are intended more to indicate the nature of these difficulties than to answer the question.

The theory of small populations, outlined in this and the preceding chapter, is essentially mathematical in nature and is unquestionably valid: given only the Mendelian mechanism of inheritance, the conclusions arrived at are a necessary consequence under the conditions specified. The question at issue, then, is whether the conditions in natural populations are often such as would allow the dispersion of gene frequencies to become detectable. The phenomena which would be expected to result from the dispersive process, if the conditions were appropriate, are differentiation between the inhabitants of different localities, and differences between successive generations. Both these phenomena are well known in subdivided or small isolated populations, and it is tempting to conclude that because they are the expected consequences of random drift, random drift must be their cause. But there are other possible causes: the environmental conditions probably differ from one locality to another and from one season to another; so the intensity, or even the direction, of selection may well vary from place to place and from year to year, and the differences observed could equally well be attributed to variation of the selection pressure. Before we can justifiably attribute these phenomena to random drift, therefore, we have to know: (1) that the effective population size is small enough; (2) that the sub-populations are well enough isolated (or the size of the 'neighbourhoods' sufficiently small); and (3) that the genes concerned are subject to very little selection.

The estimation of the present size of a population, though not technically easy, presents no difficulties of principle. But the present state of differentiation depends on the population size in the past, and this can generally only be guessed at. It is difficult to know how often the population may have been drastically reduced in size in unfavourable seasons, and the dispersion taking place in these generations of lowest numbers is permanent and cumulative. If a species colonizes a new territory, the founding members of the new sub-population may be very few in numbers, causing a substantial amount of random drift in the first generation. This is called the *founder effect*. If the sub-population then expands, its difference from the main population may seem much too great to be consistent with its present numbers. To attribute the difference to a founder effect may often be plausible but, in the absence of pedigree records, can seldom be other than a guess. One of the clearest and most interesting examples of isolated populations being differentiated as a result of founder effects is seen in the Amish communities in the USA, studied by McKusick (1978), the founder effects being established by genealogical records.

There is less difficulty in deciding whether the sub-populations are sufficiently well isolated. With a discontinuous population it is often possible to be reasonably sure that there is not too much immigration; and with a continuous population the size of the 'neighbourhoods' is, at least in principle, measurable. The greatest difficulty lies in estimating the intensity of natural selection acting on the genes concerned. Selection of an intensity far lower than could be detected experimentally is sufficient to check dispersion in all but the smallest populations. It seems rather unlikely – though this is no more than an opinion – that any gene that modifies the phenotype enough to be recognized visually would have so little effect on fitness. Many of the genes concerned with enzyme polymorphisms, however, may have selection coefficients low enough to allow populations to become differentiated. The genes concerned with quantitative differences may also be nearly enough neutral for random drift to take place. There is no doubt at all that genes of this sort do show random drift in laboratory populations, as will be shown in later chapters.

5 SMALL POPULATIONS: III. Pedigreed Populations and Close Inbreeding

In the two preceding chapters the genetic properties of small populations were described by reference to the effective number of breeding individuals; and expressions were derived, in terms of the effective number, by means of which the state of dispersion of the gene frequencies could be expressed as the coefficient of inbreeding. The coefficient of inbreeding, which is the probability of any individual being an identical homozygote, was deduced from the population size and the specified breeding structure. It expressed, therefore, the average inbreeding coefficient of all individuals of a generation. When pedigrees of the individuals are known, however, the coefficient of inbreeding can be more conveniently deduced directly from the pedigrees, instead of indirectly from the population size. This method has several advantages in practice. Knowledge is often required of the inbreeding coefficient of individuals, rather than of the generation as a whole, and this is what the calculation from pedigrees yields. In domestic animals, some individuals often appear as parents in two or more generations, and this overlapping of generations causes no trouble when the pedigrees are known. The first topic for consideration in this chapter is therefore the computation of inbreeding coefficients from pedigrees. The second topic concerns regular systems of close inbreeding. When self-fertilization is excluded, the rate of inbreeding expressed in terms of the population size is only an approximation, and the approximation is not close enough if the population size is very small. Under systems of close inbreeding, therefore, the rate of inbreeding must be deduced differently, and this is best done also by consideration of the pedigrees.

When the coefficient of inbreeding is deduced from the pedigrees of real populations, it does not necessarily describe the state of dispersion of the gene frequencies. It is essentially a statement about the pedigree relationships, and its correspondence with the state of dispersion is dependent on the absence of the processes that counteract dispersion, in particular on selection being negligible. We were able to use the coefficient of inbreeding as a measure of dispersion in the preceding chapters because the necessary conditions for its relationship with the variance of gene frequencies were specified.

Pedigreed populations

The inbreeding coefficient of an individual

This coefficient is the probability that the pair of alleles carried by the gametes that produced it were identical by descent. Computation of the inbreeding coefficient therefore requires no more than the tracing of the pedigree back to common ancestors of the parents and computing the probabilities at each segregation. Consider the simple pedigree in Fig. 5.1, representing a mating between half sibs. X is the individual whose inbreeding coefficient F_X we want to know. Its parents P and Q are related through their common parent A. They are not related in any other way, so we only have to consider the transmission of A's genes through P and Q to X, and to calculate the probability of X being an identical homozygote. Let A_1 and A_2 symbolize the genes carried by A at any particular locus. The probability that X is A_1A_1 is $1/16 = (\frac{1}{2})^4$ because the chance that A_1 is transmitted through each of the four paths AP, PX, AQ, QX, is $\frac{1}{2}$ for each path. The probability that X is A_2A_2 is similarly $(\frac{1}{2})^4$, and the probability that X is either A_1A_1 or A_2A_2 is $2(\frac{1}{2})^4 = (\frac{1}{2})^3 = \frac{1}{8}$. This probability of X being an identical homozygote represents the new inbreeding arising from A as a common ancestor of P and Q. The common ancestor A may, however, itself be an identical homozygote through previous inbreeding, in which case X will be an identical homozygote also if it gets the genotype A_1A_2 or A_2A_1 (the two being distinguished according to whether A_1 comes through P or through Q). The probability of each of these genotypes is $(\frac{1}{2})^4$ for the same reason as before, and the probability of one or the other is $(\frac{1}{2})^3$. The probability of A being an identical homozygote is its inbreeding coefficient, F_A. The additional probability of X being an identical homozygote through the previous inbreeding is then $(\frac{1}{2})^3 F_A$. Putting the two parts of the inbreeding together gives the inbreeding coefficient of X as $F_X = (\frac{1}{2})^3 + (\frac{1}{2})^3 F_A = (\frac{1}{2})^3(1 + F_A)$. Note that the index 3 is the number of individuals in the path connecting the parents through their common ancestor, i.e., individuals P, A, and Q. This makes it easy to work out the probabilities

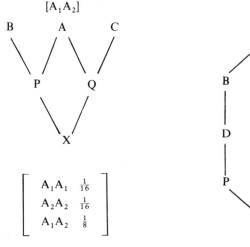

$$\begin{bmatrix} A_1A_1 & \frac{1}{16} \\ A_2A_2 & \frac{1}{16} \\ A_1A_2 & \frac{1}{8} \end{bmatrix}$$

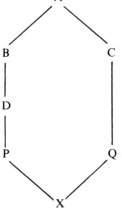

Fig. 5.1   Fig. 5.2

simply by counting individuals in the path. In Fig. 5.1 there are only the parents and the common ancestor; in Fig. 5.2 the common ancestor is further back and the individuals to be counted are P, D, B, A, C, Q, making 6. F_X in Fig. 5.2 is therefore $(\frac{1}{2})^6(1 + F_A)$. In more complicated pedigrees, the parents may be related to each other through more than one common ancestor, or from the same common ancestor through different paths, as illustrated in Example 5.1. Each common ancestor, and each path, then contributes an additional probability of the progeny being an identical homozygote, and the inbreeding coefficient is obtained by adding together the separate probabilities for each of the paths through which the parents are related.

Putting all this together gives the following general formula for the inbreeding coefficient of an individual:

$$F_X = \Sigma\,(\tfrac{1}{2})^n(1 + F_A) \qquad\qquad \ldots [5.1]$$

where n is the number of individuals in any path of relationship counting the parents of X, the common ancestor, and all individuals in the path connecting parents to common ancestor; summation is over all paths of relationship. When inbreeding coefficients are calculated in this way, it is necessary to define the base population to which the present inbreeding is referred. Individuals in the base population are assigned inbreeding coefficients of zero. In practice the individuals of the base population may be simply those at the head of the pedigree, whose ancestry further back is not known.

Example 5.1. The pedigree in Fig. 5.3 will illustrate the use of the formula [5.1]. The individual whose inbreeding coefficient is to be calculated is X. We have to look for paths through which X's parents, P and Q, are related to each other. Paths contributing nothing to the relationship are dotted. It is assumed that there are no relationships between any of the individuals other than those shown. There are four

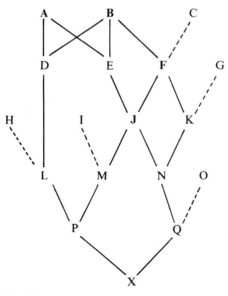

Fig. 5.3

individuals that are common ancestors, A, B, F, and J, causing relationship between P and Q. The paths of relationship and the calculation of F_X (rounded to four decimal places) are shown in the table. The inbreeding coefficient of X works out to be 0.0606. The following points should be noted: (1) D and E are full sibs. Their relationship causes some inbreeding in P, one of the parents of X, but it causes no relationship between P and Q and so contributes nothing to F_X. (2) E and F are half sibs, and the inbreeding coefficient of J, one of the common ancestors of P and Q, is therefore 1/8 as in Fig. 5.1. (3) There are four paths connecting P with Q through B as a common ancestor, and all four must be included in the calculation. (4) No individual can appear twice in the same path. For example, P M J E B F J N Q is not a valid path, because the inbreeding it produces is fully taken account of by the inbreeding coefficient of J in the shorter path P M J N Q. (5) Finally, care must be taken not to traverse paths in the wrong direction: for example, F cannot transmit genes to P through K and N.

Paths of relationship	n	F of common ancestor	Contribution to F_X
P L D A E J N Q	8	0	$(\frac{1}{2})^8 = 0.0039$
P L D B E J N Q	8	0	$(\frac{1}{2})^8 = 0.0039$
P L D B F J N Q	8	0	$(\frac{1}{2})^8 = 0.0039$
P L D B F K N Q	8	0	$(\frac{1}{2})^8 = 0.0039$
P M J E B F K N Q	9	0	$(\frac{1}{2})^9 = 0.0020$
P M J F K N Q	7	0	$(\frac{1}{2})^7 = 0.0078$
P M J N Q	5	$\frac{1}{8}$	$(\frac{1}{2})^5 \times \frac{9}{8} = 0.0352$
			$F_X = 0.0606$

When pedigrees are long and complicated, it may not be practicable to trace all the paths of relationship. A sufficiently accurate estimate of the inbreeding coefficient can, however, be got by sampling a limited number of paths (Wright and McPhee, 1925).

Coancestry or kinship

There is another method of computing inbreeding coefficients which is often more convenient and is more readily adapted to a variety of problems. It will be used in the next section to work out the inbreeding coefficients under regular systems of close inbreeding. Its chief uses in practice are for planning matings to give the least inbreeding, and for calculating the inbreeding coefficient generation by generation in a fully pedigreed population. The method does not differ in principle from the formula [5.1] given above, but instead of working from the present back to the common ancestors we work forward, keeping a running tally generation by generation, and compute the inbreeding that will result from the matings now being made. The inbreeding coefficient of an individual depends on the amount of common ancestry in its two parents. Therefore, instead of thinking about the inbreeding of the progeny, we can think of the degree of relationship by descent between the two parents. This is called the *coancestry*, or the *coefficient of*

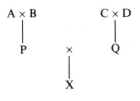

Fig. 5.4

kinship or of *consanguinity*. It will be symbolized by *f*. The coancestry of any two individuals is identical with the inbreeding coefficient of their progeny if they were mated. Thus the coancestry of two individuals is the probability that two gametes taken at random, one from each, carry alleles that are identical by descent.

Consider the generalized pedigree in Fig. 5.4. X is an individual with parents P and Q and grandparents A, B, C, and D. Now, the coancestry of P with Q is fully determined by the coancestries relating A and B with C and D, and if these are known we need go no further back in the pedigree. It can be shown that the coancestry of P with Q is simply the mean of the four coancestries AC, AD, BC, and BD. This will be clearer if stated in the form of probabilities, though the explanation is cumbersome when put into words. Take one gamete at random from P and one from Q, and repeat this many times. In half the cases, P's gamete will carry a gene from A and in half from B: similarly for Q's gamete. So the two gametes, one from P and one from Q, will carry genes from A and C in a quarter of the cases, from A and D in a quarter, from B and C in a quarter, and from B and D in a quarter of the cases. Now the probability that two gametes taken at random, one from A and the other from C, are identical by descent is the coancestry of A with C, i.e., f_{AC}, etc. So, reverting now to symbols,

$$f_{PQ} = \tfrac{1}{4}f_{AC} + \tfrac{1}{4}f_{AD} + \tfrac{1}{4}f_{BC} + \tfrac{1}{4}f_{BD}$$

This gives the basic rule relating coancestries in one generation with those in the next:

$$F_X = f_{PQ} = \tfrac{1}{4}(f_{AC} + f_{AD} + f_{BC} + f_{BD}) \qquad \ldots [5.2]$$

With this rule the experimenter can tabulate the coancestries generation by generation, and this gives a basis for planning matings and computing inbreeding coefficients. More detailed accounts of the operation are given by Plum (1954).

If there is overlapping of generations we may need to find the coancestry of individuals belonging to different generations, for which a supplementary rule is needed. Consideration of probabilities shows that the coancestry of two individuals is equivalent to the mean coancestry of one individual with the two parents of the other. Thus, referring to the same pedigree (Fig. 5.4), the rule giving the ancestry of P with C and with D is

$$\left.\begin{array}{l} f_{PC} = \tfrac{1}{2}(f_{AC} + f_{BC}) \\ f_{PD} = \tfrac{1}{2}(f_{AD} + f_{BD}) \end{array}\right\} \qquad \ldots [5.3]$$

This rule gives also

$$f_{PQ} = \tfrac{1}{2}(f_{PC} + f_{PD})$$

which by substitution from equation [5.3] reduces to the basic rule of equation [5.2].

Before we can apply the method to a pedigreed population, or to regular systems of inbreeding, we need to know the numerical values of some coancestries. The parents of the first generation have to be assumed to be all unrelated, with $f = 0$. The first non-zero coancestries are among their progeny, and when all these have been determined all subsequent generations can be calculated by the rules given above. The relationships whose coancestries may be needed in the first generation are offspring and parent, full sibs, half sibs, and self. The coancestries of these relationships are needed also in the next section for working out the consequences of continued inbreeding. The coancestries are as follows, starting with self because this appears in all the others.

Self. The coancestry of an individual with itself, f_{AA}, is the inbreeding coefficient of progeny that would be produced by self-mating. This is the probability that two gametes taken at random from A will carry identical alleles, which is $\frac{1}{2}(1 + F_A)$ for the following reason. Let A's genes be A_1 and A_2. The probability that two gametes taken at random are both A_1 or both A_2 is $\frac{1}{2}$. The probability that one is A_1 and the other A_2 is $\frac{1}{2}$, but then the probability that A_1 and A_2 are identical by descent is the inbreeding coefficient of A, F_A. Thus the total probability that the two gametes carry identical alleles is $\frac{1}{2} + \frac{1}{2}F_A$, and so

$$f_{AA} = \tfrac{1}{2}(1 + F_A) \qquad \qquad \dots [5.4]$$

If F_A is known (or assumed) to be zero, then $f_{AA} = \frac{1}{2}$.

Offspring and parent are in different generations, so the supplementary rule [5.3] is applicable. In Fig. 5.4 the coancestry of P with A is equal to the mean coancestry of P's parents with A, i.e.,

$$f_{PA} = \tfrac{1}{2}(f_{AB} + f_{AA}) \qquad \qquad \dots [5.5]$$

If it is known or assumed that A and B are not related and A is not inbred, then $f_{AB} = 0$, $f_{AA} = \frac{1}{2}$, and the coancestry reduces to $f_{PA} = \frac{1}{4}$.

Full sibs are in the same generation, so the basic rule [5.2] applies. The application of the rule is more easily understood if the pedigree is written as in Fig. 5.5. A and B are the parents of both P and Q, which are full sibs and have an

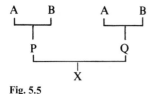

Fig. 5.5

offspring X. Applying the basic rule [5.2] and noting that $f_{AB} = f_{BA}$, we have

$$f_{PQ} = \tfrac{1}{4}(2f_{AB} + f_{AA} + f_{BB}) \qquad \qquad \dots [5.6]$$

With no previous inbreeding or relationship this reduces to $f_{PQ} = \frac{1}{4}$.

Half sibs. Figure 5.1 gives a pedigree of half sibs. Applying the basic rule

[5.2] and noting that A is a parent of both P and Q, gives

$$f_{PQ} = \tfrac{1}{4}(f_{AB} + f_{AC} + f_{BC} + f_{AA}) \qquad \ldots [5.7]$$

With no previous inbreeding or relationship this reduces to $f_{PQ} = 1/8$. This result has already been obtained as the inbreeding coefficient of X, the offspring of P and Q, in Fig. 5.1.

Regular systems of inbreeding

A regular system of inbreeding is one in which the same mating system is applied in all generations, and all individuals in the same generation have the same inbreeding coefficient. Regular systems are most often used to produce rapid inbreeding, and so the matings are between close relatives. We shall deal first with matings between the four sorts of relative already considered. (For other systems, see Wright, 1933, 1969). Then we shall deal with the inbreeding produced by backcrossing and in the generations following a cross.

Close inbreeding
The inbreeding coefficients in successive generations can be calculated from the coancestries given in equations [5.4] to [5.7]. But it is more convenient first to derive recurrence equations which relate the inbreeding coefficient in one generation to those of previous generations. The generation we are interested in is denoted by t, the previous one by $t - 1$, and the one before that by $t - 2$; $t - 3$ is as far back as we have to go with these four systems. The recurrence equations are derived as follows, and the inbreeding coefficients in successive generations are given in Table 5.1.

Self-fertilization. If X in generation t is the offspring of A in generation $t - 1$, equation [5.4] gives

$$F_X = f_{AA} = \tfrac{1}{2}(1 + F_A)$$

and the recurrence equation is therefore

$$F_t = \tfrac{1}{2}(1 + F_{t-1}) \qquad \ldots [5.8]$$

In the first generation the parents are non-inbred and $F_{t-1} = 0$, which makes $F_{(t=1)} = \tfrac{1}{2}$. In the second generation $F_{t-1} = \tfrac{1}{2}$, and $F_{(t=2)}$ becomes $\tfrac{1}{2}(1 + \tfrac{1}{2}) = \tfrac{3}{4}$. Proceeding in this way allows one to write down the inbreeding coefficients in each successive generation. Note that in this case the rate of inbreeding is constant from the beginning and it corresponds exactly with equation [3.7]: $\Delta F = 1/2N = \tfrac{1}{2}$. This is not true of the other systems. Self-fertilization gives the most rapid inbreeding possible with a normal mating system. The inbreeding coefficient reaches 99.9 per cent after 10 generations. It is possible, however, to get complete homozygosis in one step by some forms of parthenogenesis and by manipulations such as doubling the chromosome complement of haploid cells. We shall deal with full-sib mating next because it is the most often used of the other systems.

Full sibs. From the coancestry in equation [5.6], referring to Fig. 5.5, we have

$$F_X = f_{PQ} = \tfrac{1}{4}(2f_{AB} + f_{AA} + f_{BB})$$

To get the recurrence equation we have to express the coancestries as inbreeding coefficients of previous generations. First, $f_{AB} = F_P = F_{t-1}$. Since individuals in the same generation have the same inbreeding coefficient, $f_{AA} = f_{BB} = \tfrac{1}{2}(1 + F_A)$ by equation [5.4], and $F_A = F_{t-2}$. Making these substitutions leads to the recurrence equation

$$F_t = \tfrac{1}{4}(1 + 2F_{t-1} + F_{t-2}) \qquad \ldots [5.9]$$

In the first generation, F_{t-1} and F_{t-2} are both zero and so $F_{(t=1)} = 0.25$. The inbreeding coefficients in the first four generations are 0.25, 0.375, 0.50, and 0.59. The rate of inbreeding is not constant in the first few generations, as may be seen by computing ΔF from equation [3.9]. For the first four generations ΔF is 0.25, 0.17, 0.20, and 0.19. It later settles down to a constant value of 0.191.

Offspring-parent. We consider here only the mating of offspring with their younger parent; repeated backcrossing to the same parent will be considered later. Figure 5.6 shows as much of the pedigree as is needed, lettered to correspond with Fig. 5.4 and the coancestry in equation [5.5]. Each individual is an offspring in one generation and a parent in the next. The inbreeding of X is

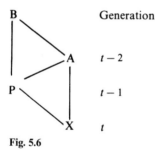

Generation

$t-2$

$t-1$

t

Fig. 5.6

given by equation [5.5] as $F_X = f_{PA} = \tfrac{1}{2}(f_{AB} + f_{AA})$. The recurrence equation is obtained by substituting $f_{AB} = F_P = F_{t-1}$, and $f_{AA} = \tfrac{1}{2}(1 + F_A) = \tfrac{1}{2} + \tfrac{1}{2}F_{t-2}$. The recurrence equation then becomes identical with that for full sibs, equation [5.9]. This is true, however, only for autosomal genes; for sex-linked genes, parent–offspring mating gives a slightly higher rate of inbreeding, with $\Delta F = 0.293$ after the first few generations (Wright, 1933).

Half sibs. Figure 5.1 gives the individuals to which the coancestry in equation [5.7] refers. To get the recurrence equation for repeated half-sib matings we have to know the relationship between individuals B and C. These could be either half sibs to each other or full sibs. With animals, B and C are usually females, both mated to the same male A. To continue half-sib mating with the equivalents of B and C always half sibs, it is necessary to mate one of the females to a second male, making 4 individuals as parents in each generation. This is difficult in practice, but if it is done the recurrence equation, obtained by substitutions in the same manner as above, becomes

$$F_t = \tfrac{1}{8}(1 + 6F_{t-1} + F_{t-2}) \qquad \ldots [5.10]$$

It is easier to continue half-sib mating with B and C being always full sibs, and the number of parents in each generation is then three. The inbreeding then goes a little faster and the recurrence equation is

$$F_t = \tfrac{1}{16}(3 + 8F_{t-1} + 4F_{t-2} + F_{t-3}) \qquad \ldots [5.11]$$

Fixation

One is often more interested in the probability of fixation as a consequence of inbreeding than in the inbreeding coefficient. The inbreeding coefficient gives the probability of an individual being a homozygote, which is $1 - 2p_0q_0(1 - F)$ from Table 3.1. But one wants to know also how soon all individuals in a line can be

Table 5.1 Inbreeding coefficients under various systems of close inbreeding, and probability of fixation under full-sib mating.

Generation (t)	A	B (1)	B (2)	C	D
0	0	0	0	0	0
1	0.500	0.250	0	0.125	0.250
2	0.750	0.375	0.063	0.219	0.375
3	0.875	0.500	0.172	0.305	0.438
4	0.938	0.594	0.293	0.381	0.469
5	0.969	0.672	0.409	0.449	0.484
6	0.984	0.734	0.512	0.509	0.492
7	0.992	0.785	0.601	0.563	0.496
8	0.996	0.826	0.675	0.611	0.498
9	0.998	0.859	0.736	0.654	0.499
10	0.999	0.886	0.785	0.691	
11		0.908	0.826	0.725	
12		0.926	0.859	0.755	
13		0.940	0.886	0.782	
14		0.951	0.908	0.806	
15		0.961	0.925	0.827	
16		0.968	0.940	0.846	
17		0.974	0.951	0.863	
18		0.979	0.960	0.878	
19		0.983	0.968	0.891	
20		0.986	0.975	0.903	

Column	System of mating	Recurrence equation
A	Self-fertilization or repeated backcrosses to highly inbred line.	$\tfrac{1}{2}(1 + F_{t-1})$
B	Full brother × sister, or offspring × younger parent:	
(1)	Inbreeding coefficient.	$\tfrac{1}{4}(1 + 2F_{t-1} + F_{t-2})$
(2)	Probability of fixation (*from Schäfer, 1937*).	
C	Half sib (females half sisters).	$\tfrac{1}{8}(1 + 6F_{t-1} + F_{t-2})$
D	Repeated backcrosses to random-bred individual.	$\tfrac{1}{4}(1 + 2F_{t-1})$

expected to be homozygous for the same allele. This is the 'purity' implied by the term 'pure line' which is often used to mean a highly inbred line. The degree of 'purity' is the probability of fixation. The probability of fixation depends on the number of alleles and their arrangement in the initial matings of the line. The probabilities of fixation over the first 20 generations of full-sib mating are given in Table 5.1, when 4 alleles were present in the initial mating. There cannot, of course, be more than 4 alleles in a sib-mated line, and when there are fewer the probability of fixation is greater (see Haldane, 1955).

Repeated backcrosses

Repeated backcrosses to an individual or to a highly inbred line are often made, for a variety of purposes. The resulting inbreeding is as follows. The pedigree (Fig. 5.7) shows an individual A, which will probably be a male, mated to

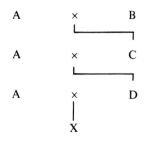

Fig. 5.7

his daughter C, his granddaughter D, etc. From the supplementary rule [5.5]

$$F_X = f_{AD} = \tfrac{1}{2}(f_{AA} + f_{AC})$$
$$= \tfrac{1}{2}\{\tfrac{1}{2}(1 + F_A) + F_D\}$$

The recurrence equation is therefore

$$F_t = \tfrac{1}{4}(1 + F_A + 2F_{t-1}) \qquad \qquad \dots [5.12]$$

where F_A is the inbreeding coefficient of the individual to which the repeated backcrosses are made. If A is an individual from the base population and $F_A = 0$, the equation becomes

$$F_t = \tfrac{1}{4}(1 + 2F_{t-1}) \qquad \qquad \dots [5.13]$$

The inbreeding coefficients over the first 9 generations are given in Table 5.1. If A is an individual from a highly inbred line and $F_A = 1$, the equation becomes

$$F_t = \tfrac{1}{2}(1 + F_{t-1}). \qquad \qquad \dots [5.14]$$

which is identical with the equation for self-fertilization. In this case A need not be the same individual in successive generations: it can be any member of the inbred line.

The chief use of repeated backcrosses is to transfer a particular gene from one strain into the genetic background of another strain. A problem then arises as to the length of foreign chromosome that will be transferred along with the desired gene. A dominant gene can be transferred by successive crosses of the heterozygote to the strain into which it is to be introduced. It can be shown (see

Crow and Kimura, 1970, p. 94) that in this case the mean length of chromosome introduced with the gene after t crosses is approximately $100/t$ cM on each side of the gene, or $200/t$ cM altogether. (1 centimorgan (cM) is the map distance corresponding to a recombination frequency of 1 per cent.) A recessive gene is commonly transferred by alternating backcrosses and intercrosses from which the homozygote is extracted. The mean length of foreign chromosome in this case is about $200/t$ cM on each side, or $400/t$ cM altogether, after t cycles (Bartlett and Haldane, 1935). From the length of linked chromosome transferred and the total map length of the organism, we can arrive at the expected proportion of the total genome that is still heterogeneous. Suppose, for example, that a dominant gene is transferred to an inbred mouse strain by five backcrosses. The gene would carry with it a length of linked chromosome amounting to $200/5 = 40$ cM. Taking the total map length of the mouse to be 2,000 cM (Slizynski, 1955), this heterogeneous segment would represent 2 per cent of the total genome. In addition, some proportion of the genome not associated with the gene being transferred is expected to be still heterogeneous. This can be taken as approximately $1 - F$ which, from column A of Table 5.1, is about 3 per cent after 5 backcrosses. So in all about 5 per cent of the genome is expected to be still heterogeneous.

Crosses and subsequent generations

A standard procedure in genetical analysis and in breeding, particularly plant breeding, is to make crosses between highly inbred lines and to raise the F_1, F_2 and subsequent generations. What is the inbreeding coefficient in the subsequent generations if these are maintained as a large random-bred population? This question is easily answered by consideration of types of gamete, but it is not difficult to verify the solution by the rules of coancestry. We shall consider populations derived from *2-way* and from *4-way* crosses, as shown in Fig. 5.8. The foundation generation of the random-bred population derived from the cross is represented by the individuals marked O, of which there are a large number to be mated at random. It is these individuals O whose inbreeding coefficient we have to find.

The inbred lines have inbreeding coefficients of $F = 1$. This means that all the gametes produced by one inbred line are identical. Consequently the F_1 individuals of the same cross have identical genotypes. Therefore, in the 2-way

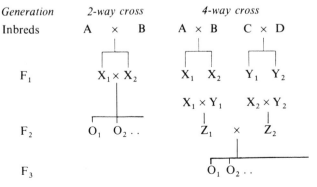

Fig. 5.8

cross all the matings of pairs of F_1 individuals to produce the F_2 are equivalent to the self-fertilization of one individual. The individuals O in F_2 are thus equivalent to the progeny of self-fertilization of one individual, and their inbreeding coefficient is $F = 0.5$. In the 4-way cross, individuals Z_1 and Z_2 are related as full sibs since X_1 and X_2 are genetically identical, and so are Y_1 and Y_2, but X and Y are not related (compare Fig. 5.5). Consequently the individuals O in the F_3 generation have an inbreeding coefficient of $F = 0.25$.

These inbreeding coefficients of the derived populations have no meaning unless the base population to which they refer is defined. The base population implicit in the reasoning above is some real or hypothetical random-breeding population from which the inbred lines were derived. The inbred lines used in the crosses are assumed to be a random sample of all possible lines produced without any change of the mean gene frequencies, i.e., with no selection. With the base population defined in this way, the meaning of the inbreeding coefficient of the derived population is as follows. If we made a large number of 2-way, or of 4-way, crosses each with a different set of inbred lines, the populations derived from the crosses would constitute a set of lines or sub-populations. The inbreeding coefficient would then indicate the expected amount of dispersion of gene frequencies among these lines. Populations derived from 2-way crosses are equivalent to progenies of one generation of self-fertilization. The gene frequencies can therefore have only three values, 0, $\frac{1}{2}$, and 1. Populations derived from 4-way crosses are equivalent to progenies of one generation of full-sib mating, and the gene frequencies can have only five values, 0, $\frac{1}{4}$, $\frac{1}{2}$, $\frac{3}{4}$, and 1.

Mixed inbreeding and crossing

Many plants are 'inbreeders', reproducing normally by self-fertilization. In many of these, however, some cross-pollination regularly occurs. The proportion of crossing varies widely, ranging, for example in lima beans and sorghum varieties, from around 5 per cent up to 50 per cent (Allard, Jain, and Workman, 1968). How much heterozygosity will the crossing generate? It is assumed that the whole population is large, and that whether an individual selfs or crosses is random, being unrelated to what its parents did. In any generation there are two sorts of progeny, those produced by self-fertilization and those produced by crossing. Let C be the proportion of individuals produced by crossing; their inbreeding coefficient is zero. The proportion produced by selfing is $(1 - C)$, and their inbreeding coefficient is $F_t = \frac{1}{2}(1 + F_{t-1})$ by equation [5.8]. The average inbreeding coefficient is therefore

$$F_t = \tfrac{1}{2}(1 + F_{t-1})(1 - C)$$

If the rate of crossing remains constant, the average inbreeding coefficient reaches an equilibrium level at which it remains. Then $F_t = F_{t-1}$, and rearrangement of the above equation gives the average inbreeding coefficient at equilibrium as

$$F = \frac{1 - C}{1 + C} \qquad \qquad \ldots [5.15]$$

where C is the proportion of individuals produced by cross-pollination. On the assumption that there is no selection for or against heterozygotes, the expected frequency of heterozygotes relative to a fully random breeding population is $1 - F$, by equation [3.15]. Application of equations [5.15] and [3.15] shows that 5 per cent of crossing generates heterozygosity amounting to 9.5 per cent of that of a random-breeding population, and 50 per cent of crossing generates 66.7 per cent. Studies of barley have shown the frequencies of heterozygotes at four esterase loci to be greater than expected from the known amount of crossing, which was 0.57 per cent, the excess being attributed to selection favouring heterozygotes (Allard, Kahler, and Weir, 1972).

The effect of crossing on the structure of a predominantly inbreeding population is more important than the generation of heterozygosity. With no crossing, an inbreeding population consists of completely homozygous lines, and natural selection operates through the elimination of the less well-adapted lines. Each local habitat is then inhabited by the line best adapted to it (Allard, Jain, and Workman, 1968), but no further adaptation can take place, nor new adaptation to different habitats. With some crossing, however, new lines are constantly generated, with genes recombined from the existing lines, and this allows continued, or new, adaptation to take place. Crossing also makes possible the elimination of deleterious genes that have arisen by mutation and been fixed by the inbreeding.

The converse problem is also of interest, namely a small amount of inbreeding in a predominantly outbreeding population. Substitution of high values of C (the proportion crossing) into equation [5.15] shows that a small amount of selfing raises the average inbreeding coefficient by very little. The reason for this is that the population does not become differentiated into permanent lines; the progeny of selfing are most likely themselves to cross-breed. If the inbreeding is by full-sib mating rather than selfing, the inbreeding coefficient expressed in a way analogous to equation [5.15] is $F = (1 - C)/(1 + 3C)$. Expressed in terms of the proportion of individuals produced by sib-mating, $S(= 1 - C)$, the formula becomes

$$F = \frac{S}{4 - 3S} \qquad \ldots [5.16]$$

(See Li, 1976, for details). For example, 5 per cent of full-sib mating in a large population would raise the average inbreeding coefficient from 0 to 1.2 per cent, and 10 per cent would raise it to 2.3 per cent. The practical implication of this is that anyone keeping a stock by random breeding, or by minimal inbreeding, need not worry about the consequences of an occasional sib mating, a point already noted in Chapter 4.

Change of base: structured population

The question to be considered here is not confined to pedigreed populations or close inbreeding. Having computed a coefficient of inbreeding with reference to a certain group of individuals as the base population, one may then want to know the inbreeding coefficient referred to a different base, either more or less remote in the ancestry. For example, an individual produced by a full-sib mating is 25 per cent inbred with reference to its parents. Its parents may themselves be inbred with reference to a more remote base. What is the inbreeding coefficient of

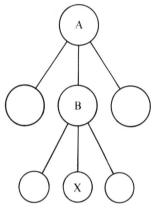

Fig. 5.9

the individual with reference to this more remote base population? This question implies a 'structured' population with a hierarchical subdivision into lines and sublines, as illustrated in Fig. 5.9. In Fig. 5.9, A represents the further-back base population with which we are concerned, B is a later stage, and X represents the individuals whose inbreeding coefficient is to be calculated. The unlettered circles contemporary with B represent the subdivision of A into lines in the manner of Fig. 3.1. In a real population only one of these lines, B, may actually exist. Line B is then further divided into sublines and X is an individual in one of these. The solution comes from a consideration of the relative frequencies of heterozygotes. Let H_X, H_B, and H_A be the frequencies of heterozygotes among the con-temporaries of X, B and A respectively. Then $H_X/H_A = (H_X/H_B)(H_B/H_A)$, and it follows from equation [3.15] that

$$P_{X \cdot A} = P_{X \cdot B} P_{B \cdot A} \qquad \qquad \ldots [5.17]$$

where $P_{X \cdot A} = 1 - F_{X \cdot A}$; $F_{X \cdot A}$ being the inbreeding coefficient of X referred to A as base, and similarly for the other subscripts. The relationship in equation [5.17] can be extended to any number of categories of subdivision, or stages of inbreeding.

Example 5.2. A strain of mice was bred for 42 generations with an effective population size of about 40, and was then inbred by full-sib mating for a further 11 generations (Falconer, 1971). What was the inbreeding coefficient at the end? The inbreeding produced by the full-sib mating, i.e., from B to X in Fig. 5.9, was 0.908, from Table 5.1. Thus $P_{X \cdot B} = 1 - 0.908 = 0.092$. The inbreeding in the line when the sib-mating was started, i.e., from A to B, was as follows: with $N_e = 40$, equation [4.1] gives $\Delta F = 0.0125$, and after 42 generations equation [3.12] gives $F_{B \cdot A} = 0.410$. Thus $P_{B \cdot A} = 1 - 0.410 = 0.590$. By equation [5.17], $P_{X \cdot A} = 0.590 \times 0.092 = 0.054$. Thus the inbreeding coefficient at the end, referred to the origin of the line as base, was $F_{X \cdot A} = 1 - P_{X \cdot A} = 0.946$.

Mutation
        After a long period of inbreeding, mutation may become an important factor in determining the frequency of heterozygotes. If u is the mutation rate of a gene

that has reached near-fixation in the line, then the frequency of heterozygotes at this locus due to mutation is $4u$ under self-fertilization, and $12u$ under full-sib mating, for autosomal loci (Haldane, 1936). These are very small frequencies if we are concerned with only one locus, but if the effects of all loci are taken together, mutation is not entirely negligible as a source of heterozygosis in long-inbred strains such as the widely used strains of mice. The practical consequences of the origin of heterogeneity by mutation are that the characteristics of a line slowly change through the fixation of mutant alleles, and that sub-lines become differentiated. An example is given in Chapter 15.

Selection favouring heterozygotes

When close inbreeding is practised, the object is generally to produce fixation, or homozygosis within the lines. It is therefore a matter of some importance to know how selection will affect the progress toward fixation. Selection against a deleterious recessive may prevent the deleterious allele from becoming fixed, but it will not delay the fixation of the more favourable allele. Selection that favours heterozygotes, however, is another matter. A consequence of inbreeding almost universally observed is a reduction of fitness, the reasons for which will be given in Chapter 14. Thus selection resists the inbreeding, since the more homozygous individuals are the less fit, and this can only mean that selection favours heterozygotes – not necessarily heterozygotes of the loci taken singly, but heterozygotes of segments of chromosome. It is only necessary to have two deleterious genes, recessive or partially recessive, linked in repulsion, to confer a selective advantage on the heterozygote of the segment of chromosome within which the genes are located. It is therefore important to find out how the opposing tendencies of inbreeding and selection in favour of heterozygotes balance each other, in order to assess the reliability of the computed inbreeding coefficient as a measure of the probability of fixation.

The outcome of the joint action of inbreeding and selection in favour of heterozygotes depends on whether there is replacement of the less fit lines by the more fit; in other words, on whether selection operates between lines or only

Table 5.2 Rate of inbreeding, ΔF, with selection favouring the heterozygote. (Except with self-fertilization, the rates are only approximate over the first few generations of inbreeding.)

Coefficient of selection against the homozygotes	ΔF (%)		
	Self-fertilization	Full sib	Half sib*
(s)			
0	50.00	19.10	13.01
0.2	44.44	14.88	9.32
0.4	37.50	10.32	5.67
0.6	28.57	5.71	2.48
0.75	20.00	2.62	0.82
0.8	16.67	1.76	0.46

*Females full sisters to each other.

within lines. Within any one line, selection against homozygotes only delays the progress toward fixation and cannot arrest it, the delay being roughly in proportion to the intensity of the selection (Reeve, 1955a). Table 5.2 shows the rates of inbreeding with various intensities of selection, when there are two alleles and selection acts equally against both homozygotes. (The rate of inbreeding, ΔF, is used here to mean the rate of dispersion of gene frequencies and, after the first few generations when the distribution of gene frequencies has become flat, it measures the rate of fixation, i.e., the proportion of unfixed loci that become fixed in each generation, as explained in Chapter 3.) The delay of fixation caused by selection is least under the closest systems of inbreeding. Thus the rate is halved under self-fertilization when the coefficient of selection is 0.67; under full-sib mating when it is 0.44; and under half-sib mating when it is 0.35. It will be seen from the table that the rate of inbreeding, though much reduced by intense selection, does not become zero until the coefficient of selection rises to 1. If there is only one line, therefore, fixation eventually goes to completion unless both homozygotes are entirely inviable or sterile.

If there are many lines, however, selection may arrest the progress of fixation and lead to a state of equilibrium, for the following reason. The amount by which the inbreeding has changed the frequency of a particular gene from its original value differs at any one time from line to line. In other words, the state of dispersion of the locus has gone further in some lines than in others. Now, if those lines in which the dispersion has gone furthest, and which are consequently most reduced in fitness, die out or are discarded, and if they are replaced by sub-lines taken from the lines in which it has gone least far, then the progress of the dispersive process will have been set back. When there is replacement of lines in this way, and the selection is sufficiently intense, a state of balance between the opposing tendencies of inbreeding and selection is reached. The intensity of selection needed to arrest the dispersive process has been worked out for regular systems of close inbreeding (Hayman and Mather, 1953). Some of the conclusions, for the case of two alleles with equal selection against the two homozygotes, are given in Table 5.3, which shows the intensity of selection against the homozygotes which will (1) just allow fixation to go eventually to completion, and (2) arrest the dispersive process at a point of balance where the frequency of heterozygotes is half its original value, i.e., where $P = 1 - F = 0.5$.

Table 5.3 Balance between inbreeding and selection in favour of heterozygotes, when selection operates between lines. The figures are the selective disadvantages of homozygotes, s, expressed as percentages. Column (a) shows the highest value of s compatible with complete fixation. Column (b) shows the value of s that leads to a steady state at $P = 1 - F = 0.5$.

Mating system	(a) ($P = 0$)	(b) ($P = 0.5$)
Self-fertilization	50.0	66.7
Full-sib	23.7	44.6
Half-sib (females half sisters)	18.8	47.2

These figures show that only a moderate advantage of heterozygotes will suffice to prevent complete fixation. Under full-sib mating, for example, loci, or segments of chromosomes that do not recombine, with a 25 per cent disadvantage in homozygotes will not all go to fixation. And, of those with a 50 per cent disadvantage, only about half will become fixed, no matter for how long the inbreeding is continued.

It must be stressed, however, that prevention of fixation in this way can only take place when there is replacement of lines and sub-lines. The following breeding methods, for example, would allow replacement of lines: if seed, set by self-fertilization, were collected in bulk and a random sample taken for planting, and this were repeated in successive generations; or, if sib pairs of mice were taken at random from all the surviving progeny, so that the same amount of breeding space was occupied in successive generations.

The conclusions outlined above refer to a single locus. If there were more than a few loci on different chromosomes all subject to selection against homozygotes of an intensity sufficient to arrest or seriously delay the progress of inbreeding, the total loss of fitness from all the loci would be very severe. Inbred lines of organisms with a high reproductive rate, such as plants and *Drosophila*, might well stand up to a total loss of fitness sufficient to keep several loci or segments of chromosome permanently unfixed. But the loss of fitness involved in preventing the fixation of more than two or three loci in an organism such as the mouse would be crippling. Under laboratory conditions the highly inbred strains of mice, after 100 or more generations of sib-mating, have a fitness not much less than half that of non-inbred strains. It is conceivable that they might have one locus permanently unfixed, but it is difficult to believe that they can have more. Complete lethality or sterility of both homozygotes at one locus means a 50 per cent loss of progeny; at two unlinked loci, a 75 per cent loss. A mouse strain with a mortality or sterility of 50 per cent can be kept going, but hardly one with 75 per cent.

6 CONTINUOUS VARIATION

It will be obvious, to biologists and layman alike, that the sort of variation discussed in the foregoing chapters embraces but a small part of the naturally occurring variation. One has only to consider one's fellow men and women to realize that they all differ in countless ways, but that these differences are nearly all matters of degree and seldom present clear-cut distinctions attributable to the segregation of single genes. If, for example, we were to classify individuals according to their height, we could not put them into groups labelled 'tall' and 'short', because there are all degrees of height and a division into classes would be purely arbitrary. Variation of this sort, without natural discontinuities, is called *continuous variation*, and characters that exhibit it are called *quantitative characters* or *metric characters*, because their study depends on measurement instead of on counting. The genetic principles underlying the inheritance of metric characters are basically those of *population genetics* outlined in the previous chapters. But since the segregation of the genes concerned cannot be followed individually, new methods of study are needed and new concepts have to be introduced. The branch of genetics concerned with metric characters is called *quantitative genetics* or *biometrical genetics*. The importance of this branch of genetics need hardly be stressed; most of the characters of economic value to plant and animal breeders are metric characters, and most of the changes concerned in micro-evolution are changes of metric characters. It is therefore in this branch that genetics has its most important application to practical problems and also its most direct bearing on evolutionary theory.

How does it come about that the intrinsically discontinuous variation caused by genetic segregation is translated into the continuous variation of metric characters? There are two reasons: one is the simultaneous segregation of many genes affecting the character, and the other is the superimposition of truly continuous variation arising from non-genetic causes. Consider, for example, a simplified situation. Suppose there is segregation at 6 unlinked loci, each with 2 alleles at frequencies of 0.5. Suppose that there is complete dominance of one allele at each locus and that the dominant alleles each add one unit to the measurement of a certain character. Then if the segregation of these genes were the only cause of variation there would be 7 discrete classes in the measurements of the character, according to whether the individual had the dominant allele present at 0, 1, 2 . . . , or 6 of the loci. The frequencies of the classes would be

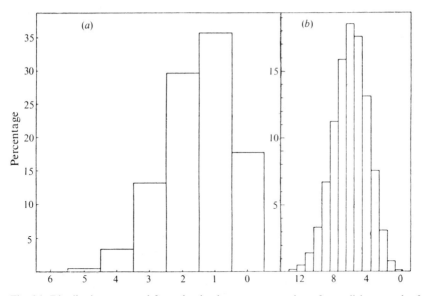

Fig. 6.1. Distributions expected from the simultaneous segregation of two alleles at each of several or many loci: (a) 6 loci, (b) 24 loci. There is complete dominance of one allele over the other at each locus, and the gene frequencies are all 0.5. Each locus, when homozygous for the recessive allele, is supposed to reduce the measurement by 1 unit in (a), and by $\frac{1}{4}$ unit in (b). The horizontal scale, representing the measurement, shows the number of loci homozygous for the recessive allele, and the vertical axis shows the probability, or the percentage of individuals expected in each class. The probabilities are derived from the binomial expansion of $(\frac{1}{4} + \frac{3}{4})^n$, where n is the number of loci.

according to the binomial expansion of $(\frac{1}{4} + \frac{3}{4})^6$, as shown in Fig. 6.1 (a). If our measurements were sufficiently accurate we should recognize these classes as being distinct and we should be able to place any individual unambiguously in its class. If there were more genes segregating but each had a smaller effect, there would be more classes with smaller differences between them, as in Fig. 6.1 (b). It would then be more difficult to distinguish the classes, and if the difference between the classes became about as small as the error of measurement we should no longer be able to recognize the discontinuities. In addition, metric characters are subject to variation from non-genetic causes, and this variation is truly continuous. Its effect is, as it were, to blur the edges of the genetic discontinuity so that the variation as we see it becomes continuous, no matter how accurate our measurements may be.

Thus the distinction between genes concerned with Mendelian characters and those concerned with metric characters lies in the magnitude of their effects relative to other sources of variation. A gene with an effect large enough to cause a recognizable discontinuity even in the presence of segregation at other loci and of non-genetic variation can be studied by Mendelian methods, whereas a gene whose effect is not large enough to cause a discontinuity cannot be studied individually. This distinction is reflected in the terms *major gene* and *minor gene*. There are, however, all intermediate grades, genes that cannot properly be classed as major or as minor. And, furthermore, as a result of pleiotropy the same gene may be classed as major with respect to one character and minor with respect to another character. The distinction, though convenient, is therefore not a fundamental one, and there is no good evidence that there are two sorts of genes with different properties. Variation caused by the simultaneous segregation of many genes may be called *polygenic* variation, and the minor genes concerned are sometimes referred to as *polygenes*.

Metric characters

The metric characters that might be studied in any higher organism are almost infinitely numerous. Any attribute that varies continuously and can be measured might in principle be studied as a metric character – anatomical dimensions and proportions, physiological functions of all sorts, and mental or psychological qualities. The essential condition is that they should be measurable. The technique of measurement, however, sets a practical limitation on what can be studied. Usually rather large numbers of individuals have to be measured and the study of any character whose measurement requires an elaborate technique therefore becomes impracticable. Consequently the characters that have been used in studies of quantitative genetics are predominantly anatomical dimensions, or physiological functions measured in terms of an end-product, such as lactation, fertility, or growth rate.

Some examples of metric characters are illustrated in Fig. 6.2. The variation is represented graphically by the frequency distribution of measurements. The measurements are grouped into equally spaced classes and the proportion of individuals falling in each class is plotted on the vertical scale. The resulting histogram is discontinuous only for the sake of convenience in plotting. If the class ranges were made smaller and the number of individuals measured were increased indefinitely, the histogram would become a smooth curve. The variation of some metric characters, such as bristle number or litter size, is not strictly speaking continuous because, being measured by counting, their values can only be whole numbers. Nevertheless, one can regard the measurements in such cases as referring to an underlying character whose variation is truly continuous though expressible only in whole numbers, in a manner analogous to the grouping of measurements into classes. For example, litter size may be regarded as a measure of the underlying, continuously varying character, fertility. For practical purposes such characters can be treated as continuously varying, provided the number of classes is not too small. When there are too few classes, as for example when susceptibility to disease is expressed as death or survival, different methods have to be employed, as will be explained in Chapter 18.

The frequency distributions of most metric characters approximate more or less closely to normal curves. This can be seen in Fig. 6.2, where the smooth curves drawn through the histograms are normal curves having means and variances calculated from the data. In the study of metric characters it is therefore possible to make use of the properties of the normal distribution and to apply the appropriate statistical techniques. Sometimes, however, the scale of measurement must be modified if a distribution approximating to the normal is to be obtained. The distribution in Fig. 6.2 (*d*), for example, would be skewed if measured and plotted simply as the number of facets. But it becomes symmetrical, and

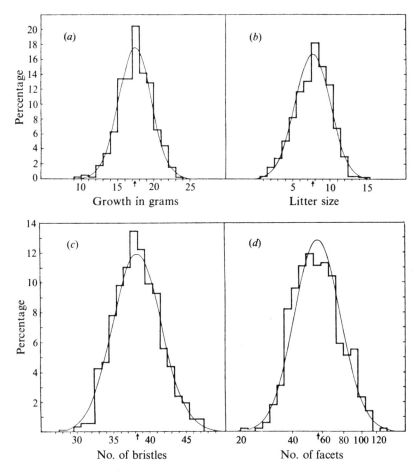

Fig. 6.2. Frequency distributions of four metric characters, with normal curves superimposed. The means are indicated by arrows. The characters are as follows, the number of observations on which each histogram is based being given in brackets:

(*a*) Mouse (♂♂): growth from 3 to 6 weeks of age. (380)

(*b*) Mouse: litter size (number of live young in 1st litters). (689)

(*c*) *Drosophila melanogaster* (♀♀): number of bristles on ventral surface of 4th and 5th abdominal segments, together. (900)

(*d*) *Drosophila melanogaster* (♀♀): number of facets in the eye of the mutant "Bar". (488)

(*a*), (*b*), and (*c*) are from original data: (*d*) is from data of Zeleny (1922).

approximates to a normal distribution, if measured and plotted in logarithmic units. The criteria on which the choice of a scale of measurement rests cannot be fully appreciated at this stage, and will be explained in Chapter 17. Meantime it will be assumed that any metric character under discussion is measured on an appropriate scale and has a distribution that is approximately normal.

General survey of the subject-matter

There are two basic genetic phenomena concerned with metric characters, both more or less familiar to all biologists, and each forms the basis of a breeding method. The first is the resemblance between relatives. Everyone is familiar with the fact that relatives tend to resemble each other, and the closer the relationship, in general the closer the resemblance. Though it is only in our own species that resemblances are readily discernible without measurement, the phenomenon is equally present in other species. The degree of resemblance varies with the character, some showing more, some less. The resemblance between offspring and parents provides the basis for selective breeding. Use of the more desirable individuals as parents brings about an improvement of the mean level of the next generation, and just as some characters show more resemblance than others, so some are more responsive to selection than others. The degree of resemblance between relatives is one of the properties of a population that can be readily observed, and it is one of the aims of quantitative genetics to show how the degree of resemblance between different sorts of relatives can be used to predict the outcome of selective breeding and to point to the best method of carrying out the selection. This problem will form the central theme of the next seven chapters, the resemblance between relatives being dealt with in Chapters 9–10, and the effects of selection in Chapters 11–13.

The second basic genetic phenomenon is inbreeding depression, with its converse hybrid vigour, or heterosis. This phenomenon is less familiar to the layman than the first, since the laws against incest prevent its more obvious manifestations in our own species; but it is well known to animal and plant breeders. Inbreeding tends to reduce the mean level of all characters closely connected with fitness in animals and in naturally outbreeding plants, and to lead in consequence to loss of general vigour and fertility. Since most characters of economic value in domestic animals and plants are aspects of vigour or fertility, inbreeding is generally deleterious. The reduced vigour and fertility of inbred lines is restored on crossing, and in certain circumstances this hybrid vigour can be made use of as a means of improvement. The enormous improvement of the yield of commercially grown maize has been achieved by this means and represents probably the greatest practical achievement of genetics (see Mangelsdorf, 1951). The effects of inbreeding and crossing will be described in Chapters 14–16.

The properties of a population that we can observe in connection with a metric character are means, variances, and covariances. The natural subdivision of the population into families allows us to analyse the variance into components which form the basis for the measurement of the degree of resemblance between relatives. We can in addition observe the consequences of experimentally applied breeding methods, such as selection, inbreeding, or cross-breeding. The practical objective of quantitative genetics is to find out how we can use the observations made on the population as it stands to predict the outcome of any particular breeding method. The more general aim is to find out how the observable properties of the population are influenced by the properties of the genes concerned and by the various non-genetic circumstances that may influence a metric character. The chief properties of genes that have to be taken into account are the degree of dominance, the manner in which genes at different loci combine their effects, pleiotropy, linkage, and fitness under natural selection. To take account of all these properties simultaneously, in addition to a variety of non-genetic circumstances, would make the problems unmanageably complex. We therefore have to simplify matters by dealing with one thing at a time, starting with the simpler situations.

The plan to be followed in the succeeding chapters is this: we shall first show what determines the population mean, and then introduce two new concepts – average effect and breeding value – which are necessary to an understanding of the variance. Then we shall discuss the variance, its analysis into components, and the covariance of relatives, which will lead us to the degree of resemblance between relatives. In all this we shall take full account of dominance from the beginning: the other complicating factors will be more briefly discussed when they become relevant. The most important simplification that we shall make concerns the effect of genes on fitness: we shall assume that Mendelian segregation is undisturbed by differential fitness of the genotypes. The description of means, variances, and covariances will refer to a random-breeding population, with Hardy–Weinberg equilibrium genotype frequencies, with no selection and no inbreeding. That is to say, we shall describe the population before any special breeding method is applied to it. Then in Chapters 11–13 we shall describe the effects of selection, and in Chapters 14–16 the effects of inbreeding. This will cover the fundamentals of quantitative genetics, and in the final chapters we shall discuss some special topics.

7 VALUES AND MEANS

We have seen in the early chapters that the genetic properties of a population are expressible in terms of the gene frequencies and genotype frequencies. In order to deduce the connection between these on the one hand and the quantitative differences exhibited in a metric character on the other, we must introduce a new concept, the concept of *value*, expressible in the metric units by which the character is measured. The value observed when the character is measured on an individual is the *phenotypic value* of that individual. All observations, whether of means, variances, or covariances, must clearly be based on measurements of phenotypic values. In order to analyse the genetic properties of the population we have to divide the phenotypic value into component parts attributable to different causes. Explanation of the meanings of these components is our chief concern in this chapter, though we shall also be able to find out how the population mean is influenced by the array of gene frequencies.

The first division of phenotypic value is into components attributable to the influence of genotype and environment. The *genotype* is the particular assemblage of genes possessed by the individual, and the *environment* is all the non-genetic circumstances that influence the phenotypic value. Inclusion of all non-genetic circumstances under the term 'environment' means that the genotype and the environment are by definition the only determinants of phenotypic value. The two components of value associated with genotype and environment are the *genotypic value* and the *environmental deviation*. We may think of the genotype conferring a certain value on the individual and the environment causing a deviation from this, in one direction or the other. Or, symbolically,

$$P = G + E \qquad \qquad \ldots [7.1]$$

where P is the phenotypic value, G is the genotypic value, and E is the environmental deviation. The mean environmental deviation in the population as a whole is taken to be zero, so that the mean phenotypic value is equal to the mean genotypic value. The term *population mean* then refers equally to phenotypic or to genotypic values. When dealing with successive generations we shall assume for simplicity that the environment remains constant from generation to generation, so that the population mean is constant in the absence of genetic change. If we could replicate a particular genotype in a number of individuals and measure them under environmental conditions normal for the

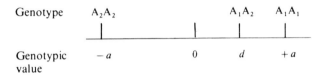

Fig. 7.1. Arbitrarily assigned genotypic values.

population, their mean environmental deviations would be zero, and their mean phenotypic value would consequently be equal to the genotypic value of that particular genotype. This is the meaning of the genotypic value of an individual. In principle it is measurable, but in practice it is not, except when we are concerned with a single locus where the genotypes are phenotypically distinguishable, or with the genotypes represented in highly inbred lines.

For the purposes of deduction we must assign arbitrary values to the genotypes under discussion. This is done in the following way. Considering a single locus with two alleles, A_1 and A_2, we call the genotypic value of one homozygote $+a$, that of the other homozygote $-a$, and that of the heterozygote d. (We shall adopt the convention that A_1 is the allele that increases the value.) We thus have a scale of genotypic values as in Fig. 7.1. The origin, or point of zero value, on this scale is mid-way between the values of the two homozygotes. The value d of the heterozygote depends on the degree of dominance. If there is no dominance, $d = 0$; if A_1 is dominant over A_2, d is positive, and if A_2 is dominant over A_1, d is negative. If dominance is complete, d is equal to $+a$ or $-a$, and if there is overdominance, d is greater than $+a$ or less than $-a$. The degree of dominance may be expressed as d/a.

Example 7.1 For the purposes of illustration in this chapter, and also later on, we shall refer to a dwarfing gene in the mouse, known as 'pygmy' (symbol pg), described by King (1950, 1955), and by Warwick and Lewis (1954). This gene reduces body-size and is nearly, but not quite, recessive in its effect on size. It was present in a strain of small mice (MacArthur's) at the time the studies cited above were made. The weights of mice of the three genotypes at 6 weeks of age were approximately as follows (sexes averaged):

	Genotypes		
	+ +	+ pg	pg pg
Weight in grams	14	12	6

(The weight of heterozygotes given here is to some extent conjectural, but it is unlikely to be more than 1 g in error.) These are average weights obtained under normal environmental conditions, and they are therefore the genotypic values. The mid-point in genotypic value between the two homozygotes is 10 g, and this is the origin, or zero-point, on the scale of values assigned as in Fig. 7.1. The value of a on this scale is therefore 4 g, and that of d is 2 g.

Population mean

We can now see how the gene frequencies influence the mean of the character in the population as a whole. Let the gene frequencies of A_1 and A_2 be p and q

Table 7.1

Genotype	Frequency	Value	Freq. × Val.
A_1A_1	p^2	$+a$	p^2a
A_1A_2	$2pq$	d	$2pqd$
A_2A_2	q^2	$-a$	$-q^2a$
		Sum =	$a(p-q)+2dpq$

respectively. Then the first two columns of Table 7.1 show the three genotypes and their frequencies in a random breeding population, from formula [1.2]. The third column shows the genotypic values as specified above. The mean value in the whole population is obtained by multiplying the value of each genotype by its frequency and summing over the three genotypes. The reason why this yields the mean value may be understood by converting frequencies to numbers of individuals. Multiplying the value by the number of individuals in each genotype and summing over genotypes gives the sum of values of all individuals. The mean value would then be this sum of values divided by the total number of individuals. The procedure in working with frequencies is the same, but since the sum of the frequencies is 1, the sum of values × frequencies is the mean value. In other words, the division by the total number has already been made in obtaining the frequencies. Multiplication of values by frequencies to obtain the mean value is a procedure that will be often used in this chapter and subsequent ones. Returning to the population mean, multiplication of the value by the frequency of each genotype is shown in the last column of Table 7.1. Summation of this column is simplified by noting that $p^2 - q^2 = (p+q)(p-q) = p - q$. The population mean, which is the sum of this column, is thus

$$M = a(p-q) + 2dpq \qquad \ldots [7.2]$$

This is both the mean genotypic value and the mean phenotypic value of the population with respect to the character.

The contribution of any locus to the population mean thus has two terms: $a(p-q)$ attributable to homozygotes, and $2dpq$ attributable to heterozygotes. If there is no dominance $(d = 0)$, the second term is zero, and the mean is proportional to the gene frequency: $M = a(1 - 2q)$. If there is complete dominance $(d = a)$, the mean is proportional to the square of the gene frequency: $M = a(1 - 2q^2)$. The *total range* of values attributable to the locus is $2a$, in the absence of overdominance. That is to say, if A_1 were fixed in the population $(p = 1)$ the population mean would be a, and if A_2 were fixed $(q = 1)$ it would be $-a$. If the locus shows overdominance, however, the mean of an unfixed population may be outside this range.

The genotypic values a and d are deviations from the mean value of the two homozygotes, as shown in Fig. 7.1. It follows that the population mean expressed in equation [7.2] is a deviation from the mid-homozygote value, which is the origin or zero-point of the scale. If the mean is to be expressed as a deviation from some other value, an appropriate constant must be added or subtracted. For example, one might want to express the mean as a deviation from the value of the lower homozygote. This would require the addition of a, and the mean would become, after some simplification, $M = 2p(a + dq)$. Or, expressed as a deviation from the upper homozygote, it would be $M = 2q(-a + dp)$.

Example 7.2 Let us take again the pygmy gene in mice, as described in Example 7.1, and see what effect this gene would have on the population mean when present at two particular frequencies. First, the total range is from 6 g to 14 g: a population consisting entirely of pygmy homozygotes would have a mean of 6 g, and one from which the gene was entirely absent would have a mean of 14 g. (These values refer specifically to MacArthur's Small Strain at the time the observations were made.) Now suppose the gene were present at a frequency of 0.1, so that under random mating homozygotes would appear with a frequency of 1 per cent. The values to be substituted in equation [7.2] are $p = 0.9$, $q = 0.1$, and $a = 4$ g, $d = 2$ g, as shown in Example 7.1. The population mean, by equation [7.2], is therefore: $M = 4 \times 0.8 + 2 \times 0.18 = 3.56$. This value of the mean, however, is measured from the mid-homozygote point, which is 10 g, as origin. Therefore the actual value of the population mean is 13.56 g. Next suppose the gene were present at a frequency of 0.45. Substituting in the same way, we find $M = 1.76$, to which must be added 10 g for the origin, giving a value of 11.76 g. Rough corroboration of these figures is given by the records of the strain carrying the gene. When the gene was present at a frequency of about 0.4 the mean weight was about 12 g. Two generations, later, when the pygmy gene had been deliberately eliminated, the mean weight rose to about 14 g.

Now we have to put together the contributions of genes at several loci and find their joint effect on the mean. This introduces the question of how genes at different loci combine to produce a joint effect on the character. For the moment we shall suppose that combination is by addition, which means that the value of a genotype with respect to several loci is the sum of the values attributable to the separate loci. For example, if the genotypic value of A_1A_1 is a_A and that of B_1B_1 is a_B, then the genotypic value of $A_1A_1B_1B_1$ is $a_A + a_B$. The consequences of non-additive combination will be explained at the end of this chapter. With additive combination, then, the population mean resulting from the joint effects of several loci is the sum of the contributions of each of the separate loci, thus:

$$M = \Sigma a(p - q) + 2 \Sigma dpq \qquad \dots [7.3]$$

This is again both the genotypic and the phenotypic mean value. The total range in the absence of overdominance is now $2\Sigma a$. If all alleles that increase the value were fixed, the mean would be $+ \Sigma a$, and if all alleles that decrease the value were fixed, it would be $- \Sigma a$. These are the theoretical limits to the range of potential variation in the population. The origin from which the mean value in equation [7.3] is measured is the mid-point of the total range. This is equivalent to the average mid-homozygote point of all the loci separately.

Example 7.3 As an example of two loci that combine additively, we shall refer to two colour genes in mice, whose effects on the number of pigment granules have been described by Russell (1949). This is a metric character which reflects the intensity of pigmentation in the coat. The two genes are 'brown' (b) and 'extreme dilution' (c^e), an allele of the albino series. Measurements were made of the number of melanin granules per unit volume of hair, in wild-type homozygotes, in the two single mutant homozygotes, and in the double mutant homozygote. We shall assume both wild-type alleles to be completely dominant, so that only these four genotypes need be considered. The mean numbers of granules in the four genotypes were as shown in the table.

	B –	bb	$2a_B$
C –	95	90	5
$c^e c^e$	38	34	4
$2a_C$	57	56	

The difference between the two figures in each row and in each column measures the homozygote difference, or $2a$ on the scale of values assigned as in Fig. 7.1. Apart from the trivial discrepancy of 1 unit, these differences are independent of the genotype at the other locus. In other words, the difference of value between B– and bb is the same among C– genotypes as it is among $c^e c^e$ genotypes; and similarly the difference between C– and $c^e c^e$ is the same in B– as it is in bb. Thus the two loci combine additively, and the value of a composite genotype can be rightly predicted from knowledge of the values of the single genotypes. For example, the bb genotype is 5 units less than the wild-type, and the $c^e c^e$ is 57 units less; therefore bb $c^e c^e$ should be 62 units less than the wild-type value of 95, namely 33, which is almost identical with the observed value of 34.

Average effect

In order to deduce the properties of a population connected with its family structure, we have to deal with the transmission of value from parent to offspring, and this cannot be done by means of genotypic values alone, because parents pass on their genes and not their genotypes to the next generation, genotypes being created afresh in each generation. A new measure of value is therefore needed which will refer to genes and not to genotypes. This will enable us to assign a 'breeding value' to individuals, a value associated with the genes carried by the individual and transmitted to its offspring. The new measure is the 'average effect'. We can assign an average effect to a gene in the population, or to the difference between one gene and another of an allelic pair. The *average effect of a gene* is the mean deviation from the population mean of individuals which received that gene from one parent, the gene received from the other parent having come at random from the population. This may be stated in another way. Let a number of gametes all carrying A_1 unite at random with gametes from the population; then the mean deviation from the population mean of the genotypes so produced is equal to the average effect of the gene A_1. The concept of average effect is perhaps easier to grasp in the form of the *average effect of a gene-substitution*, which can more conveniently be used when only two alleles at a locus are under consideration. If we could change, say, A_2 genes into A_1 at random in the population, and could then note the resulting change of value, this would be the average effect of the gene-substitution. It is equal to the difference between the average effects of the two genes involved in the substitution. A graphical representation of the average effect of a gene-substitution is given later in Fig. 7.2.

It is important to realize that the average effect of a gene or a gene-substitution depends on the gene frequency, and that the average effect is therefore a property of the population as well as of the gene. The reason for this can be seen in the words 'taken at random' in the definitions, because the content of the random sample depends on the gene frequency in the population. The point may perhaps be more easily understood from a specific example. Consider the substitution of a recessive gene, a, for its dominant allele, A. The substitution will change the value only when the individual already carries one recessive allele, in other words in heterozygotes. Changing AA into Aa will not affect the value, but changing Aa into aa will. Now, when the frequency of the recessive allele, a, is low there will be many AA individuals, which the substitution will not affect; but when the recessive is at high frequency there will be very few AA individuals, and most of the individuals in which a substitution can be made will be affected by it. Therefore the average effect of the substitution will be small when the frequency of the recessive allele is low, and large when it is high.

Let us see how the average effect is related to the genotypic values a and d, in terms of which the population mean was expressed. This will help to make the concept clearer. The reasoning is set out in Table 7.2. Consider a locus with two alleles, A_1 and A_2, at frequencies p and q respectively, and take first the average effect of the gene A_1, for which we shall use the symbol α_1. If gametes carrying A_1 unite at random with gametes from the population, the frequencies of the genotypes produced will be p of A_1A_1 and q of A_1A_2. The genotypic value of A_1A_1 is $+a$ and that of A_1A_2 is d, and the mean of these, taking account of the proportions in which they occur, is $pa + qd$. The difference between this mean value and the population mean is the average effect of the gene A_1. Taking the value of the population mean from equation [7.2], we get

$$\begin{aligned}\alpha_1 &= pa + qd - [a(p - q) + 2dpq]\\ &= q[a + d(q - p)]\end{aligned} \qquad \ldots [7.4a]$$

Similarly, the average effect of the gene A_2 is

$$\alpha_2 = -p[a + d(q - p)] \qquad \ldots [7.4b]$$

Now consider the average effect of the gene-substitution, letting A_2 be changed into A_1. Of the A_2 genes taken at random from the population, a proportion p will be found in A_1A_2 genotypes and a proportion q in A_2A_2 genotypes. Changing A_1A_2 into A_1A_1 will change the value from d to $+a$, and the effect will therefore

Table 7.2

Type of gamete	Values and frequencies of genotypes produced			Mean value of genotypes produced	Population mean to be deducted	Average effect of gene
	A_1A_1 a	A_1A_2 d	A_2A_2 $-a$			
A_1	p	q		$pa + qd$	$-[a(p-q) + 2dpq]$	$q[a + d(q - p)]$
A_2		p	q	$-qa + pd$	$-[a(p-q) + 2dpq]$	$-p[a + d(q - p)]$

be $(a - d)$. Changing A_2A_2 into A_1A_2 will change the value from $-a$ to d, and the effect will be $(d + a)$. The average change is therefore $p(a - d) + q(d + a)$, which on rearrangement becomes $a + d(q - p)$. Thus the average effect of the gene-substitution (written as α, without subscript) is

$$\alpha = a + d(q - p) \qquad \qquad \text{...[7.5]}$$

The relation of α to α_1 and α_2 can be seen by comparing equations [7.5] and [7.4], whence

$$\alpha = \alpha_1 - \alpha_2 \qquad \qquad \text{...[7.6]}$$

and

$$\left. \begin{aligned} \alpha_1 &= q\alpha \\ \alpha_2 &= -p\alpha \end{aligned} \right\} \qquad \qquad \text{...[7.7]}$$

Example 7.4 Consider again the pygmy gene and its effect on body weight, for which $a = 4$ g and $d = 2$ g. If the frequency of the pg gene were $q = 0.1$, the average effect of substituting + for pg would be, by equation [7.5], $\alpha = 4 + 2 \times -0.8 = 2.4$ g. And if the frequency were $q = 0.4$, the average effect of the gene-substitution would be: $\alpha = 4 + 2 \times -0.2 = 3.6$ g. The average effects of the genes seperately are, by equation [7.7]:

	$q = 0.1$	$q = 0.4$
Average effect of + : $\alpha_1 =$	+ 0.24	+ 1.44
Average effect of pg: $\alpha_2 =$	− 2.16	− 2.16
$\alpha = \alpha_1 - \alpha_2 =$	2.40	3.60

Thus, the average effect of the gene substitution, α, is greater when the gene frequency is greater. The identity of the average effects of pg at the two gene frequencies is only a coincidence.

Breeding value

The usefulness of the concept of average effect arises from the fact, already noted, that parents pass on their genes and not their genotypes to their progeny. It is therefore the average effects of the parents' genes that determine the mean genotypic value of its progeny. The value of an individual, judged by the mean value of its progeny, is called the *breeding value* of the individual. Breeding value, unlike average effect, can therefore be measured. If an individual is mated to a number of individuals taken at random from the population, then its breeding value is twice the mean deviation of the progeny from the population mean. The deviation has to be doubled because the parent in question provides only half the genes in the progeny, the other half coming at random from the population. Breeding values can be expressed in absolute units, but are usually more conveniently expressed in the form of deviations from the population mean, as defined above. Just as the average effect is a property of the gene and the population, so is the breeding value a property of the individual and the population from which its mates are drawn. One cannot speak of an individual's breeding value without specifying the population in which it is to be mated.

Defined in terms of average effects, the breeding value of an individual is equal to the sum of the average effects of the genes it carries, the summation being made over the pair of alleles at each locus and over all loci. Thus, for a single locus with two alleles, the breeding values of the genotypes are as follows:

Genotype	Breeding value
A_1A_1	$2\alpha_1 = 2q\alpha$
A_1A_2	$\alpha_1 + \alpha_2 = (q-p)\alpha$
A_2A_2	$2\alpha_2 = -2p\alpha$

Example 7.5. Let us illustrate breeding values by reference to the pygmy gene in mice. The average effects of the + and pg genes were given in the last example. From these we may find the breeding values of the three genotypes as explained above. These breeding values, which are given below, are deviations from the population mean. The population means with gene frequencies of 0.1 and 0.4 were found in Example 7.2 and are shown again below in the column headed M.

	M	Breeding values		
		+ +	+ pg	pg pg
$q = 0.1$	13.56	+ 0.48	− 1.92	− 4.32
$q = 0.4$	11.76	+ 2.88	− 0.72	− 4.32

(The breeding values of pygmy homozygotes are only hypothetical because in fact pygmy homozygotes are nearly all sterile: but this complication may be overlooked in the present context.)

Extension to a locus with more than two alleles is straightforward, the breeding value of any genotype being the sum of the average effects of the two alleles present. If all loci are to be taken into account, the breeding value of a particular genotype is the sum of the breeding values attributable to each of the separate loci. If there is non-additive combination of genotypic values, a slight complication arises. We have given two definitions of breeding value, a practical one in terms of the measured value of the progeny and a theoretical one in terms of average effects. Non-additive combination renders these two definitions not quite equivalent. This point will be more fully explained in Chapter 9.

Consideration of the definition of breeding value will show that in a population in Hardy–Weinberg equilibrium the mean breeding value must be zero; or if breeding values are expressed in absolute units the mean breeding value must be equal to the mean genotypic value and to the mean phenotypic value. This can be verified from the breeding values listed above. Multiplying the breeding value by the frequency of each genotype and summing gives the mean breeding value (expressed as a deviation from the population mean) as

$$2p^2q\alpha + 2pq(q-p)\alpha - 2q^2p\alpha = 2pq\alpha(p + q - p - q) = 0$$

The breeding value is sometimes referred to as the 'additive genotype', and variation in breeding value ascribed to the 'additive effects' of genes. Though we shall not use these terms we shall follow custom in using the term 'additive' in connection with the variation of breeding values to be discussed in the next chapter, and we shall use the symbol A to designate the breeding value of an individual.

Dominance deviation

We have separated off the breeding value as a component part of the genotypic value of an individual. Let us consider now what makes up the remainder. When a single locus only is under consideration, the difference between the genotypic value G and the breeding value A of a particular genotype is known as the *dominance deviation D*, so that

$$G = A + D \qquad \qquad \dots [7.8]$$

The dominance deviation arises from the property of dominance among the alleles at a locus, since in the absence of dominance, breeding values and genotypic values coincide. From the statistical point of view the dominance deviations are interactions between alleles, or within-locus interactions. They represent the effect of putting genes together in pairs to make genotypes; the effect not accounted for by the effects of the two genes taken singly. Since the average effects of genes, and the breeding values of genotypes, depend on the gene frequency in the population, the dominance deviations are also dependent on gene frequency. They are therefore partly properties of the population and are not simply measures of the degree of dominance.

Example 7.6. Continuing with the example of pygmy gene, we may now list the genotypic values and the breeding values, and so obtain the dominance deviations of the three genotypes, by equation [7.8]. These values, all now expressed as deviations from the population mean M, are given in the table.

	$q = 0.1 : M = 13.56$			$q = 0.4 : M = 11.76$		
	+ +	+ pg	pg pg	+ +	+ pg	pg pg
Frequency	0.81	0.18	0.01	0.36	0.48	0.16
Genotypic value, G	+ 0.44	− 1.56	− 7.56	+ 2.24	+ 0.24	− 5.76
Breeding value, A	+ 0.48	− 1.92	− 4.32	+ 2.88	− 0.72	− 4.32
Dominance dev., D	− 0.04	+ 0.36	− 3.24	− 0.64	+ 0.96	− 1.44

The relations between genotypic values, breeding values and dominance deviations can be illustrated graphically, as in Fig. 7.2, and the meaning of the dominance deviation is perhaps more easily understood in this way. In the figure the genotypic values (closed circles) are plotted against the number of A_1 genes in the genotype. A straight regression line is fitted by least squares to these points, each point being weighted by the frequency of the genotype it represents. The position of this line gives the breeding values of each genotype, as shown by the open circles. The differences between the breeding values and the genotypic values are the dominance deviations, indicated by vertical dotted lines. The cross

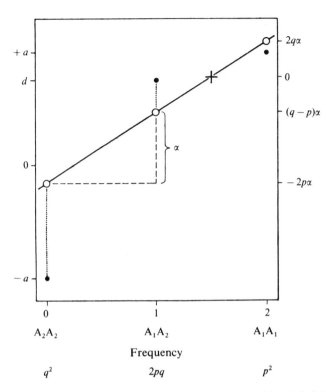

Fig. 7.2. Graphical representation of genotypic values (closed circles), and breeding values (open circles), of the genotypes for a locus with two alleles, A_1 and A_2, at frequencies p and q, as explained in the text. Horizontal scale: number of A_1 genes in the genotype. Vertical scales of value: on left – arbitrary values assigned as in Fig. 7.1; on right – deviations from the population mean. The figure is drawn to scale for the values: $d = \frac{3}{4}a$, and $q = \frac{1}{4}$.

marks the population mean. The average effect α of the gene-substitution is given by the difference in breeding value between A_2A_2 and A_1A_2, or between A_1A_2 and A_1A_1, as indicated. The original definition of the average effect of a gene-substitution was given by Fisher (1918, 1941) in terms of this linear regression of genotypic value on number of genes.

The dominance deviation can be expressed in terms of the arbitrarily assigned genotypic values a and d, by subtraction of the breeding value from the genotypic value, as shown in Table 7.3. The genotypic values must first be converted to deviations from the population mean, because the breeding values have been expressed in this way. The genotypic values, so converted, are given in two forms: in terms of a and in terms of α. Let us take the genotype A_1A_1 to show how these are obtained and how the dominance deviation is obtained by subtraction of the breeding value. The arbitrarily assigned genotypic value of A_1A_1 is $+a$, and the population mean is $a(p-q) + 2dpq$. Expressed as a deviation from the population mean, the genotypic value is therefore

$$a - [a(p-q) + 2dpq] = a(1-p+q) - 2dpq = 2qa - 2dpq = 2q(a - dp)$$

Table 7.3 Values of genotypes in a two-allele system, measured as deviations from the population mean.
Population mean : $M = a(p - q) + 2dpq$
Average effect of gene-substitution: $\alpha = a + d(q - p)$

	Genotypes		
	A_1A_1	A_1A_2	A_2A_2
Frequencies	p^2	$2pq$	q^2
Assigned values	a	d	$-a$
Deviations from population mean:			
Genotypic value	$\begin{cases} 2q(a - pd) \\ 2q(\alpha - qd) \end{cases}$	$\begin{matrix} a(q - p) + d(1 - 2pq) \\ (q - p)\alpha + 2pqd \end{matrix}$	$\begin{matrix} -2p(a + qd) \\ -2p(\alpha + pd) \end{matrix}$
Breeding value	$2q\alpha$	$(q - p)\alpha$	$-2p\alpha$
Dominance deviation	$-2q^2d$	$2pqd$	$-2p^2d$

This may be expressed in terms of the average effect α by substituting $a = \alpha - d(q - p)$ (from equation [7.5]), and the genotypic value then becomes $2q(\alpha - qd)$. Subtraction of the breeding value, $2q\alpha$, gives the dominance deviation as $-2q^2d$. By similar reasoning the dominance deviation of A_1A_2 is $2pqd$, and that of A_2A_2 is $-2p^2d$. Thus all the dominance deviations are functions of d. If there is no dominance, d is zero and the dominance deviations are also all zero. Therefore in the absence of dominance, breeding values and genotypic values are the same. Genes that show no dominance ($d = 0$) are sometimes called 'additive genes', or are said to 'act additively'.

Since the mean breeding value and the mean genotypic value are equal, it follows that the mean dominance deviation is zero. This can be verified by multiplying the dominance deviation by the frequency of each genotype and summing. The mean dominance deviation is thus

$$-2p^2q^2d + 4p^2q^2d - 2p^2q^2d = 0$$

Interaction deviation

When only a single locus is under consideration, the genotypic value is made up of the breeding value and the dominance deviation only. But when the genotype refers to more than one locus, the genotypic value may contain an additional deviation due to non-additive combination. Let G_A be the genotypic value of an individual attributable to one locus, G_B that attributable to a second locus, and G the aggregate genotypic value attributable to both loci together. Then

$$G = G_A + G_B + I_{AB} \qquad \ldots [7.9]$$

where I_{AB} is the deviation from additive combination of these genotypic values. In dealing with the population mean, earlier in this chapter, we assumed that I was zero for all combinations of genotypes. If I is not zero for any combination of genes at different loci, those genes are said to 'interact' or to exhibit 'epistasis', the term epistasis being given a wider meaning in quantitative genetics than in Mendelian genetics. The deviation I is called the *interaction deviation* or *epistatic deviation*. If the interaction deviation is zero the genes concerned are said to 'act additively' between loci. Thus 'additive action' may mean two different things. Referred to genes at one locus it means the absence of dominance, and referred to genes at different loci it means the absence of epistasis.

Loci may interact in pairs or in threes or higher numbers, and the interactions may be of several different sorts, as the behaviour of major genes shows. The complex nature of the interactions, however, need not concern us, because in the aggregate genotypic value interactions of all sorts are treated together as a single interaction deviation. So for all loci together we can write

$$G = A + D + I \qquad \ldots [7.10]$$

where A is the sum of the breeding values attributable to the separate loci, and D is the sum of the dominance deviations.

The mean interaction deviation of all the genotypes in a population is zero, when values are expressed as deviations from the population mean. That this must be so can be seen from equation [7.10], remembering that the mean G, A, and D are all zero. The interaction deviation is not just a property of the interacting genotypes, but depends also on the frequencies of the genotypes in the population, and so on the gene frequencies.

Example 7.7 As an example of non-additive combination of two loci we shall take the same two colour genes in mice that were used in Example 7.3 to illustrate additive combination; but this time we refer to their effects on the size of the pigment granules, instead of their number (Russell, 1949). The mean size (diameter in μ) of the granules in each of the four genotypes was as shown in the table. This time the differences are not independent of the other genotype: the c^e gene, for example, has quite a large effect on the B– genotype, but none at all on the bb genotype. Thus the two loci show epistatic interaction and do not combine additively. The interaction deviations of the four genotypes in any particular population would depend on the gene frequencies at both loci.

	B –	bb	*Diff.*
C –	1.44	0.77	0.67
$c^e c^e$	0.94	0.77	0.17
Diff.	0.50	0.00	

8 VARIANCE

The genetics of a metric character centres round the study of its variation, for it is in terms of variation that the primary genetic questions are formulated. The basic idea in the study of variation is its partitioning into components attributable to different causes. The relative magnitude of these components determines the genetic properties of the population, in particular the degree of resemblance between relatives. In this chapter we shall consider the nature of these components and how the genetic components depend on the gene frequency. Then, in the next chapter, we shall show how the degree of resemblance between relatives is determined by the magnitudes of the components.

Components of variance

The amount of variation is measured and expressed as the variance: when values are expressed as deviations from the population mean the variance is simply the mean of the squared values. The components into which the variance is partitioned are the same as the components of value described in the last chapter; so that, for example, the genotypic variance is the variance of genotypic values, and the environmental variance is the variance of environmental deviations. The total variance is the phenotypic variance, or the variance of phenotypic values, and is the sum of the separate components. The components of variance and the values whose variance they measure are listed in Table 8.1.

The total variance is then, with certain qualifications, the sum of the

Table 8.1 Components of variance

Variance component	Symbol	Value whose variance is measured
Phenotypic	V_P	Phenotypic value
Genotypic	V_G	Genotypic value
Additive	V_A	Breeding value
Dominance	V_D	Dominance deviation
Interaction	V_I	Interaction deviation
Environmental	V_E	Environmental deviation

components, thus:

$$V_P = V_G + V_E \qquad \qquad \ldots [8.1a]$$
$$= V_A + V_D + V_I + V_E \qquad \ldots [8.1b]$$

The qualifications are, first, that genotypic values and environmental deviations may be correlated, in which case V_P will be increased by twice the covariance of G with E; and, second, there may be interaction between genotypes and environments, in which case there will be an additional component of variance attributable to the interaction. These two complications will be dealt with later in this chapter; meantime it will be assumed that they do not apply.

Components as proportions of the total

The partitioning of the variance into its components allows us to estimate the relative importance of the various determinants of the phenotype, in particular the role of heredity versus environment, or nature and nurture. The question of 'relative importance' can be answered only if it is expressed in terms of the variance attributable to the different sources of variation. The relative importance of a source of variation is the variance due to that source, as a proportion of the total phenotypic variance. The relative importance of heredity in determining phenotypic values is called the *heritability* of the character. There are, however, two distinctly different meanings of 'heredity', and heritability, according to whether they refer to genotypic values or to breeding values. A character can be 'hereditary' in the sense of being determined by the genotype or in the sense of being transmitted from parents to offspring, and the extent to which it is hereditary in the two senses may not be the same. The ratio V_G/V_P expresses the extent to which individuals' phenotypes are determined by their genotypes. This is called the *heritability in the broad sense*, or the *degree of genetic determination*. The ratio V_A/V_P expresses the extent to which phenotypes are determined by the genes transmitted from the parents. This is called the *heritability in the narrow sense*, or simply the *heritability*. In all that follows, the term 'heritability' will be restricted to mean the narrow-sense heritability, V_A/V_P. The heritability V_A/V_P determines the degree of resemblance between relatives and is therefore of the greatest importance in breeding programmes. The degree of genetic determination V_G/V_P is of more theoretical interest than practical importance. It can be estimated in the following way.

Estimation of the degree of genetic determination, V_G/V_P

Estimation of the genotypic variance V_G is simple in theory though not so easy in practice. Neither the genotypic nor the environmental components of variance, V_G and V_E, can be estimated directly from observations on a single population, but in certain circumstances they can be estimated in experimental populations. If one or other component could be completely eliminated, the remaining phenotypic variance would provide an estimate of the remaining component. Environmental variance cannot be removed because it includes by definition all non-genetic variance, and much of this is beyond experimental control. Elimination of genotypic variance can, however, be achieved experimentally. Individuals with identical genotypes can be obtained from a highly inbred line or the F_1 of a cross between two such lines, or from a clone propagated from a single individual. If a group of such individuals is raised under the normal range of environmental circumstances, their phenotypic variance provides an estimate of the environmental variance V_E. Subtraction of this from the phenotypic variance of a genetically mixed population then gives an estimate of the genotypic variance of this population. This estimation is illustrated in the following example.

Example 8.1 Partitioning of the phenotypic variance into its genotypic and environmental components has been done for several characters in *Drosophila melanogaster*. The results are given later, in Table 8.2, but here we may describe the results for one character in more detail in order to show how the partitioning is made. The character is the length of the thorax (in units of 1/100 mm), which may be regarded as a measure of body-size. The phenotypic variance was measured in a genetically mixed, i.e., a random-bred, population, and in a genetically uniform population, consisting of the F_1 generation of three crosses between highly inbred lines. The first estimates the genotypic and environmental variance together, and the second estimates the environmental variance alone, as shown in the table. So, by subtraction, an estimate of the genotypic variance is obtained. This shows that 49 per cent of the total variation of thorax length in the genetically mixed population is attributable to genetic differences between individuals and 51 per cent to non-genetic differences. (Data from F. W. Robertson, 1957b).

Population	Components	Observed variance
Mixed	$V_G + V_E$	0.366
Uniform	V_E	0.186
Difference	V_G	0.180
	$V_G/V_P =$	$0.180/0.366 = 49\%$

The estimation of the genotypic variance in the manner described above is not quite as straightforward as it may seem. It rests on the assumption that the environmental variance is the same in all genotypes, and this is certainly not always true. The environmental variance measured in one inbred line or cross is that shown by this one particular genotype, and other genotypes may be more or less sensitive to environmental influences and may therefore show more or less environmental variance. The environmental variance of the mixed population may therefore not be the same as that measured in the genotypically uniform group. Furthermore, some characters have been found to be more variable among inbred, homozygous, individuals than among cross-bred, heterozygous, individuals, the homozygotes being more sensitive to environmental differences. For these reasons it is necessary to estimate the environmental variance from several different genetically uniform groups, and preferably from F_1 crosses rather than inbred lines. Different sensitivities to the environment are an aspect of genotype–environment interaction which is discussed more fully later in this chapter. Individuals of identical genotype are also provided by identical twins in man and cattle, but their use in partitioning the variance is very limited: they will be discussed in Chapter 10 when the problems that they raise will be better understood.

The difficulties arising from the use of unrepresentative genotypes can be overcome with plants that can be propagated clonally. It is then possible to have a large number of genotypes each represented by many clonally produced individuals. If the individuals are grown with genotypes randomized with respect to the environment, an analysis of variance provides an estimate of the component of variance between clones. The variance between clones is due mainly to differences of genotype, and can be regarded as an estimate of V_G. But there may be environmental effects included in it. That is to say, some part of the environmental differences between individuals may be transmitted to all their clonal descendants. For this reason the ratio V_G/V_P estimated in this way is liable to be an overestimate. Strictly speaking, it should be called the 'clonal repeatability'; the meaning of 'repeatability' is explained later in this chapter.

Genetic components of variance

The partition into genotypic and environmental variance does not take us far toward an understanding of the genetic properties of a population, and in particular it does not reveal the cause of resemblance between relatives. The genotypic variance must be further divided according to the division of genotypic value into breeding value, dominance deviation, and interaction deviation. Thus we have:

Values	G	$=$	A	$+$	D	$+$	I	
Variance components	V_G	$=$	V_A	$+$	V_D	$+$	V_I	$\dots$ [8.2]
	(genotypic)		(additive)		(dominance)		(interaction)	

The *additive variance*, which is the variance of breeding values, is the important component since, as already mentioned, it is the chief cause of resemblance between relatives and therefore the chief determinant of the observable genetic properties of the population and of the response of the population to selection. Moreover, it is the only component that can be readily estimated from observations made on the population. In practice, therefore, the important partition is into additive genetic variance versus all the rest, the rest being non-additive genetic and environmental variance. This partitioning yields the ratio V_A/V_P, which is the heritability of the character.

Estimation of the additive variance rests on observation of the degree of resemblance between relatives, and will be described later when we have discussed the causes of resemblance between relatives. Our immediate concern here is to show how the genetic components of variance are influenced by the gene frequency. To do this we have to express the variance in terms of the gene frequency and the assigned genotypic values a and d. We shall consider first a single locus with two alleles, thus excluding interaction variance for the moment.

Additive and dominance variance

The information needed to obtain expressions for the variance of breeding values and the variance of dominance deviations was given in the last chapter in Table 7.3. This table gives the breeding values and dominance deviations of the three genotypes, expressed as deviations from the population mean. It will be remembered that the means of both breeding values and dominance deviations are zero. Therefore no correction for an assumed mean is needed, and the variance is simply the mean of the squared values. The variances are thus obtained by squaring the values in the table, multiplying by the frequency of the genotype concerned, and summing over the three genotypes. The additive variance, which is the variance of breeding values, is obtained as follows:

$$\begin{aligned}
V_A &= 4p^2q^2\alpha^2 + 2pq(q-p)^2\alpha^2 + 4p^2q^2\alpha^2 \\
&= 2pq\alpha^2(2pq + q^2 - 2pq + p^2 + 2pq) \\
&= 2pq\alpha^2(p^2 + 2pq + q^2) \\
&= 2pq\alpha^2 & \dots [8.3a] \\
&= 2pq[a + d(q-p)]^2 & \dots [8.3b]
\end{aligned}$$

Similarly, the variance of dominance deviations is

$$\begin{aligned}
V_D &= d^2(4q^4p^2 + 8p^3q^3 + 4p^4q^2) \\
&= 4p^2q^2d^2(q^2 + 2pq + p^2) \\
&= (2pqd)^2 & \dots [8.4]
\end{aligned}$$

If there is no dominance at the locus under consideration ($d = 0$), the expression for the additive variance simplifies to

$$V_A = 2pqa^2 \qquad \dots [8.5]$$

If there is complete dominance ($d = a$), the additive variance becomes

$$V_A = 8pq^3a^2 \qquad \dots [8.6]$$

With any degree of dominance, the expressions for both the additive and the dominance variances become much simpler if the gene frequencies are one-half, $p = q = 0.5$. Equations [8.3b] and [8.4] then reduce to

$$\left.\begin{aligned}
V_A &= \tfrac{1}{2}a^2 \\
V_D &= \tfrac{1}{4}d^2
\end{aligned}\right\} \qquad \dots [8.7]$$

Populations in which all segregating genes have frequencies of one-half are the F_2 and subsequent generations derived from a cross of two highly inbred, homozygous, lines. For a full account of the analysis of such populations, see Mather and Jinks (1971, 1977).

Total genetic variance

The total genetic variance, V_G, arising from one locus can be calculated directly from Table 7.3 in the same way as was done above for V_A and V_D. But to do so requires a lengthy algebraic reduction, and it is simpler to get V_G from the values of V_A and V_D calculated above. Since $G = A + D$, the variance of G is given by $V_G = V_A + V_D + 2\text{cov}_{AD}$, where cov_{AD} is the covariance of breeding values with dominance deviations. This covariance can be shown to be zero as follows. The breeding values, dominance deviations, and frequencies of the three genotypes were given in Table 7.3. Multiplying breeding value by dominance deviation by frequency, and summing over the three genotypes, gives the covariance as

$$-4p^2q^3\alpha d + 4p^2q^2(q-p)\alpha d + 4p^3q^2\alpha d = 4p^2q^2\alpha d(-q + q - p + p) = 0$$

Thus

$$V_G = V_A + V_D$$
$$= 2pq[a + d(q - p)]^2 + [2pqd]^2 \qquad \ldots [8.8]$$

Example 8.2 To illustrate the genetic components of variance arising from a single locus, let us return to the pygmy gene in mice, used for several examples in the last chapter. From the values tabulated in Example 7.6, we may compute the components of variance directly. Since the values are expressed as deviations from the population mean, the variance is obtained by multiplying the frequency of each genotype by the square of its value, and summing over the three genotypes. For example, the genotypic variance when $q = 0.1$ is $0.81(0.44)^2 + 0.18(-1.56)^2 + 0.01(-7.56)^2 = 1.1664$. The additive variance is obtained in the same way from the variance of breeding values, and the dominance variance from the variance of dominance deviations. The variances obtained are as follows:

	$q = 0.1$	$q = 0.4$
Genotypic, V_G	1.1664	7.1424
Additive, V_A	1.0368	6.2208
Dominance, V_D	0.1296	0.9216

The variances may be obtained also, and with less trouble, by use of the formulae given above in equations [8.3], [8.4], and [8.8]. The values to be substituted were given in Example 7.1; namely, $a = 4$ and $d = 2$. Notice that the dominance variance is quite small in comparison with the additive.

The ways in which the gene frequency and the degree of dominance influence the magnitude of the genetic components of variance can best be appreciated from graphical representations of the relationships derived above in equations [8.3], [8.4], and [8.8]. The graphs in Fig. 8.1 show the amounts of genotypic, additive, and dominance variance arising from a single locus with two alleles, plotted against the gene frequency. Three cases are shown to illustrate the effect of different degrees of dominance: in graph (a) there is no dominance $(d = 0)$; in graph (b) there is complete dominance $(d = a)$; and in graph (c) there is 'pure' overdominance $(a = 0)$. In the first case the genotypic variance is all additive, and it is greatest when $p = q = 0.5$. In the second case the dominance variance is maximal when $p = q = 0.5$, the additive variance is maximal when the frequency of the recessive allele is $q = 0.75$, and the genotypic variance is maximal when $q^2 = 0.5$, i.e., $q = 0.71$. In the third case the dominance variance is the same as in the second and is maximal when $p = q = 0.5$. The additive variance, however, is zero when $p = q = 0.5$, and has two maxima, one at $q = 0.15$ and the other at $q = 0.85$. The genotypic variance, in this case, remains practically constant over a wide range of gene frequency, though its composition changes profoundly. The general conclusion to be drawn from these graphs is that genes contribute much more variance when at intermediate frequencies than when at high or low frequencies: recessives at low frequency, in particular, contribute very little variance.

The foregoing account of the genetic variance is mainly theoretical. In practice we are not concerned with gene frequencies or gene effects because these are not known except in specially constructed populations. In practice, therefore,

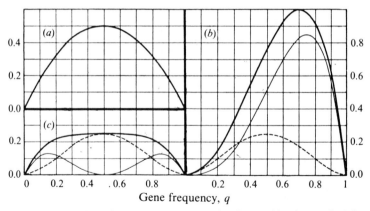

Fig. 8.1. Magnitude of the genetic components of variance arising from a single locus with two alleles, in relation to the gene frequency. Genotypic variance – thick lines; additive variance – thin lines; dominance variance – broken lines. The gene frequency, q, is that of the recessive allele. The degrees of dominance are: in (a) no dominance ($d = 0$); in (b) complete dominance ($d = a$); and in (c) 'pure' overdominance ($a = 0$). The figures on the vertical scale, showing the amount of variance, are to be multiplied by a^2 in graphs (a) and (b), and by d^2 in graph (c).

we are concerned only with the estimation of the components. It should be noted, however, that all the components of genetic variance are dependent on the gene frequencies, so any estimates of them are valid only for the population from which they are estimated. When observations on the resemblance between relatives are available, the additive genetic variance can be estimated. (The way in which this is done is the subject of the next two chapters.) The total phenotypic variance can then be partitioned into $V_A : (V_D + V_I + V_E)$, the non-additive genetic variance being included with the environmental variance. If inbred lines are available the environmental variance can be estimated and so the phenotypic variance can be partitioned into $V_G : V_E$.

If both these partitions are made, we can separate the additive genetic from the rest of the genetic variance, and so make the three-fold partition into additive genetic, non-additive genetic, and environmental variance, $V_A : (V_D + V_I) : V_E$, the dominance and interaction components being lumped together as non-additive genetic variance. Examples of this partitioning are given in Table 8.2.

A possible misunderstanding about the concept of additive genetic variance, to which the terminology may give rise, should be mentioned here. The concept of additive variance does not carry with it the assumption of additive gene action; and the existence of additive variance is not an indication that any of the genes act additively (i.e., show neither dominance nor epistasis). No assumption is made about the mode of action of the genes concerned. Additive variance can arise from genes with any degree of dominance or epistasis, and only if we find that all the genotypic variance is additive can we conclude that the genes show neither dominance nor epistasis.

The existence of more than two alleles at a locus introduces no new principle, though it complicates the theoretical description of the effect of the locus. Expressions for the additive and dominance variances are given by Kempthorne (1955a). The locus contributes additive variance arising from the average effects of

Table 8.2 Partitioning of the variance of four characters in *Drosophila melanogaster*. Components as percentages of the total, phenotypic, variance.

		Character			
		(1) Bristles	(2) Thorax	(3) Ovary	(4) Eggs
Phenotypic	V_P	100	100	100	100
Additive genetic	V_A	52	43	30	18
Non-additive genetic	$V_D + V_I$	9	6	35	44
Environmental	V_E	39	51	35	38

Characters and sources of data:
(1) Number of bristles on 4th + 5th abdominal segments (Clayton, Morris, and Robertson, 1957; Reeve and Robertson, 1954).
(2) Length of thorax (Robertson, 1957b).
(3) Size of ovaries, i.e. number of ovarioles in both ovaries. (Robertson, 1957a).
(4) Number of eggs laid in 4 days (4th to 8th after emergence) (Robertson, 1957b).

its several alleles, and dominance variance arising from the several dominance deviations.

To arrive at the variance components expressed in the population, the separate effects of all loci that contribute variance have to be combined. When a random-mating population is in equilibrium, the additive variance arising from all loci together is the sum of the additive variances attributable to each locus separately; and the dominance variance is similarly the sum of the separate contributions. But when more than one locus is under consideration then the interaction deviations, if present, give rise to another component of variance, the interaction variance, which is the variance of the interaction deviations.

Interaction variance

If the genotypes at different loci show epistatic interaction, in the manner described in the previous chapter, then the interactions give rise to a component of variance V_I, which is the variance of the interaction deviations. Theoretical description of the properties of interaction variance rests on its further subdivision into components. It is first subdivided according to the number of loci involved: two-factor interaction arises from the interaction of two loci, three-factor from three loci, etc. Interactions involving larger numbers of loci contribute so little variance that they can be ignored, and we shall confine our attention to two-factor interactions since these suffice to illustrate the principles involved. The next subdivision of the interaction variance is according to whether the interaction involves breeding values or dominance deviations. There are thus three sorts of two-factor interactions. Interaction between the two breeding values gives rise to additive × additive variance, V_{AA}; interaction between the breeding value of one locus and the dominance deviation of the other gives rise to additive × dominance variance, V_{AD}; and interaction between the two dominance deviations gives rise to dominance × dominance variance, V_{DD}. So the interaction variance is broken down into components thus:

$$V_I = V_{AA} + V_{AD} + V_{DD} + \text{etc.} \qquad \dots [8.9]$$

the terms designated 'etc.' being similar components arising from interactions between more than two loci.

There is no doubt that interaction between loci controlling quantitative characters is a frequent occurrence: it has been demonstrated in many studies of *Drosophila*, for example (see Kearsey and Kojima, 1967). It is not easy, however, to estimate the amount of variance that it generates, and little is known about its relative importance as a source of variation. For further details of the interaction variance and its estimation, see Cockerham (1954, 1963), Kempthorne (1954, 1955a, b), and Crow and Kimura (1970).

In the partitioning of the variance by relatively simple experiments, such as are considered here, most of the interaction variance is included with the dominance component, which is then referred to as *non-additive genetic* variance. This is as far as we can go here in the description of the interaction variance, but we shall see in the next chapter how it contributes to the resemblance between relatives.

Variance due to disequilibrium

There is one additional source of genetic variance that must be mentioned at this point, although it will concern us only at a few places in later chapters. It arises when a population is not in equilibrium under random mating. Disequilibrium exists when the genotype frequencies at two or more loci considered jointly are not what would be expected from the gene frequencies. The disequilibrium introduces an additional source of genetic variance for the following reason. For simplicity, consider just two loci which do not interact in the manner described above. Let G' and G'' be genotypic values of individuals with respect to each locus separately, and let G be the genotypic value with respect to both jointly, i.e., $G = G' + G''$. The total genotypic variance caused by the two loci together is then

$$V_G = V_{G'} + V_{G''} + 2\text{cov}_{G'G''} \qquad \ldots [8.10]$$

The covariance term represents correlation between the genotypic values at the two loci in different individuals. The correlation can be positive or negative, so disequilibrium can either increase or decrease the variance. When more than two loci are to be considered, there will be a covariance term for each pair of loci. When there is no disequilibrium, all the covariance terms are zero and the variance is as described in the previous sections.

There are two forms of non-random mating that generate disequilibrium, and they differ in the way they produce the covariance in equation [8.10]. The first occurs when parents are not a random sample of the individuals in their generation. Selection of parents, which is the subject of later chapters, constitutes non-random mating of this sort. The second form of non-random mating is assortative mating, as described in Chapter 1. The two sorts of covariance produced represent different correlations of gene effects. First, there is a correlation between genes at different loci in the same gamete. This is gametic phase, or linkage, disequilibrium, which was explained in Chapter 1. The second sort of covariance represents correlation between the genes in uniting pairs of gametes, i.e., between the genes an individual receives from its two parents. The first form of non-random mating alone, i.e., selection of parents which are then random-mated, generates the first sort of covariance, that due to gametic phase disequilibrium. The second form of non-random mating, i.e., assortative mating, generates both sorts of covariance. If the source of the disequilibrium ceases to operate, the covariance that is not due to gametic phase disequilibrium disappears immediately. The gametic phase disequilibrium of unlinked loci is halved in each subsequent generation, but with linked loci it persists for longer.

Correlation and interaction between genotype and environment

Two complications arise in connection with the partitioning of the variance into genotypic and environmental components as expressed in equation [8.1]. These can both normally be neglected without seriously affecting the conclusions drawn from partitioning the variance, but it is important to know what the consequences of neglecting them are.

Correlation

In the foregoing account of the variance components it has been assumed that environmental deviations and genotypic values are independent of each other; in other words, that there is no correlation between genotypic value and environmental deviation, such as would arise if the better genotypes were given better environments. Correlation between genotype and environment is seldom an important complication, and can usually be neglected in experimental populations, where randomization of environment is one of the chief objects of experimental design. There are some situations, however, in which the correlation exists. Milk-yield in dairy cattle provides an example. The normal practice of dairy husbandry is to feed cows according to their yield, the better phenotypes being given more food. This introduces a correlation between phenotypic value and environmental deviation; and, since genotypic and phenotypic values are correlated, there is also a correlation between genotypic value and environmental deviation. Another example is human intelligence. The phenotypic values of the parents affect the environment in which the children grow up; so, to the extent that intelligence is inherited, this introduces a correlation between the genotype and the environment of the children. Equation [8.1a] is true only if environmental deviations and genotypic values are uncorrelated. When a correlation is present the phenotypic variance is increased by twice the covariance of genotypic values and environmental deviations, and equation [8,1a] becomes

$$V_P = V_G + V_E + 2\text{cov}_{GE} \qquad \ldots [8.11]$$

The consequence of neglecting the correlation can be seen by consideration of the estimation of V_G illustrated in Example 8.1. The genetically uniform group, being subject to only one source of variance, can have no covariance. Its phenotypic variance therefore still estimates V_E. The genetically mixed group, however, has its phenotypic variance as in equation [8.11]. Subtraction of the two phenotypic variances therefore yields an estimate of $V_G + 2\text{cov}_{GE}$. Thus, neglect of a correlation between genotype and environment leads to the covariance being included with the genotypic variance. This is not unreasonable with cows' milk-yield, because the environment is a direct consequence of the genotypic value, and the non-random aspects of the environment can be thought of as part of the individual's genotype. It is less satisfactory with human intelligence because the environmental effects on the children are not a consequence of their own genotypes but of their parents' genotypes.

Interaction

Another assumption that has been made, which is not always justifiable, is that a specific difference of environment has the same effect on different genotypes; or, in other words, that we can associate a certain environmental deviation with a specific difference of environment, irrespective of the genotype on which it acts. When this is not so there is an interaction, in the statistical sense, between genotypes and environments. There are several forms which this interaction may take. For example, a specific difference of environment may have a greater effect on some genotypes than on others; or there may be a change in the order of merit of a series of genotypes when measured under different environments. That is to say, genotype A may be superior to genotype B in environment X, but inferior in environment Y.

When interaction between genotypes and environments is present, the phenotypic value of an individual is not simply $P = G + E$, as in equation [7.1], but includes also an interaction component: $P = G + E + I_{GE}$. The interaction component gives rise to an additional source of variation and equation [8.1a] becomes $V_P = V_G + V_E + V_{GE}$.

In an experiment of the sort illustrated in Example 8.1 the genetically uniform group is a single genotype. Its variance is due entirely to environmental differences among individuals, and depends on the way in which the particular genotype responds to the environmental differences. Therefore the variance due to interaction is included with the environmental variance estimated from the phenotypic variance of that genotype. Some genotypes, as already noted, may be more sensitive than others to environmental differences. So, to some extent, the environmental variance is a property of the genotype. But the source of the variation is environmental and not genetic. It is therefore logical, as well as experimentally necessary, to regard any variance due to genotype–environment interaction as being part of the environmental variance included in any estimate of V_E.

Genotype–environment interaction becomes very important if individuals of a particular population are to be reared under different conditions. For example, a breed of livestock may be used by different farmers who treat it differently; and varieties of plants are grown in different seasons, at different places, and under different conditions. This situation is different in one respect from what we have been considering hitherto. The different farms, seasons, or locations are 'specific environments', shared by all the individuals in them, and are more in the nature of 'treatments'. In the situation considered hitherto, each individual has its own particular environment, and individuals cannot be grouped according to any particular aspect of their environments, such as nutrition, temperature, or crowding. When individuals are reared in specific environments the genotype–environment interaction can be studied in more detail. If genotypes can be replicated, and more than one individual of each of several genotypes are reared in different specific environments, then an analysis of variance in a two-way classification of genotypes × environments will yield estimates of the variance between genotypes, the variance between the specific environments, and the variance attributable to interaction of genotypes with environments. If there is no interaction, then the best genotype in one environment will be the best in all. But if there is much interaction then particular genotypes must be sought for particular environments. The specialization of breeds or varieties for specific environments will be taken up again later, in Chapter 19, because it can be discussed more usefully from a different viewpoint. We shall next consider the idea of environmental sensitivity and how it can be measured.

Environmental sensitivity. Some of the genotype–environment interaction can be ascribed to differences of sensitivity of different genotypes. In other words, a given environmental difference has more effect on some genotypes than it has on others. To measure environmental sensitivities, and to see how much of the interaction variance is ascribable to differences of sensitivity, different genotypes are reared or grown in a range of specific environments. The specific environments have to be quantified as more or less favourable for expression of the character under study. The only way in which environments can be quantified is by the mean performance of all the genotypes. In other words, the measure of an environment is the mean of all genotypes in that environment. This will be called the *environmental value.* Each genotype has its own mean value in each of the specific environments. The genotype's environmental sensitivity is then the regression of its own value on the environmental value. The procedure will be made clearer by Example 8.3 below. The variance due to interaction of genotypes with the specific environments is estimated from an analysis of variance, and the amount attributable to differences of sensitivity is obtained from the heterogeneity of regression slopes. For details, see Perkins and Jinks (1968); and for another example, see Zuberi and Gale (1976).

Example 8.3 (*From data kindly supplied by Professor J.L Jinks*) Ten inbred lines of the tobacco plant *Nicotiana rustica* were grown in each of eight specific environments created by different dates of sowing and different densities of planting. The final heights of 8 plants of each line in each environment were measured. An analysis of variance showed that the differences between the genotypes (lines) and between the environments were significant and that there was a significant genotype × environment interaction. The environmental sensitivities of four of the genotypes are depicted in Fig. 8.2. To estimate the environmental sensitivity of a genotype, the general effect of each environment is first evaluated as the mean of all 10 genotypes in that environment. Then the value of each genotype is plotted against the environmental mean. The slope of the regression line measures the environmental sensitivity of the genotype. A regression coefficient of 1.0 represents the average sensitivity of all genotypes. The sensitivities of the four genotypes in the graphs are shown at the right-hand margin. Of particular interest are the two genotypes in the middle, which had the highest and lowest sensitivities of the 10 lines. They had nearly equal means over all environments, but in consequence of their different sensitivities one was taller in good environments and the other was taller in poor environments, a reversal of the order of merit.

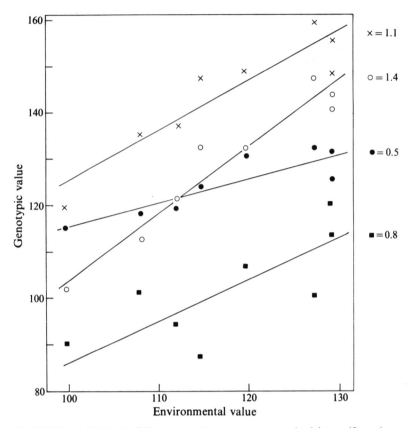

Fig. 8.2. Plant height (cm) of *Nicotiana rustica* genotypes grown in eight specific environments as explained in Example 8.3. (*Drawn from data kindly supplied by Professor J. L. Jinks.*)

Environmental variance

Environmental variance, which by definition embraces all variation of non-genetic origin, can have a great variety of causes and its nature depends very much on the character and the organism studied. Generally speaking, environmental variance is a source of error that reduces precision in genetic studies and the aim of the experimenter or breeder is therefore to reduce it as much as possible by careful management or proper design of experiments. Nutritional and climatic factors are the commonest external causes of environmental variation, and they are at least partly under experimental control. Maternal effects form another source of environmental variation that is sometimes important, particularly in mammals, but is less susceptible to control. Maternal effects are prenatal and postnatal influences, mainly nutritional, of the mother on her young: we shall have more to say about them in the next chapter in connection with resemblance between relatives. Error of measurement is another source of variation, though it is usually quite trivial. When a character can be measured in units of length or weight it is usually measured so accurately that the variance attributable to measurement is negligible in comparison with the rest of the variance. Some characters, however, cannot strictly speaking be measured, but have to be graded by judgement into classes. Carcass qualities of livestock are an example. With such characters the variance due to measurement may be considerable.

In addition to the variation arising from recognizable causes, such as those mentioned, there is usually also a substantial amount of non-genetic variation whose cause is unknown, and which therefore cannot be eliminated by experimental design. This is generally referred to as 'intangible' variation. Some of the intangible variation may be caused by 'environmental' circumstances, in the common meaning of the word – that is, by circumstances external to the individual – even though their nature is not known. Some, however, may arise from 'developmental' variation: variation, that is, which cannot be attributed to external circumstances, but is attributed, in ignorance of its exact nature, to 'accidents' or 'errors' of development as a general cause. Characters whose intangible variation is predominantly developmental are those connected with anatomical structure, which do not change after development is complete, such as skeletal form, pigmentation, or bristle number in *Drosophila*. Characters more susceptible to the influences of the external environment, in contrast, are those connected with metabolic processes, such as growth, fertility, and lactation.

Example 8.4 Human birth weight provides an example of a character subject to much environmental variation whose nature has been analysed in detail (Penrose, 1954; Robson, 1955). The partitioning of the phenotypic variance given in the table shows the relative importance of all the identified sources of variation, birth weight being regarded as a character of the child. All the environmental variation is 'maternal' in the sense that it is connected with the prenatal environment, but several distinct components of the maternal environment are distinguished. 'Maternal genotype', which accounts for 20 per cent of the total phenotypic variance, reflects genetic variation (chiefly additive) between mothers in the birth weight of their children; i.e., birth weight regarded as a character of the mother. 'Maternal environment, general', which accounts for another 18 per cent, reflects non-genetic

Partitioning of variance of human birth-weight. Components as percentages of the total, phenotypic, variance.

Cause of variation	% of total
Genetic	
Additive	15
Non-additive (approx)	1
Sex	2
Total genotypic	18
Environmental	
Maternal genotype	20
Maternal environment, general	18
Maternal environment, immediate	6
Age of mother	1
Parity	7
Intangible	30
Total environmental	82

variation between mothers in the same way. These two components, totalling 38 per cent, are maternal causes of variation in birth weight that affect all children of the same mother alike. 'Maternal environment, immediate' means causes attributable to the mother but differing in successive pregnancies. Two causes of the same nature – 'Age of mother' and 'Parity' (i.e., whether the child is the first, second, etc.) – are separately identifiable. Finally, the 'Intangible' variation is all the remainder, of which the cause cannot be identified. To explain how these various components were estimated would take too much space, and could not properly be done until the end of Chapter 10. It must suffice to say that the estimates all come from comparisons of the degree of resemblance between identical twins, fraternal twins, full sibs, children of sisters, and other sorts of cousins. It should be noted that the estimates are not very precise and must not be taken as definitive parameters of human populations. In another study (Morton, 1955) with different data, no effect of genetic factors in the foetus was found, all the variation of birth weight being environmental.

Multiple measurements: repeatability

When more than one measurement of the character can be made on each individual, the phenotypic variance can be partitioned into variance within individuals and variance between individuals. This partitioning leads to a ratio of variance components called the repeatability which has three main uses: to show how much is to be gained by the repetition of measurements, to set upper limits to the ratios V_G/V_P or V_A/V_P, and to predict future performance from past records. It may also throw light on the nature of the environmental variance. The partitioning of the variance corresponding to the repeatability is not a part of genetic theory, because it is the environmental, not the genetic, variance that is partitioned. It does, however, have some practical implications for genetical analysis and breeding programmes, as we shall see.

There are two ways by which the repetition of a character may provide multiple measurements: by temporal repetition and by spatial repetition. Milk-yield and litter size are examples of characters repeated in time. Milk-yield can be measured in successive lactations, and litter size in successive pregnancies. Several measurements of each individual can thus be obtained. The variance of yield per lactation, or of the number of young per litter, can then be analysed into a component within individuals, measuring the differences between the performances of the same individual, and a component between individuals, measuring the permanent differences between individuals. The within-individual component is entirely environmental in origin, caused by temporary differences of environment between successive performances. The between-individual component is partly environmental and partly genetic, the environmental part being caused by circumstances that affect the individuals permanently. By this analysis, therefore, the variance due to temporary environmental circumstances is separated from the rest, and can be measured.

Characters repeated in space are chiefly structural or anatomical, and are found more often in plants than in animals. For example, plants that bear more than one fruit yield more than one measurement of any character of the fruit, such as its shape or seed content. Spatial repetition in animals is chiefly found in characters that can be measured on the two sides of the body or on serially repeated parts, such as the number of bristles on the abdominal segments of

Drosophila. With spatially repeated characters the within-individual variance is again entirely environmental in origin but, unlike that of temporally repeated characters, it represents the 'developmental' variation arising from localized circumstances operating during development.

In order that we may discuss both temporal and spatial repetition together, we shall use the terms *special environmental variance*, V_{Es}, to refer to the within-individual variance arising from temporary or localized circumstances; and *general environmental variance*, V_{Eg}, to refer to the environmental variance contributing to the between-individual component and arising from permanent or non-localized circumstances. The ratio of the between-individual component to the total phenotypic variance is the intraclass correlation r. It is the correlation between repeated measurements of the same individual, and is known as the *repeatability* of the character. The partitioning of the phenotypic variance expressed by the repeatability is thus into two components, V_{Es} versus $(V_G + V_{Eg})$, so that the repeatability is

$$r = \frac{V_G + V_{Eg}}{V_P} \qquad \qquad \ldots [8.12]$$

The repeatability therefore expresses the proportion of the variance of single measurements that is due to permanent, or non-localized, differences between individuals, both genetic and environmental. It allows the separate estimation of the component V_{Es} due to the special environment which, as a proportion of the total, is given by

$$\frac{V_{Es}}{V_P} = 1 - r \qquad \qquad \ldots [8.13]$$

From equation [8.12] it can be seen that the repeatability sets an upper limit to the degree of genetic determination V_G/V_P, and to the heritability V_A/V_P. The repeatability is usually much easier to determine than either of these two ratios and it may often be known when they are not. The heritability, which is the ratio of practical importance, may of course be much less than the repeatability, but it cannot be greater, and this knowledge is better than no knowledge at all of the heritability. The repeatability differs very much according to the nature of the character, and also, of course, according to the genetic properties of the population and the environmental conditions under which the individuals are kept. The estimates in Table 8.3 give some idea of the sort of values that may be found with various characters, and two cases are described in more detail in Example 8.5.

There are two assumptions implicit in the idea of repeatability. The first is that the variances of the different measurements are equal, and have their components in the same proportions. The second is that the different measurements reflect what is genetically the same character – a point that will be explained in Chapter 19. If these assumptions are not valid, the repeatability becomes a somewhat vague concept, without precise meaning in relation to the components of variance. Some of the characters in Table 8.3 do not conform strictly with the assumptions.

Table 8.3 Some examples of repeatability

	Repeatability
Drosophila melanogaster:	
Abdominal bristle number (see Example 8.5)	0.42
Ovary size (see Table 8.4)	0.54
Mouse: (original data)	
Weight at 6 weeks	0.96
Litter size in 1st and 2nd litters	0.45
Sheep: (Morley, 1951)	
Weight of fleece in different years	0.74
Cattle (British Friesians): (Barker and Robertson, 1966)	
Milk yield in 1st and 2nd lactations	0.40
Percent fat in 1st and 2nd lactations	0.67

Example 8.5 The number of bristles on the ventral surfaces of the abdominal segments is a character that has been much studied in *Drosophila melanogaster*, because it is technically convenient and its genetic properties are relatively simple. We have already mentioned it several times but have not yet used it as an example. There are about 20 bristles on each of 3 segments in males, and each of 4 segments in females. The number of bristles per segment can therefore be treated as a spatially repeated character. The sources of variation in this character have been studied in detail by Reeve and Robertson (1954), and the components of variance found are given in the table.

		♂ ♂	♀ ♀
Total phenotypic	V_P	4.24	5.44
Between flies	$V_G + V_{Eg}$	1.82	2.19
Within flies	V_{Es}	2.42	3.25
Repeatability		0.429	0.403

Estimation of the repeatability of a character separates off the component of variance due to special environment, V_{Es}, but it leaves the other component of environmental variance – that due to general environment, V_{Eg} – confounded with the genotypic variance, as shown in the above example. To complete the partitioning we need to separate V_{Eg} from V_G. This can be done in two ways: either by estimating the genotypic variance V_G, in the manner of Example 8.1, or by calculating the repeatability in a genetically uniform group such as an inbred line or F_1 cross. Where there is no genetic variation, the between-individual component of variance consists only of the general environment component V_{Eg}, and the repeatability measures the ratio $V_{Eg}/(V_{Eg} + V_{Es})$. The environmental variance has been partitioned in this way for two characters in *Drosophila* and the full partitioning that this leads to is shown in Table 8.4. The main point of interest is the very small amount of variation arising from the general environment. These characters are therefore very little influenced by the external environment; or perhaps it would be more accurate to say that the technique of rearing the flies has been very successful in eliminating unwanted sources of environmental variation.

Table 8.4 Partitioning of the phenotypic variance of two characters in *Drosophila melanogaster*. Each component is given as a percentage of the total variance of single measurements.

Component		(1) *Bristle number*	(2) *Ovary size*
Additive genetic,	V_A	33	23
Non-additive genetic,	V_{NA}	6	27
General environment,	V_{Eg}	3	4
Special environment,	V_{Es}	58	46
Total, phenotypic,	V_P	100	100

Characters and sources of data:
(1) Counts on one abdominal segment (Reeve and Robertson, 1954). The results for males and females were calculated separately and then averaged.
(2) Number of ovarioles in one ovary (Robertson, 1957*a*).

The proportions of the genetic components here are lower than those in Table 8.2 because Table 8.2 refers to the variance of the sum of two measurements; see Example 8.6.

Under the conditions of the experiment, virtually all the non-genetic variation is due to strictly localized causes that influence the segments or the ovaries independently. Because the V_{Eg} component is so small, the repeatability gives a good estimate of the degree of genetic determination, V_G/V_P. Furthermore, the non-additive genetic variance of bristle-number is small, so the repeatability is not very different from the heritability, V_A/V_P.

Gain from multiple measurements. One way in which knowledge of the repeatability is useful is to indicate the gain in accuracy expected from multiple measurements. If the repeatability is high, little will be gained; if it is low, more will be gained. The question is: how is the gain in accuracy related to the repeatability? The only component of variance that is reduced by repeated measurements is that due to the special environment, V_{Es}, and the amount by which it is reduced depends on the number of measurements made. Suppose that each individual is measured n times, and that the mean of these n measurements is taken to be the phenotypic value of the individual, say $P_{(n)}$. Then the phenotypic variance is made up of the genotypic variance, the general environmental variance, and one nth of the special environmental variance:

$$V_{P(n)} = V_G + V_{Eg} + \frac{1}{n} V_{Es} \qquad \dots [8.14]$$

Thus, increasing the number of measurements reduces the amount of variance due to special environment that appears in the phenotypic variance, and this reduction of the phenotypic variance represents the gain in accuracy. The variance of the mean of n measurements as a proportion of the variance of one measurement can be expressed in terms of the repeatability, as follows. Writing the components in terms of r and $1 - r$, from equations [8.12] and [8.13], and substituting into equation [8.14], gives

$$V_{P(n)} = \left(r + \frac{1-r}{n}\right)V_P$$

and rearrangement leads to

$$\frac{V_{P(n)}}{V_P} = \frac{1+r(n-1)}{n} \qquad \ldots [8.15]$$

This ratio is plotted in Fig. 8.3 to show how the phenotypic variance is reduced by multiple measurements, with characters of different repeatabilities. When the repeatability is high, and there is therefore little special environmental variance, multiple measurements give little gain in accuracy. When the repeatability is low, multiple measurements may lead to a worthwhile gain in accuracy. The gain in accuracy, however, falls off rapidly as the number of measurements increases, and it is seldom worth while to make more than two or three measurements. In practice it does not make any difference whether one works with the mean or with

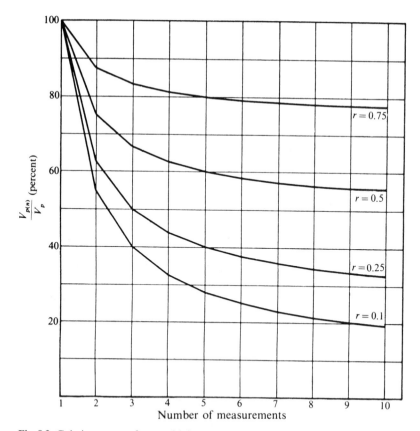

Fig. 8.3. Gain in accuracy from multiple measurements of each individual. The vertical scale gives the variance of the mean of n measurements as a percentage of the variance of one measurement. The horizontal scale gives the number of measurements, up to 10. The four graphs refer to characters of different repeatability as indicated.

the sum of the measurements: though the actual variances will be different, the relative magnitudes of the components are not affected.

Example 8.6   Most of the studies of abdominal bristle number in *Drosophila* have been based on counts of the bristles on two segments. The table shows how the percentage composition of the variance is affected by counting two segments instead of one. Column (1) gives the percentage composition of the phenotypic variance when only one segment is counted, as in Table 8.4. If two segments are counted and the mean of the two counts is taken as the phenotype of the individual, V_{Es} is halved but

	One segment	Two segments	
	(1)	(2)	(3)
V_A	33	33	46.4
V_{NA}	6	6	8.2
V_{Eg}	3	3	4.1
V_{Es}	58	29	41.3
V_P	100	71	100.0

the other components are unaltered, giving the figures in column (2). The total variance V_P is now reduced to 71. Dividing each component by 71 gives, in column (3), the percentage composition of the variance when phenotypic values are based on counts of two segments. The point of practical importance is that the additive genetic variance has been increased from 33 to 46 per cent. (The reason why V_A is 46 per cent and not 52 per cent as in Table 8.2 is that the two estimates are derived from different strains).

The advantage for breeding programmes from the gain in accuracy is the increased proportion of additive genetic variance. This increase in the proportion of additive variance, however, cannot be relied on unless the two assumptions mentioned earlier are valid, namely that the different measurements have equal variances and represent the same character genetically. These conditions are met by the number of bristles on the abdominal segments of *Drosophila*, and the conclusions reached in Example 8.6 are valid. A character for which the assumptions do not hold is milk-yield of cows in successive lactations (Rendel *et al.*, 1957). In this case the proportion of additive genetic variance is actually less for the mean of several lactations than it is for first lactations only.

Prediction of future performance.   The prediction of future performance is a problem that occurs in many contexts. It has no genetic connotation but rests on the partitioning of the variance into components due to permanent and temporary effects, i.e., the partitioning made by the repeatability. Performances, both past and future, must be thought of in terms of deviations from the population means, past and future. A good past performance is partly due to the temporary environmental effects on the individual and these are not carried through to the subsequent performance, so the future performance tends to 'regress' toward the population mean. No prediction can be made without a knowledge of the characteristics of the population with respect to the two performances, for example milk-yield in first and second lactations. The repeatability, which is the correlation between the two performances, tells us how accurately we can predict the second from a knowledge of the first. The prediction itself is made from the regression coefficient of second on first performance. If x and y are first and second performances respectively, $\bar{x}$ and $\bar{y}$ are the population means, and b is the regression coefficient of y on x, then the prediction is given by $(y - \bar{y}) = b(x - \bar{x})$. The relationship between the regression and correlation coefficients is $b = r\sigma_y/\sigma_x$, where σ_x and σ_y are the standard deviations.

Example 8.7 The prediction of future performance can be illustrated from the data of Barker and Robertson (1966) on the milk-yield of British Friesian cows. The data refer to 3,764 cows with records of yields in first, second, and third lactations. The means and standard deviations are given in the table, yields being here converted to kilograms. Both mean and standard deviation increased in successive lactations. Let us predict the mean yield in second and third lactations of a heifer with a yield of 5,000 kg in her first lactation. The repeatabilities required are the correlations of second with first and of third with first; both were 0.40. The regression, for example of second on first, is then calculated as $b = r\sigma_2/\sigma_1$. The predicted yield in the second lactation is obtained from $Y - \bar{Y} = b(X - \bar{X})$ where Y is the prediction, $\bar{Y}$ is the mean yield in second lactations, X is observed yield of 5,000 kg and $\bar{X}$ is the mean yield in first lactations. The calculations for second and third lactations and the predicted mean are set out in the table.

	Lactation		
	1st	*2nd*	*3rd*
Mean, kg	4,096	4,232	4,731
Standard deviation (σ)	696	934	960
Correlation with 1st (r)	—	0.40	0.40
Regression on 1st (b)	—	0.536	0.552

Observed yield in 1st lactation = 5,000
Deviation from mean = + 904
Predicted yield in 2nd = 4,232 + (0.536 × 904) = 4,716.5
Predicted yield in 3rd = 4,731 + (0.552 × 904) = 5,230.0

Predicted mean in 2nd and 3rd	4,973

Summary of variance partitioning

This chapter has shown how the phenotypic variance of a genetically variable population can be partitioned into four components, two genetic and two environmental. The data needed to do this are of three different kinds, each making a partition into two parts, but in different ways. Table 8.5 summarizes the different partitions that can be made.

Table 8.5 Summary of variance partitioning.

Data needed	Partition made	Ratio estimated
Resemblance between relatives	$(V_A):(V_{NA} + V_{Eg} + V_{Es})$	heritability, V_A/V_P
Genetically uniform group	$(V_A + V_{NA}):(V_{Eg} + V_{Es})$	degree of genetic
	$= (V_G):(V_E)$	determination, V_G/V_P
Multiple measurements	$(V_G + V_{Eg}):V_{Es}$	repeatability $(V_G + V_{Eg})/V_P$
All three	$V_A:V_{NA}:V_{Eg}:V_{Es}$	

9 RESEMBLANCE BETWEEN RELATIVES

The resemblance between relatives is one of the basic genetic phenomena displayed by metric characters, and the degree of resemblance is a property of the character that can be determined by relatively simple measurements made on the population without special experimental techniques. The degree of resemblance provides the means of estimating the amount of additive genetic variance, and it is the proportionate amount of additive variance (i.e., the heritability) that chiefly determines the best breeding method to be used for improvement. An understanding of the causes of resemblance between relatives is therefore fundamental to the practical study of metric characters and to its application in animal and plant improvement. In this chapter, therefore, we shall examine the causes of resemblance between relatives, and show in principle how the amount of additive variance can be estimated from the observed degree of resemblance, leaving the more practical aspects of the estimation of the heritability for consideration in the next chapter.

In the last chapter we saw how the phenotypic variance can be partitioned into components attributable to different causes. These components we shall call *causal components* of variance, and denote them as before by the symbol V. The measurement of the degree of resemblance between relatives rests on the partitioning of the phenotypic variance in a different way, into components corresponding to the grouping of the individuals into families. These components can be estimated directly from the phenotypic values and for this reason we shall call them *observational components* of phenotypic variance, and denote them by the symbol σ^2 in order to keep the distinction clear. Consider, for example, the grouping of individuals into families of full sibs. By the analysis of variance we can partition the total observed variance into two components, between (or among) groups and within groups. The between-group component is the variance of the 'true' means of the groups about the population mean, and the within-group component is the variance of individuals about the true mean of their group. The true mean of a group is the mean that would be found if it were estimated without error from a very large number of individuals. An explanation of the estimation of these two components will be given, with examples, in the next chapter. Now, the resemblance between related individuals, i.e., between full sibs in the case under discussion, can be looked at either as similarity of individuals in the same group, or as difference between individuals in different groups. The greater the similarity within the groups, the greater will be the difference between the groups. The degree of resemblance can therefore be expressed as the between-group component as a proportion of the total variance. This is the intraclass correlation coefficient and is given by

$$t = \frac{\sigma_B^2}{\sigma_B^2 + \sigma_W^2}$$

where σ_B^2 is the between-group component and σ_W^2 the within-group component. (It is customary to use the symbol t for the intraclass correlation of phenotypic values in order to avoid confusion with other correlations for which the symbol r is used.) The between-group component expresses the amount of variation that is common to members of the same group, and it can equally well be referred to as the covariance of members of the groups. In the case of the resemblance between offspring and parents, the grouping of the observations is into pairs rather than groups; one parent, or the mean of two parents, paired with one offspring or the mean of several offspring. The between-pair component of variance is then meaningful only if the parental values and the offspring values have the same variance, which they often do not. The covariance of offspring with parents is therefore calculated from the sum of cross-products, and the degree of resemblance is expressed as the regression of offspring on parents. The reason why the correlation is often inappropriate will become apparent later. The regression is given by

$$b_{OP} = \frac{\text{cov}_{OP}}{\sigma_P^2}$$

where cov_{OP} is the covariance of offspring and parents, and σ_P^2 is the variance of parents.

Thus, the covariance of related individuals is the new property of the population that we have to deduce in seeking the cause of resemblance between relatives, whether sibs or offspring and parents. The covariance, being simply a portion of the total phenotypic variance, is composed of the causal components described in the last chapter, but in amounts and proportions differing according to the sort of relationship. By finding out how the causal components contribute to the covariance, we shall see how an observed covariance can be used to estimate the causal components of which it is composed.

The commonest and most useful relationships are offspring with parents, half sibs, full sibs, and (in human studies) twins. The covariances in these relationships will be explained fully, and those of other relationships summarized afterwards. Twins have their special problems which will be dealt with in the next chapter.

Both genetic and environmental sources of variance contribute to the covariance of relatives, the covariance of phenotypic values being the sum of the genetic and environmental covariances. The genetic covariances will be described first, with the regressions or correlations that they give rise to, and the environmental causes of resemblance will be commented on later.

Genetic covariance

Our object now is to deduce from theoretical considerations the covariance of relatives arising from genetic causes, neglecting for the time being any non-genetic causes of resemblance that there may be. This means that we have to deduce the covariance of the genotypic values of the related individuals. The population will be assumed to be in Hardy–Weinberg equilibrium, all parents mating at random with respect to the character under consideration. The effects of assortative mating will be considered in the next chapter. Any variance arising from epistatic interaction between loci will at first be neglected, its effects being described briefly later. For each relationship, two ways of deducing the covariance will be described, the first being more concise and the second more explicit.

Offspring and one parent
The covariance to be deduced is that of the genotypic values of individuals with the mean genotypic values of their offspring produced by mating at random in the population. If values are expressed as deviations from the population mean, then the mean value of the offspring is by definition half the breeding value of the parent, as explained in Chapter 7. Therefore the covariance to be computed is that of an individual's genotypic value with half its breeding value, i.e., the covariance of G with $\frac{1}{2}A$. Since $G = A + D$ (D being the dominance deviation), the covariance is that of $(A + D)$ with $\frac{1}{2}A$. Multiplying these terms together and summing over all parents gives

$$\text{Sum of cross-products} = \Sigma \tfrac{1}{2}A(A + D)$$
$$= \tfrac{1}{2}\Sigma\, A^2 + \tfrac{1}{2}\Sigma\, AD$$

Dividing both sides by the number of parents gives the covariance as $\frac{1}{2}V_A + \frac{1}{2}\text{cov}_{AD}$. It was shown in the previous chapter that cov_{AD} is zero, so

$$\text{cov}_{OP} = \tfrac{1}{2}V_A \qquad\qquad \dots [9.1]$$

The genetic covariance of offspring and one parent is therefore half the additive genetic variance of the parents.

The second, more explicit, way of deriving the covariance is by consideration of the effects of single loci. This will be done by reference to a locus with two alleles but the conclusions are equally valid for loci with any number of alleles. Table 9.1 gives the genotypes of the parents, their frequencies in the population, and their genotypic values expressed as deviations from the population mean (from Table 7.3). The right-hand column gives the mean genotypic values of the offspring, which are half the breeding values of the parents as given in Table 7.3.

Table 9.1

Parents			Offspring
Genotype	Frequency	Genotypic value	Mean genotypic value
A_1A_1	p^2	$2q(\alpha - qd)$	$q\alpha$
A_1A_2	$2pq$	$(q - p)\alpha + 2pqd$	$\frac{1}{2}(q - p)\alpha$
A_2A_2	q^2	$-2p(\alpha + pd)$	$-p\alpha$

The covariance of offspring and parent is then the mean cross-product, and is obtained by multiplying together the three columns – frequency × genotypic value of parent × genotypic value of offspring – and summing over the three genotypes of the parents. After collecting together the terms in α^2 and the terms in αd we obtain

$$\begin{aligned} \text{cov}_{\text{OP}} &= pq\alpha^2(p^2 + 2pq + q^2) + 2p^2q^2\alpha d(-q + q - p + p) \\ &= pq\alpha^2 \\ &= \tfrac{1}{2}V_A \end{aligned}$$

since, from equation [8.3a], $V_A = 2pq\alpha^2$. Summing over all loci we again reach the conclusion that the covariance of offspring and one parent is equal to half the additive variance.

The regression of offspring on one parent is got by dividing the covariance by the variance of the parents, which is the phenotypic variance of the population. Thus the regression is

$$b_{\text{OP}} = \tfrac{1}{2}\frac{V_A}{V_P} \qquad\qquad \ldots [9.2]$$

The covariance was deduced above by considering the mean value of the offspring of each parent, without specifying the number of offspring on which the mean is based. In fact, the covariance is the same whatever the number of offspring, even if only one is used, for the following reason. The mean of n offspring is $(1/n)\Sigma\,\text{O}$, where $\Sigma\,\text{O}$ is the sum of the values of the offspring. The covariance of one parent with the mean of n offspring is $\text{cov}(\text{P},(1/n)\Sigma\,\text{O}) = (1/n)\Sigma\,\text{cov}(\text{P}, \text{O}) = \text{cov}(\text{P}, \text{O})$, which is the covariance of parents with any one offspring. This conclusion is applicable to relatives of any kind. In general, therefore, the covariance of any individual with the mean value of a number of relatives is equal to its covariance with any one of those relatives. The regression of offspring on parents is also unaffected by the number of offspring used, because the variance of offspring does not enter into the calculation of the regression.

Offspring and mid-parent.

The covariance of the mean of the offspring and the mean of both parents (commonly called the 'mid-parent') may be deduced in the following way. Let O be the mean of the offspring, and P and P' be the values of the two parents. The mid-parent value is $\bar{\text{P}} = \tfrac{1}{2}(\text{P} + \text{P}')$. The sum of cross-products is $\Sigma\,\text{O}\bar{\text{P}} = \tfrac{1}{2}(\Sigma\,\text{OP} + \Sigma\,\text{OP}')$, and the covariance is $\text{cov}_{\text{O}\bar{\text{P}}} = \tfrac{1}{2}(\text{cov}_{\text{OP}} + \text{cov}_{\text{OP}'})$. If P and P' have the same variance, then $\text{cov}_{\text{OP}} = \text{cov}_{\text{OP}'}$ and consequently

$$\text{cov}_{\text{O}\bar{\text{P}}} = \text{cov}_{\text{OP}} = \tfrac{1}{2}V_A \qquad\qquad \ldots [9.3]$$

Thus, provided the two sexes have equal variances, the covariance of offspring and mid-parent is the same as that of offspring with one parent, which we have seen is equal to half the additive variance.

The longer method of demonstrating the covariance of offspring with mid-parent is rather laborious, but it must be given since it will be needed for arriving at the covariance of full sibs. We shall, however, omit some of the steps of algebraic reduction. A table (Table 9.2) is made in the same manner as for offspring and one

Table 9.2

Genotype of parents		Frequencies of matings	Mid-parent value	Progeny			Mean value of progeny	Progeny mean × mid-parent	Square of progeny mean
				A_1A_1 a	A_1A_2 d	A_2A_2 $-a$			
A_1A_1	A_1A_1	p^4	a	1	—	—	a	a^2	a^2
A_1A_1	A_1A_2	$4p^3q$	$\frac{1}{2}(a+d)$	$\frac{1}{2}$	$\frac{1}{2}$	—	$\frac{1}{2}(a+d)$	$\frac{1}{4}(a^2+2ad+d^2)$	$\frac{1}{4}(a^2+2ad+d^2)$
A_1A_1	A_2A_2	$2p^2q^2$	0	—	1	—	d	0	d^2
A_1A_2	A_1A_2	$4p^2q^2$	d	$\frac{1}{4}$	$\frac{1}{2}$	$\frac{1}{4}$	$\frac{1}{2}d$	$\frac{1}{2}d^2$	$\frac{1}{4}d^2$
A_1A_2	A_2A_2	$4pq^3$	$\frac{1}{2}(-a+d)$	—	$\frac{1}{2}$	$\frac{1}{2}$	$\frac{1}{2}(-a+d)$	$\frac{1}{4}(a^2-2ad+d^2)$	$\frac{1}{4}(a^2-2ad+d^2)$
A_2A_2	A_2A_2	q^4	$-a$	—	—	1	$-a$	a^2	a^2

parent, but now we have to tabulate types of mating and their frequencies, instead of single parents. Against each type of mating we put the mean genotypic value of the two parents, i.e., the mid-parent value; then the proportions of the three genotypes among the progeny, and the mean genotypic value of the progeny. The working is made easier by writing the genotypic values in terms of a and d instead of as deviations from the population mean. The last two columns of the table give the product of progeny-mean × mid-parent, and the square of the progeny for later use. To get the covariance of progeny-mean and mid-parent value, we take the product of progeny-mean × mid-parent and multiply it by the frequency of the mating type, and then sum over mating types. This gives the mean product (MP) from which we have to deduct a correction for the population mean, since values are not here expressed as deviations from the mean. The correction is simply the square of the population mean (M^2) since the means of parents and of progeny are equal. Both MP and M^2 contain terms in a^2, in ad, and in d^2. By collecting together these terms and simplifying a little we obtain

$$MP = a^2[p^3(p+q)+q^3(p+q)] + 2adpq(p^2-q^2) + d^2pq(p^2+2pq+q^2)$$
$$M^2 = a^2(p^2-2pq+q^2) \qquad\quad + 4adpq(p-q) \quad + 4d^2p^2q^2$$

Then,

$$\text{cov}_{O\bar{P}} = MP - M^2$$
$$= a^2pq - 2adpq(p-q) + d^2pq(p-q)^2$$
$$= pq[a + d(q-p)]^2$$
$$= pq\alpha^2$$
$$= \tfrac{1}{2}V_A$$

when summed over all loci.

Though the covariance of offspring with the mean of both parents is the same as the covariance with a single parent, the degree of resemblance is not the same. The regression of offspring on mid-parent values is $b = \text{cov}_{O\bar{P}}/\sigma_{\bar{P}}^2$, where $\sigma_{\bar{P}}^2$ is the variance of mid-parent values. If the variances of the two sexes are equal, then $\sigma_{\bar{P}}^2 = \tfrac{1}{2}V_P$ because, in general, the variance of the mean of n individuals is one nth of the variance of single individuals. The regression of offspring on mid-parent values is therefore

$$b_{O\bar{P}} = \frac{\tfrac{1}{2}V_A}{\tfrac{1}{2}V_P} = \frac{V_A}{V_P} \qquad\qquad \dots [9.4]$$

which is twice the regression on single parents. As with single parents, the number of offspring used does not affect the covariance or the regression.

The regression of offspring on parents is a useful measure of the degree of resemblance because it is simply related to the causal components of variance. The correlation between offspring and parents, however, does not have this useful feature. The correlation is calculated as $\text{cov}_{OP}/\sigma_O\sigma_P$, where σ_O and σ_P are the square roots of the variances of offspring and parents respectively, whether single or the mean of more than one. So the correlation is affected by the number of offspring as well as by the number of parents. If there is only one offspring, the correlation with a single parent is the same as the regression, but under all other circumstances it is different. The correlation of one offspring with mid-parent values is $(\sqrt{\frac{1}{2}})V_A/V_P$. When there are more than one offspring, the correlation depends on the variance of the observed means of the offspring and has no simple relationship with the causal components of variance.

Half sibs

Half sibs are individuals that have one parent in common and the other parent different. A group of half sibs is therefore the progeny of one individual mated to a random group of the other sex and having one offspring by each mate. Thus the mean genotypic value of the group of half sibs is by definition half the breeding value of the common parent. The covariance is the variance of the true means of the half-sib groups, and is therefore the variance of half the breeding values of the common parents, which is a quarter of the additive variance:

$$\text{cov}_{(HS)} = V_{\frac{1}{2}A} = \tfrac{1}{4}V_A \qquad \qquad \ldots [9.5]$$

This covariance also can be demonstrated by the longer method, from the values already given in Table 9.1. The covariance is the variance of the means of the groups of offspring listed in the right-hand column. Squaring the offspring values and multiplying by their frequencies gives:

Variance of means of half-sib families

$$\begin{aligned}
&= p^2q^2\alpha^2 + 2pq.\tfrac{1}{4}(q-p)^2\alpha^2 + q^2p^2\alpha^2 \\
&= pq\alpha^2[pq + \tfrac{1}{2}(q-p)^2 + pq] \\
&= pq\alpha^2[\tfrac{1}{2}(p+q)^2] \\
&= \tfrac{1}{2}pq\alpha^2
\end{aligned}$$

Therefore, since $2pq\alpha^2 = V_A$ (from equation [8.3a]),

$$\text{cov}_{(HS)} = \tfrac{1}{4}V_A$$

summation being made over all loci.

The degree of resemblance between sibs is expressed as the intraclass correlation, which is the between-group variance, i.e., the covariance, as a proportion of the total variance. So the correlation of half sibs is

$$t = \tfrac{1}{4}\frac{V_A}{V_P} \qquad \qquad \ldots [9.6]$$

Full sibs

The covariance of full sibs is less simple than those of the relationships so far considered because the dominance variance contributes to it. Consider first the covariance due to the additive variance alone. Full sibs have both parents in common and the mean genotypic value of a group of full sibs is then equal to the mean breeding value of the two parents. Let A and A' be the breeding values of the two parents. Then the covariance is the variance of $\frac{1}{2}(A+A')$ which is $\frac{1}{4}(V_A+V_{A'})= \frac{1}{2}V_A$ if the additive variance is the same in the two sexes. Now consider the contribution of dominance. It is easier to think of the covariance being calculated from the sum of cross-products in pairs of sibs taken at random. Let the parents have genotypes A_1A_2 and A_3A_4. There are then four genotypes among the progeny, A_1A_3, A_1A_4, A_2A_3, and A_2A_4, each with a frequency of $\frac{1}{4}$. Let the first sib chosen have any one of these genotypes. Then the probability that the second sib has the same genotype is $\frac{1}{4}$. Thus, one-quarter of all sib-pairs have the same genotype and consequently the same dominance deviation, D. For these pairs having the same dominance deviation, the cross-product of the dominance deviations is D^2; other pairs, with different dominance deviations, have a mean cross-product of zero. Over all pairs, therefore, the sum of cross-product is $\frac{1}{4}\Sigma D^2$, and so the mean cross-product is $\frac{1}{4}V_D$. This is the covariance due to dominance deviations, and it adds to the covariance due to breeding values. The genetic covariance of full sibs is therefore

$$\text{cov}_{(FS)} = \tfrac{1}{2}V_A + \tfrac{1}{4}V_D \qquad \qquad \dots [9.7]$$

The second way of deriving the covariance of full sibs comes from Table 9.2 with little additional work. The covariance is the variance of the means of families. The right-hand column shows the squares of the progeny means, and it will be seen that these are all exactly the same as the products of progeny mean $\times$ mid-parent, except for the two entries in the middle involving terms in d^2. The mean square (MS) can therefore be got from the mean product (MP) already calculated; thus

$$MS = MP + d^2 2p^2q^2 - \tfrac{1}{4}d^2 4p^2q^2$$
$$= MP + d^2 p^2 q^2$$

The correction for the mean is the same as before, so we have

$$\text{cov}_{(FS)} = \text{cov}_{O\bar{P}} + d^2 p^2 q^2$$
$$= pq\alpha^2 + d^2 p^2 q^2$$

Since $2pq\alpha^2 = V_A$ (from equation [8.3a]) and $4d^2p^2q^2 = V_D$ (from equation [8.4]) the covariance of full sibs is

$$\text{cov}_{(FS)} = \tfrac{1}{2}V_A + \tfrac{1}{4}V_D$$

summing over all loci.

The correlation of full sibs is

$$t = \frac{\tfrac{1}{2}V_A + \tfrac{1}{4}V_D}{V_P} \qquad \qquad \dots [9.8]$$

In principle the difference between the covariances of full sibs and of half sibs provides a way of estimating the dominance variance, since $\text{cov}_{(FS)} - 2\text{cov}_{(HS)} = \tfrac{1}{4}V_D$. In practice, however, this can be done only if there are no environmental contributions to the phenotypic covariances.

Twins

The genetic covariances of twins are very simple. Dizygotic (fraternal) twins are related as full sibs and their genetic covariance is that of full sibs. Monozygotic (identical) twins have identical genotypes, so there is no genetic variance within pairs and the whole of the genetic variance appears in the between-pair component. The genetic covariance is therefore

$$\text{cov}_{(MZ)} = V_G \qquad \qquad \dots [9.9]$$

General

From the relationships explained above, it will have been seen that the covariance is made up of simple fractions of the causal components of variance, V_A and V_D, the fractions in the cases dealt with being $\frac{1}{2}$ or $\frac{1}{4}$. These fractions, or coefficients, are related in a simple manner to the coancestries of the relatives and their parents, so that the covariance of any sort of relatives can be easily deduced from a consideration of the appropriate pedigree. Let r be the fraction of the additive genetic variance, and u that of the dominance variance, appearing in the covariance. Then the generalized covariance for any sort of relationship is

$$\text{cov} = rV_A + uV_D \qquad \dots [9.10]$$

Let P and Q be two individuals representing the relationship whose covariance is required, and let A, B and C, D be their parents respectively, as shown in Fig. 5.4. Then, letting f stand for the coancestry as explained in Chapter 5, the values of r and u are obtained as

$$r = 2f_{PQ} \qquad \dots [9.11]$$

$$u = f_{AC}f_{BD} + f_{AD}f_{BC} \qquad \dots [9.12]$$

For the derivation of these two equations, see Crow and Kimura (1970, p. 134). In applying equations [9.11] and [9.12], the condition of random mating can be relaxed to the extent that any individual in the pedigree can be inbred except the relatives P and Q themselves, which must not be inbred. Table 9.3 gives the values of r and u, summarizing the relationships already described and adding some other, more distant, relationships.

The coefficient u of the dominance variance represents the probability of the relatives having the same genotype through identity by descent, and it is zero unless the related individuals have paths of coancestry through both of their respective parents. The only relatives in Table 9.3 having u non-zero are full sibs and double first cousins. The coefficient r of the additive variance is sometimes called the *coefficient of relationship*, or the *theoretical correlation*, between the relatives in question. It is the correlation between their breeding values, and it represents the correlation that would be found if all the phenotypic variance were additive genetic. We shall return to this point when considering the estimation of the heritability in the next chapter.

Table 9.3 Coefficients of the variance components in the covariances of relatives.

Relationship	Component Coefficient	V_A r	V_D u
MZ twins		1	1
First-degree. Offspring : parent		$\frac{1}{2}$	0
Full sib		$\frac{1}{2}$	$\frac{1}{4}$
Second-degree. Half sib, Offspring : grandparent, Uncle (aunt) : Nephew (niece)		$\frac{1}{4}$	0
Double first cousin		$\frac{1}{4}$	$\frac{1}{16}$
Third-degree. Offspring : great-grandparent Single first cousin		$\frac{1}{8}$	0

Epistatic interaction

The variance arising from epistatic interaction between loci contributes small fractions to the covariances of relatives. In Chapter 8 it was noted that the interaction variance V_I is subdivided into components according to the number of interacting loci, and according to whether the interaction is between breeding values or dominance deviations. The generalized covariance of any sort of relatives is as follows when the interaction components are included (for details, see Kempthorne, 1955a, b):

$$\left.\begin{aligned}
\mathrm{cov} = rV_A + uV_D + r^2 V_{AA} + ru V_{AD} + u^2 V_{DD} \\
+ r^3 V_{AAA} + r^2 u V_{AAD} + ru^2 V_{ADD} + u^3 V_{DDD} \\
\text{etc.}
\end{aligned}\right\} \quad \dots [9.13]$$

Table 9.4 gives the coefficients of the variance components in the covariances with two-factor interactions included. The offspring–parent covariance refers equally to one parent and to mid-parent values. The conclusions that come from consideration of the interaction components are the following. First, only small fractions of the interaction components contribute to any covariance, the most being $\frac{1}{4}$; the contributions of interactions between more than two loci are even smaller. Second, interaction components involving dominance variance do not contribute unless the dominance variance itself contributes. Third, and this is the most important point, the interactions of breeding values, V_{AA}, V_{AAA}, etc., contribute to all covariances of relatives. Any estimate of V_A made from half-sib correlations will contain also $\frac{1}{4}V_{AA} + \frac{1}{16}V_{AAA}$, etc.; and any estimate of V_D obtained from a full-sib correlation will contain also portions of the $A \times D$ and $D \times D$ interaction components. It was noted in Chapter 7 that the two definitions of breeding value given there are not equivalent if there is interaction between loci. We can now see how this comes about. Defined in terms of the measured values of progeny – the practical definition – breeding value includes additive $\times$ additive interaction deviations as well as the average effects of the genes carried by the parents; whereas, defined in terms of the average effects of genes – the theoretical definition – it does not.

From the coefficients in Table 9.4 one can see how in principle the interaction components can be estimated. For example, $\mathrm{cov_{OP}} - 2\mathrm{cov_{(HS)}} = \frac{1}{8}V_{AA}$. To estimate

Table 9.4 Covariances of relatives including the contributions of two-factor interactions.

| Relatives | Variance components and the coefficients of their contributions | | | | |
	V_A	V_D	V_{AA}	V_{AD}	V_{DD}
Offspring–parent : $\mathrm{cov_{OP}} =$	$\frac{1}{2}$	—	$\frac{1}{4}$	—	—
Half sibs: $\mathrm{cov_{(HS)}} =$	$\frac{1}{4}$	—	$\frac{1}{16}$	—	—
Full sibs: $\mathrm{cov_{(FS)}} =$	$\frac{1}{2}$	$\frac{1}{4}$	$\frac{1}{4}$	$\frac{1}{8}$	$\frac{1}{16}$
General: $\mathrm{cov} =$	r	u	r^2	ru	u^2

the interaction components, however, requires complex experiments of great precision, providing comparisons of the covariances of many different sorts of relatives free of environmental covariance. This is hardly practicable with animals but it can be done with plants: see, for example, Pooni *et al.* (1978), and Chi *et al.* (1969), who found negligible interaction variances in seven characters of maize.

Effects of linkage.   In the derivations of all the covariances given above, the effects of linkage have been ignored, the summation over loci carrying the implicit assumption that all the loci segregate independently. Linkage can increase the covariances of sibs, even in a random-mating population with the coupling and repulsion phases in equilibrium. The covariance of offspring and parents is not affected. The covariances of full and half sibs are affected through an increase in the components due to epistatic interaction; the closer the linkage the greater the increase. If the linked genes do not interact, there is no increase of covariance (see Cockerham, 1956).

Environmental covariance

Genetic causes are not the only reasons for resemblance between relatives; there are also environmental circumstances that tend to make relatives resemble each other; some sorts of relatives more than others. If members of a family are reared together, as with human families or litters of pigs or mice, they share a common environment. This means that some environmental circumstances that cause differences between unrelated individuals are not a cause of difference between members of the same family. In other words, there is a component of environmental variance that contributes to the variance between means of families but not to the variance within the families, and it therefore contributes to the covariance of the related individuals. This between-group environmental component, for which we shall use the symbol V_{Ec}, is usually called the *common environment*, a term that seems more appropriate when we think of the component as a cause of similarity between members of a group than when we think of it as a cause of difference between members of different groups. The remainder of the environmental variance, which we shall denote by V_{Ew}, arises from causes of difference that are unconnected with whether the individuals are related or not. It therefore appears in the within-group component of variance, but does not contribute to the between-group component, which is the variance of the true means of the groups. In considerations of the resemblance between relatives, therefore, the environmental variance must be divided into two components:

$$V_E = V_{Ec} + V_{Ew} \qquad \ldots [9.14]$$

one of the components, V_{Ec}, contributing to the covariance of the related individuals.

The sources of common environmental variance are many and varied, and arise from environmental factors such as nutrition, climatic conditions or, in man, cultural influences. Whenever families differ in respect of these factors there will be, or may be, environmentally caused differences between the means of families, which appear in the covariance as the V_{Ec} component. What we designate as the V_{Ec} component depends on the way in which individuals are grouped when we estimate the observational components of phenotypic variance. Whatever the form of the analysis, the part of the variance between the means of groups that can be ascribed to environmental causes is called the V_{Ec} component. The nature of this component thus depends on the form of the analysis applied. If the groups in the analysis are full-sib families then the V_{Ec} component represents environmental causes of similarity between full sibs; if the groups are half sibs it represents causes of similarity between half sibs. And in parent–offspring relationships a comparable covariance term represents environmental causes of resemblance between offspring and parent. Thus, whenever we measure a phenotypic covariance with the object of using it to estimate a causal component of variance, we have to decide whether it includes an appreciable component due to common environment, and this is often a matter of judgement based on a biological understanding of the organism and the character. In experiments, much of the V_{Ec} component can often be eliminated by suitable design. For example, members of the same family need not always be reared in the same vial, cage, or plot: they can be randomized over the rearing environments. Or, by dividing families into two or more groups, the V_{Ec} component can be measured and deducted from the covariance.

Maternal effects are a frequent, and often troublesome, source of environmental resemblance, particularly with mammals. The young are subject to a maternal environment during the first stages of their life, and this influences the phenotypic values of many metric characters even when measured on the adult, causing offspring of the same mother to resemble each other. Two sorts of maternal effect need to be distinguished. First, the phenotypic value of the mother for the character in question may influence the value of the offspring for the same character. For example, large mice give more milk than small mice and consequently their young grow better. This leads to an environmentally caused resemblance between the weight of the offspring and the weight of their mother. Furthermore, offspring of the same mother resemble each other in weight because they have shared the same milk supply. This sort of maternal effect therefore contributes an environmental component to the covariance of offspring with mothers, and to the covariance of full sibs or maternal half sibs. The second sort of maternal effect causes resemblance between offspring of the same mother, but not between the offspring and their mother. This arises when the character in the mother that gives rise to the maternal effect is not the character whose covariance is being studied. For example, the growth of the tails of young mice is influenced by the temperature in the nest. Mothers differ in the assiduity with which they nurse their young, and consequently there are differences in nest temperature between families. This produces an environmental component in the covariance of sibs in respect of tail length. But the nest temperature is not related to the mother's tail length, so there is no environmental covariance of offspring and mothers in respect of tail length. The variation among offspring due to a maternal effect results from variation among the mothers in the character that gives rise to the maternal effect, such as milk-yield. The maternal character is, to a greater or lesser degree, determined by the mother's genotype. Therefore the environmental variance V_{Ec} seen in the offspring is to some extent the consequence of genetic variation of some other character in the mothers. The resemblance between relatives becomes very complicated when the genetic basis of a maternal effect is taken into account. For details, see Willham (1963), Thompson (1976).

Relatives of all sorts may be subject to an environmental source of resemblance. In what follows, however, we shall make the simplification of disregarding the V_{Ec} component for all relatives except full sibs. The common maternal environment of full sibs is often the most troublesome source of environmental resemblance to overcome by experimental design. Consequently, a V_{Ec} component contributes more often and in greater amount to the covariance of full sibs than to that of any other sort of relative.

Competition. Brief mention must be made of a way by which resemblance between relatives can be reduced, instead of increased, for environmental reasons. This occurs when members of the same family complete for limited resources. Suppose, for example, that sib-families of animals are reared with each family in a separate pen, and that all pens are given the same fixed amount of food, growth rate being the character of interest. There would then be little or no variation in growth rate between families, and the covariance would consequently be reduced. There would, however, still be variation within families; indeed, there might be more variation than with unrestricted feeding because the competition is an additional source of variation. The intraclass correlation would therefore be reduced by the competition for fixed resources. The correlation could even be negative, because if one individual gets more to eat, another must of necessity get less. Competition is an important factor in plants, often making sib-correlations largely meaningless, particularly with characters related to yield.

Phenotypic resemblance

The covariance of phenotypic values is the sum of the covariances arising from genetic and from environmental causes. Thus by putting together the conclusions of the two preceding sections we arrive at the phenotypic covariances

Table 9.5 Phenotypic resemblance between relatives

Relatives	Covariance	Regression (b) or correlation (t)
Offspring and one parent	$\frac{1}{2}V_A$	$b = \frac{1}{2}\dfrac{V_A}{V_P}$
Offspring and mid-parent	$\frac{1}{2}V_A$	$b = \dfrac{V_A}{V_P}$
Half sibs	$\frac{1}{4}V_A$	$t = \frac{1}{4}\dfrac{V_A}{V_P}$
Full sibs	$\frac{1}{2}V_A + \frac{1}{4}V_D + V_{Ec}$	$t = \dfrac{\frac{1}{2}V_A + \frac{1}{4}V_D + V_{Ec}}{V_P}$

summarized in Table 9.5. (It will be remembered that some possible sources of environmental covariance are being neglected, particularly in offspring–parent relationships involving the mother.) In all these relationships except that of full sibs, the covariance is either a half or a quarter of the additive genetic variance. By observing the phenotypic covariance of relatives, we can thus estimate the amount of additive genetic variance. Similarly, the regression or correlation provides a means of estimating the proportionate amount of additive genetic variance, V_A/V_P, which is the heritability, and this is the chief use of measurements of the degree of resemblance between relatives. The methods of estimating the heritability will be considered more fully in the next chapter.

10 HERITABILITY

The heritability of a metric character is one of its most important properties. It expresses, as we have seen, the proportion of the total variance that is attributable to the average effects of genes, and this is what determines the degree of resemblance between relatives. But the most important function of the heritability in the genetic study of metric characters has not yet been mentioned, namely its predictive role, expressing the reliability of the phenotypic value as a guide to the breeding value. Only the phenotypic values of individuals can be directly measured, but it is the breeding value that determines their influence on the next generation. Therefore if the breeder or experimenter chooses individuals to be parents according to their phenotypic values, his success in changing the characteristics of the population can be predicted only from a knowledge of the degree of correspondence between phenotypic values and breeding values. This degree of correspondence is measured by the heritability, as the following considerations will show.

The heritability is defined as the ratio of additive genetic variance to phenotypic variance:

$$h^2 = \frac{V_A}{V_P} \qquad \qquad \dots [10.1]$$

(The customary symbol h^2 stands for the heritability itself and not for its square. The symbol derives from Wright's (1921) terminology, where h stands for the corresponding ratio of standard deviations.) An equivalent meaning of the heritability is the regression of breeding value on phenotypic value:

$$h^2 = b_{AP} \qquad \qquad \dots [10.2]$$

The equivalence of these meanings can be seen from reasoning similar to that by which the genetic covariance of offspring and one parent was derived in the previous chapter. If we split the phenotypic value into breeding value and a remainder (R) consisting of the environmental, dominance, and interaction deviations, then $P = A + R$. Since A and R are uncorrelated, $\text{cov}_{AP} = V_A$ and so $b_{AP} = V_A/V_P = h^2$.

We may note also that the correlation between breeding values and phenotypic values, r_{AP}, is equal to the square-root of the heritability. This follows from the general relationship between correlation and regression coefficients, which gives

$$r_{AP} = b_{AP}\frac{\sigma_P}{\sigma_A} = h^2\frac{1}{h} = h \qquad \dots [10.3]$$

By regarding the heritability as the regression of breeding value on phenotypic value we see that an individual's estimated breeding value is the product of its phenotypic value and the heritability:

$$A_{(expected)} = h^2 P \qquad \dots [10.4]$$

breeding values and phenotypic values both being reckoned as deviations from the population mean. In other words, the heritability expresses the reliability of the phenotypic value as a guide to the breeding value, or the degree of correspondence between phenotypic value and breeding value. For this reason the heritability enters into almost every formula connected with breeding methods, and many practical decisions about procedure depend on its magnitude. These matters, however, will be considered in the next chapters; here we are concerned only to point out that the determination of the heritability is one of the first objectives in the genetic study of a metric character.

It is important to realize that the heritability is a property not only of a character but also of the population and of the environmental circumstance to which the individuals are subjected. Since the value of the heritability depends on the magnitude of all the components of variance, a change in any one of these will affect it. All the genetic components are influenced by gene frequencies and may therefore differ from one population to another, according to the past history of the population. In particular, small populations maintained long enough for an appreciable amount of fixation to have taken place are expected to show lower heritabilities than large populations. The environmental variance is dependent on the conditions of culture or management: more variable conditions reduce the heritability; more uniform conditions increase it. So, whenever a value is stated for the heritability of a given character it must be understood to refer to a particular population under particular conditions. Values found in other populations under other circumstances will be more or less the same according to whether the structure of the population and the environmental conditions are more or less alike.

Very many determinations of heritabilities have been made for a great variety of characters in animals and plants. Some examples are given in Table 10.1. Heritabilities cannot easily be estimated with any great precision, and most estimates have rather large standard errors. Different estimates for the same character in the same organism show a wide range of variation, some of which may reflect real differences between populations or the conditions under which they are studied. Nevertheless, within the range of sampling errors, estimates tend to be similar in different populations. Because of the large sampling errors, the estimates in Table 10.1 are given to the nearest 5 per cent. Despite the lack of precision, it is very clear that heritabilities differ greatly according to the character. There is, moreover, some connection between the magnitude of the heritability and the nature of the character. This can be seen in Table 10.1. On the whole, the characters with the lowest heritabilities are those

Table 10.1 Approximate values of the heritability of various characters in various animal species. The estimates are rounded to the nearest 5 percent; their standard errors range from about 2 per cent to about 10 per cent.

	$h^2(\%)$	Ref.
Man		
Stature	65	(1)
Serum immunoglobulin (IgG) level	45	(2)
Cattle		
Body weight (adult)	65	(3)
Butterfat, %	40	(4)
Milk-yield	35	(4)
Pigs		
Back-fat thickness	70	(5)
Efficiency of food conversion	50	(5)
Weight gain per day	40	(5)
Litter size	5	(6)
Poultry		
Body weight (at 32 wks)	55	(7)
Egg weight (at 32 wks)	50	(7)
Egg production (to 72 wks)	10	(7)
Mice		
Tail length (at 6 wks)	40	(8)
Body weight (at 6 wks)	35	(8)
Litter size (1st litters)	20	(9)
Drosophila melanogaster		
Abdominal bristle number	50	(10)
Body size	40	(11)
Ovary size	30	(12)
Egg production	20	(11)

(1) West African population. Roberts, Billewicz, and McGregor (1978).
(2) US whites. Grundbacher (1974).
(3) Beef cattle; average of many estimates. Preston and Willis (1970).
(4) British Friesians, 1st lactations. Barker and Robertson (1966).
(5) British Large White. Smith, King, and Gilbert (1962).
(6) British Large White. Strang and Smith (1979).
(7) White Leghorn strain-crosses. Emsley, Dickerson, and Kashyap (1977).
(8) Rutledge, Eisen, and Legates (1973).
(9) Falconer (1965b).
(10) Clayton, Morris, and Robertson (1957).
(11) Robertson (1957b)
(12) Robertson (1957a)

most closely connected with reproductive fitness, while the characters with the highest heritabilities are those that might be judged on biological grounds to be the least important as determinants of natural fitness. This is well seen in the gradation of the four characters of *Drosophila*.

Some care is needed in applying the concept of heritability to plants. When defined as $h^2 = V_A/V_P$, the variances are those of individual values. Individual values of plants, particularly of their yields, are often not available or, if available, are rendered largely meaningless by competition, which was mentioned as an environmental factor in the previous chapter. Yields are usually expressed per unit area of plot in which the plants are grown. The unit of measurement is therefore the plot yield, not the individual yield. If the individuals in a plot are members of one family – full or half sibs – the 'heritability' is the heritability of differences between families, the meaning of which will be explained in Chapter 13. The rest of this chapter refers only to the heritability of individual values.

Estimation of heritability

The heritability is estimated from the degree of resemblance between relatives. Table 10.2 shows again the composition of the phenotypic covariances derived in the previous chapter. The right-hand column gives the regression or correlation expressed in terms of the heritability, from which it can be seen that with any relationship

$$h^2 = b/r \qquad \text{or} \qquad t/r \qquad \qquad \dots [10.5]$$

where r is the coefficient of the additive variance in the covariance, or the 'theoretical' correlation. Thus, when expressed in terms of the correlation (or regression) between relatives, the heritability is the observed correlation as a proportion of the correlation that would be found if the character were completely inherited, i.e., if all the variance were additive genetic.

The choice of what sort of relatives to use for the estimation of the heritability depends on the circumstances. In addition to the practical matter of which sorts of relatives are in fact available, there are two points to consider: precision and bias. In general, the closer the relationship, the more precise is the estimate. The reason for this is that the observed regression or correlation must be multiplied by a larger factor $(1/r)$ with more distant relatives, and the standard error of the regression or correlation must be multiplied by the same factor to give the standard error of the estimated heritability. The statistical precision will be considered more fully later in this chapter. Bias in the estimate of the heritability is usually a more important consideration than precision. It is introduced by

Table 10.2

Relatives	Covariance*	Regression (b) or correlation (t)
Offspring and one parent	$\frac{1}{2}V_A$	$b = \frac{1}{2}h^2$
Offspring and mid-parent	$\frac{1}{2}V_A$	$b = h^2$
Half sibs	$\frac{1}{4}V_A$	$t = \frac{1}{4}h^2$
Full sibs	$\frac{1}{2}V_A + \frac{1}{4}V_D + V_{Ec}$	$t \geqslant \frac{1}{2}h^2$

*The contributions of epistatic interactions are ignored, and so are the possible environmental contributions to relatives other than full sibs.

environmental sources of covariance and, in the case of full sibs, by dominance. From considerations of the biology of the character and the experimental design, we have to decide which covariance is least likely to be augmented by an environmental component, a matter already discussed in the last chapter. Generally speaking, the half-sib correlation and the regression of offspring on father are the most reliable from this point of view. The regression of offspring on mother is sometimes liable to give too high an estimate on account of maternal effects, as it would, for example, with body size in most mammals. (Example 10.3 illustrates the bias due to a maternal effect.) The full-sib correlation, which is the only relationship for which an environmental component of covariance is shown in the table, is the least reliable of all. The component due to common environment is often present in large amount and is difficult to overcome by experimental design; and the full-sib covariance is further augmented by the dominance variance. The full-sib correlation can therefore seldom do more than set an upper limit to the heritability.

Example 10.1 The heritability of abdominal bristle number in *Drosophila melanogaster* has been determined by three different methods applied to the same population (Clayton, Morris, and Robertson, 1957), with the following results:

Method of estimation	Heritability
Offspring–parent regression	0.51 ± 0.07
Half-sib correlation	0.48 ± 0.11
Full-sib correlation	0.53 ± 0.07
Combined estimate	0.52

The estimates obtained by the three methods are in very satisfactory agreement. In this case, the character – bristle number – is free of complications arising from maternal effects and common environment.

Let us now consider briefly some technical matters concerning the translation of observational data into estimates of heritability. For the moment it will be assumed that all observations are made on a random-mating population with no selection of the parents. Later, the effects of assortative mating and of selection will be described.

Offspring–parent regression
    The estimation of the heritability from the regression of offspring on parents is comparatively straightforward. The data are obtained in the form of measurements of parents – one or the mean of both – and the mean of their offspring. The covariance is then computed from the cross-products of the paired values. The following example illustrates the regression on mid-parent values.

Example 10.2. Figure 10.1 illustrates the regression of offspring on mid-parent values for wing length in *Drosophila melanogaster* (Reeve and Robertson, 1953). There are 37 pairs of parents and a mean of 2.73 offspring were measured from each pair of parents. The parents were mated assortatively, with the result that the variance of mid-parent values is greater than it would be if mating had been at random, as will be explained in a later section. Each point on the graph represents the mean value of one pair of parents (measured along the horizontal axis), and the mean value of their

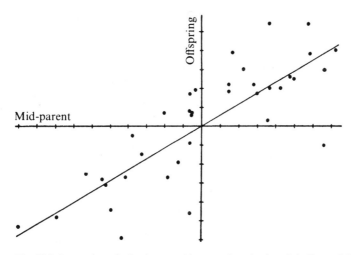

Fig. 10.1. Regression of offspring on mid-parent for wing length in *Drosophila*, as explained in Example 10.2. Mid-parent values are shown along the horizontal axis, and mean value of offspring along the vertical axis. (*Drawn from data kindly supplied by Dr E. C. R. Reeve.*)

offspring (measured along the vertical axis). The axes are marked at intervals of 1/100 mm, and they intersect at the mean value of all parents and all offspring. The sloping line is the linear regression of offspring on mid-parent. The slope of this line estimates the heritability, and has the value ($\pm$ standard error):

$$h^2 = b_{O\bar{P}} = 0.58 \pm 0.07$$

A complication in the use of the regression of offspring on parents arises if the variance is not equal in the two sexes. It was noted in the previous chapter that the covariance of offspring and mid-parent values is equal to the additive genetic variance on condition that the sexes are equal in phenotypic variance. If the variances are not equal, the regression on mid-parent cannot, strictly speaking, be used, and the heritability must be calculated separately for each sex. The heritability in males, for example, is estimated from the regression of sons on fathers, and of daughters on fathers. The regression of daughters on fathers, however, must be adjusted for the difference in variance, multiplying it by the ratio of phenotypic standard deviations of males to females. Thus if b is the regression of daughters on fathers, the adjusted regression is $b' = b\sigma_{\male}/\sigma_{\female}$. Similarly, the heritability in females is estimated from the regression of daughters on mothers, and of sons on mothers adjusted by $\sigma_{\female}/\sigma_{\male}$. Estimations from the four regressions, and the adjustments for unequal variances, are illustrated in the following example.

Example 10.3 The heritability of the body weight at 6 weeks of age was estimated in a random-bred strain of mice by offspring–parent regression (Falconer, 1973). The variances of males and females were not equal, and so the regressions were calculated separately for each sex of offspring and of parent. The phenotypic standard deviations and their ratios were as follows:

$$\sigma_{\male} = 3.786, \ \sigma_{\female} = 2.675, \ \sigma_{\male}/\sigma_{\female} = 1.415, \ \sigma_{\female}/\sigma_{\male} = 0.707$$

Table (i) Regression coefficients with standard errors and adjustment factors.

	Parents	
	Male	
Offspring	Male	Female
Male	0.110 ± 0.040	$(0.324 \pm 0.064) \times 0.707$ $= 0.229 \pm 0.045$ 0.237 ± 0.043
Female	$(0.111 \pm 0.029) \times 1.415$ $= 0.157 \pm 0.041$	

Table (ii) Heritabilities, per cent, with standard errors.

	Parents	
Offspring	Male	Female
Male	22 ± 8	46 ± 9
Female	31 ± 8	47 ± 9
Both	27 ± 6	47 ± 6

Table (i) gives the regression coefficients and their standard errors, with the factors by which both must be multiplied to adjust for the difference in variance. The regressions are all of offspring on one parent, so the regressions and their standard errors must be multiplied by two to obtain the heritabilities given in table (ii). The estimates do not differ significantly according to the sex of offspring, so male and female offspring are averaged in the third line of table (ii). The estimates do, however, differ significantly between male and female parents. The much higher estimate from females is attributable to bias from a maternal effect.

Sib analysis

The estimation of heritability from half sibs is more complicated than appears at first sight and needs more detailed comment. A common form in which data are obtained with animals is the following. A number of males (sires) are each mated to several females (dams), the males and females being randomly chosen and randomly mated. A number of offspring from each female are measured to provide the data. The individuals measured thus form a population of half-sib and full-sib families. An analysis of variance is then made by which the phenotypic variance is divided into observational components attributable to differences between the progeny of different males (the between-sire component, σ_S^2), to differences between the progeny of females mated to the same male (between-dam, within-sires, component, σ_D^2); and to differences between individual offspring of the same female (within-progenies component, σ_W^2). The form of the analysis is shown in Table 10.3. There are supposed to be s sires, each mated to d dams, which produce k offspring each. The values of the mean squares are denoted by MS_S, MS_D, and MS_W. The mean square within progenies is itself the estimate of the within-progeny variance component, σ_W^2; but the other mean squares are not the variance components. The compositions of the mean squares in terms of the observational components of variance are shown in the right-hand column of the table, consideration of which will show how the variance

Table 10.3 Form of analysis of half-sib and full-sib families.

Source	d.f.	Mean square	Composition of mean square
Between sires	$s - 1$	MS_S	$= \sigma_W^2 + k\sigma_D^2 + dk\sigma_S^2$
Between dams (within sires)	$s(d - 1)$	MS_D	$= \sigma_W^2 + k\sigma_D^2$
Within progenies	$sd(k - 1)$	MS_W	$= \sigma_W^2$

s = number of sires
d = number of dams per sire
k = number of offspring per dam

components are to be estimated. The between-dam mean square, for example, is made up of the within-progeny component together with k times the be-tween-dam component; so the between-dam component is estimated as $\sigma_D^2 = (1/k)(MS_D - MS_W)$. Similarly, the between-sire component is estimated as $\sigma_S^2 = (1/dk)(MS_S - MS_D)$, where dk is the number of offspring per sire. If there are unequal numbers of offspring from the dams, or of dams in the sire-groups, the mean values of k and d can be used with little error, provided the inequality of numbers is not very great. The exact solution, which is too complicated for description here, can be found in Snedecor and Cochran (1967, Section 10.19), Turner and Young (1969), or Searle (1971).

The next step is to deduce the connections between the observational components that have been estimated from the data and the causal components, in particular the additive genetic variance, the estimation of which is the main purpose of the analysis. Though all the information needed has already been given, the interpretation of the observational components, which is given in Table 10.4, is not immediately apparent without explanation. The first point to note is that the estimate of the phenotypic variance is given by the sum (σ_T^2) of the three observational components: $V_P = \sigma_T^2 = \sigma_S^2 + \sigma_D^2 + \sigma_W^2$. This is not necessarily equal to the observed variance as estimated from the total sum of squares, though the two seldom differ by much. Now consider the interpretation of the between-sire component, σ_S^2. This is the variance between the means of half-sib families and it therefore estimates the phenotypic covariance of half sibs, $\text{cov}_{(HS)}$, which is $\frac{1}{4}V_A$. Thus $\sigma_S^2 = \frac{1}{4}V_A$. Next consider the within-progeny component, σ_W^2. Since any between-group variance component is equal to the covariance of the members of the groups, it follows that a within-group component is equal to the total variance minus the covariance of members of the groups. The progenies of the dams are full-sib families and so the within-progeny variance estimates $V_P - \text{cov}_{(FS)}$. This leads to the interpretation $\sigma_W^2 = \frac{1}{2}V_A + \frac{3}{4}V_D + V_{Ew}$. Finally, there remains the between-dam component, and what it estimates can be found by subtraction as follows:

$$\sigma_D^2 = \sigma_T^2 - \sigma_S^2 - \sigma_W^2 = \text{cov}_{(FS)} - \text{cov}_{(HS)} = \frac{1}{4}V_A + \frac{1}{4}V_D + V_{Ec}$$

Consideration of the between-sire and between-dam components will show that their sum gives an estimate of the full-sib covariance, $\text{cov}_{(FS)}$, but this provides no

Table 10.4 Interpretation of the observational components of variance in a sib analysis.

Observational component			Covariance and causal components estimated	
Sires:	$\sigma_S^2 =$	$\mathrm{cov}_{(HS)}$	$= \frac{1}{4} V_A$	
Dams:	$\sigma_D^2 =$	$\mathrm{cov}_{(FS)} - \mathrm{cov}_{(HS)}$	$= \frac{1}{4} V_A + \frac{1}{4} V_D + V_{Ec}$	
Progenies:	$\sigma_W^2 =$	$V_P - \mathrm{cov}_{(FS)}$	$= \frac{1}{2} V_A + \frac{3}{4} V_D + V_{Ew}$	
Total: $\sigma_T^2 = \sigma_S^2 + \sigma_D^2 + \sigma_W^2 =$		V_P	$= V_A + V_D + V_{Ec} + V_{Ew}$	
Sires + Dams:	$\sigma_S^2 + \sigma_D^2 =$	$\mathrm{cov}_{(FS)}$	$= \frac{1}{2} V_A + \frac{1}{4} V_D + V_{Ec}$	

new information for estimating the causal components. These conclusions about the connection between observational and causal components of variance are summarized in Table 10.4. The contributions of the interaction variance to the observational components can be deduced from the contributions to the covariances given in Table 9.4.

The estimation of the heritability from sib analyses is illustrated in the two following examples. The calculation of the standard error of the estimate, which is complicated, is described by Turner and Young (1969).

Example 10.4 As an illustration of the estimation of heritability from a sib analysis, we refer to the study of Danish Landrace pigs based on the records of the Danish Pig Progeny Testing Stations (Fredeen and Jonsson, 1957). The data came from 468 sires

Sib analysis of body length in Danish Landrace pigs; data for male offspring only.

Source	d.f.	Mean square	Component of variance	
Between sires	432	6.03	$\sigma_S^2 = \frac{1}{4}(6.03 - 3.81) = 0.555$	
Between dams, within sires	468	3.81	$\sigma_D^2 = \frac{1}{2}(3.81 - 2.87) = 0.47$	
Within progenies	936	2.87	$\sigma_W^2 =$	2.87
			$\sigma_T^2 =$	3.895

Interpretation of analysis

Sib correlations		Estimates of heritability		
Half sibs: $t_{(HS)} = \dfrac{\sigma_S^2}{\sigma_T^2} = 0.142$		Sire-component: $h^2 = \dfrac{4\sigma_S^2}{\sigma_T^2}$	$= 0.57$	
		Dam-component: $h^2 = \dfrac{4\sigma_D^2}{\sigma_T^2}$	$= 0.48$	
Full sibs: $t_{(FS)} = \dfrac{\sigma_S^2 + \sigma_D^2}{\sigma_T^2} = 0.263$		Sire + Dam: $h^2 = \dfrac{2(\sigma_S^2 + \sigma_D^2)}{\sigma_T^2}$	$= 0.53$	

each mated to 2 dams, the analysis being made on the records of 2 male and 2 female offspring from each dam. Only one such analysis is given here: that of body length in the male offspring. The analysis, shown in the table, was made within stations and within years, and this accounts for the degrees of freedom being fewer than would appear appropriate from the numbers stated above. The interpretation of the analysis, shown at the foot of the table, has been slightly simplified by the omission of some minor adjustments not relevant for us at this stage. The between-dam component is not greater than the between-sire component, so there cannot be much non-additive genetic variance or variance due to common environment. The two estimates of the heritability, from the sire and dam components respectively, can therefore be regarded as equally reliable, and their combination based on the resemblance between full sibs may be taken as the best estimate.

Example 10.5 We have not yet had an example to illustrate the effect of common environment in augmenting the full-sib correlation. This is provided by body size in mice. The analysis given in table (i) refers to the weight of female mice at 6 weeks of age (J. C. Bowman, unpublished). There were 719 offspring from 74 sires and 192 dams, each with one litter. These were spread over 4 generations and the analysis was made within generations. The analysis is complicated by the inequality of the number of offspring per dam and of dams per sire. The adjustments made for these inequalities are shown, without explanation, in the compositions of the mean squares from which the components are estimated. The dam component is much larger than the sire component, indicating a substantial bias due to common environment or dominance. Therefore only the sire component can be used to estimate the heritability. The estimate obtained is $h^2 = 4 \times (0.48/5.14) = 0.37$. (This estimate has a standard error of 0.26, so that it is not significantly different from zero. The experiment was on too small a scale to be of much practical use, though it serves to illustrate the method.) The causal components can now be estimated from the analysis according to the interpretation given in Table 10.4. It is not possible to discriminate between common environment and dominance as the cause of the difference between the dam and sire components. The estimates in table (ii) are based on the assumption that the difference

Table (i)

Source	d.f.	Mean square	Composition of M.S.	Components
Sires	70	17.10	$\sigma_W^2 + k'\sigma_D^2 + dk'\sigma_S^2$	$\sigma_S^2 = 0.48$
Dams	118	10.79	$\sigma_W^2 + k\sigma_D^2$	$\sigma_D^2 = 2.47$
Progenies	527	2.19	σ_W^2	$\sigma_W^2 = 2.19$

$$k = 3.48; \quad k' = 4.16; \quad d = 2.33 \qquad \sigma_T^2 = 5.14$$

Table (ii)

$V_P = \sigma_T^2$	$= 5.14 = 100\%$	
$V_A = 4\sigma_S^2$	$= 1.92 = 37\%$	
$V_{Ec} = \sigma_D^2 - \sigma_S^2$	$= 1.99 = 39\%$	
$V_{Ew} = \sigma_W^2 - 2\sigma_S^2$	$= 1.23 = 24\%$	

is all due to common environment, and that $V_D = 0$. We can go a little further than this in the interpretation of the analysis and put an upper limit on the dominance variance as follows. The maximum possible value for V_D is set by the within-progenies component σ_W^2: it is possible, though very unlikely, that $V_{Ew} = 0$, which would make $\sigma_W^2 - \sigma_S^2 = \frac{3}{4}V_D$, from which $V_D = 1.64 = 32$ per cent as an upper limit. V_{Ec} would then be $\sigma_D^2 - \sigma_S^2 - \frac{1}{4}V_D = 1.58 = 31$ per cent as a lower limit. The true values of the causal components are, however, likely to be much nearer those in table (ii).

Intra-sire regression of offspring on dam

The heritability can be estimated from the offspring–parent relationship in a population with the structure described in the foregoing section, but a slight modification is necessary. Since each male is mated to several females, the regression of offspring on mid-parent is inappropriate; and, since there are usually rather few male parents, the simple regressions on one or other parent are both unsuitable. The heritability can, however, be satisfactorily estimated from the average regression of offspring on dams, calculated within sire groups. That is to say, the regression of offspring on dam is calculated separately for each set of dams mated to one sire, and the regression from each set pooled in a weighted average. This method is commonly used for the estimation of heritabilities in farm animals. The intra-sire regression of offspring on dam estimates half the heritability, as the following consideration will show. The progeny of one sire has a mean deviation from the population mean equal to half the breeding value of the sire, provided the females he is mated to are a random sample from the population. The progeny of one dam deviates from the mean of the sire group by half the breeding value of the dam. Therefore the within-sire covariance of offspring and dam is equal to half the additive variance of the population as a whole; and the within-sire regression of offspring on dam is equal to half the heritability, just like the simple regression of offspring on one parent. The validity of the estimate is, of course, dependent on the absence of maternal effects contributing to the resemblance between offspring and dams. Inequality of the variance of males and females calls for an adjustment if the heritability is to be estimated from the intra-sire regression of male offspring on dams. The regression coefficient should then be multiplied by the ratio of the phenotypic standard deviation of females to that of males.

Twins and human data

Identical twins seem at first sight to provide, for man and cattle, a means of estimating the genotypic variance. They provide individuals of identical genotype, just as inbred lines, or crosses between lines, do for laboratory animals or for plants. Many studies of human twins have been made, and have shown the members of the pairs to be extremely alike in most characters, even when reared apart from childhood. Studies of cattle twins, though on a much smaller scale, show the same thing (see Hancock, 1954; Brumby, 1958). Taken at their face value, these studies seem to indicate a very high degree of genetic determination – up to 90 per cent or even more – for many characters. The use of identical twins in this way is, however, vitiated by the additional similarity due to common environment. Twins share a common environment from conception to birth and

Table 10.5 Composition of the components of variance between and within pairs of twins, omitting interaction components.

	Between pairs, σ_b^2	Within pairs, σ_w^2
Identical (MZ)	$V_A + V_D + V_{Ec}$	V_{Ew}
Fraternal (DZ)	$\frac{1}{2}V_A + \frac{1}{4}V_D + V_{Ec}$	$\frac{1}{2}V_A + \frac{3}{4}V_D + V_{Ew}$
Difference	$\frac{1}{2}V_A + \frac{3}{4}V_D$	$\frac{1}{2}V_A + \frac{3}{4}V_D$

over the period during which they are reared together, so that the between-pair variance contains the variance due to common environment, V_{Ec}, confounded with the genetic variance, V_G. This difficulty may be partly overcome by comparison of the two sorts of twins, identical or monozygotic (MZ) and fraternal or dizygotic (DZ). Dizygotic twins are full sibs that share a common environment to approximately the same extent as monozygotic twins. To estimate the amount of genetic variance, we ask how much less alike are DZ than MZ twins. Table 10.5 shows the composition of the components of variance between and within pairs, on the assumption that both components of the environmental variance, V_{Ec} and V_{Ew}, are the same in MZ as in DZ twins. The contributions from the interaction variance, which are omitted for simplicity, can be added from Table 9.4. The difference between MZ and DZ twins in both components estimates half of the additive variance together with three-quarters of the dominance variance. It may be noted that the difference between the between-pairs mean squares has the same expectation as the difference between the components, i.e., $(MS_{MZ} - MS_{DZ}) = (\sigma_{b(MZ)}^2 - \sigma_{b(DZ)}^2) = \frac{1}{2}V_A + \frac{1}{4}V_D$. The correlation between co-twins is the between-pair component divided by the phenotypic variance, so twice the difference between the MZ correlation and the DZ correlation estimates $(V_A + 1\frac{1}{2}V_D)/V_P$. This is nearer to the degree of genetic determination (broad-sense heritability) than it is to the heritability (narrow-sense), which is perhaps what is wanted from human data. It should be noted, however, that the twin analysis does not provide a strictly valid estimate of either V_A/V_P or of V_G/V_P, even with the assumption of equality of the environmental components. If the epistatic components are added to the co-variances it will be seen that the bias is increased. Example 10.6 illustrates the twin-analysis applied to four human characters.

The analysis of twin data outlined above rests critically on two assumptions. The first, already mentioned, is that the environmental components of variance are the same in the two types of twins. The second, not yet mentioned, is that the total genetic variance is the same in the two types. Furthermore, the object of the analysis is to estimate parameters of the population, most of whom are not twins, so for these estimates to be valid the environmental components of variance of twins must be the same as those of single-born individuals. There are many possible causes of differences in the environmental components, of which the following are some (Stern, 1973, explains and discusses these more fully).

(1) Genotype–environment interaction: as explained in Chapter 8, this is formally included with the environmental variance. It will contribute different amounts to the MZ and DZ environmental components.

(2) MZ twins are of three types according to the arrangement of the foetal membranes – a single amnion and single chorion, a single chorion, or separate amnions and chorions; all DZ twins are of the last type.

(3) Competition between co-twins *in utero*, which is probably more severe in MZ than in DZ pairs.

(4) Exact contemporaneity of twins as opposed to singletons.

(5) Parental treatment of twins, which may either enhance or diminish the similarity, and may affect MZ and DZ twins differently.

(6) Errors in the diagnosis of zygosity.

(7) The inclusion of unlike-sexed pairs among the DZ twins.

Differences between the total genetic variance of MZ and DZ twins can arise in the following way (Nance, 1976). The frequency of DZ twinning is influenced by genetic factors including racial differences, whereas the frequency of MZ twinning is little, if at all, influenced by genetic or racial factors. Therefore the different sections or strata of the population may be differently represented among samples of MZ and DZ twins, and the genetic variances may differ in consequence. The requirement of equality of variances may be tested by comparison of the total variances estimated as $\sigma_b^2 + \sigma_w^2$ though counterbalancing differences of genetic and environmental components cannot be ruled out. If the total variances prove to be equal and there is no obvious preference for the between-pair or within-pair comparison, the information from the two can be combined by averaging them (Christian *et al.*, 1974), i.e., $(\sigma_{bMZ}^2 - \sigma_{bDZ}^2) + (\sigma_{wDZ}^2 - \sigma_{wMZ}^2) = V_A + 1\frac{1}{2}V_D$. To estimate the 'heritability', the value obtained for $V_A + 1\frac{1}{2}V_D$ is divided by the total variance V_P. If the total variance of twins is not the same as that of singletons, the 'heritability' applicable to the population would be obtained by taking V_P from singletons.

Despite all the difficulties in twin analyses, there is probably less bias from inequality of environmental variances than there is from the common-environment component V_{Ec} in full-sib correlations. The following example illustrates the point.

Example 10.6 The table gives the correlations of MZ twins, like-sexed DZ twins, and full sibs for four characters, from Huntley (1966). The characters, all measured on children, are the total ridge count on ten fingers, height adjusted for age, a verbal IQ test, and a social-maturity score which 'assesses the individual's ability to look after his practical needs and to take responsibility in relation to his age'. These were chosen to represent characters that would be expected to be, in the order stated, increasingly subject to environmental influences. The 'heritabilities' estimated from the twin-differences, shown at the foot of the table, are consistent with this expectation. The

	Finger-ridge count	Height	IQ score	Social-maturity score
MZ twins	0.96	0.90	0.83	0.97
DZ twins	0.47	0.57	0.66	0.89
Full sibs	0.51	0.50	0.58	0.32
$2(t_{MZ} - t_{DZ})$	0.98	0.66	0.34	0.16
$2t_{FS}$	1.02	1.00	1.16	0.64

estimates from doubling the full-sib correlation are obviously too high, except for the ridge count, being biased upwards by common environment V_{Ec}. The twin analyses have, at least partially, removed this bias. The heritability of the finger-ridge count has been estimated from offspring–parent regressions as about 0.8 (Mi and Rashad, 1975). The heritabilities of the counts of single fingers are lower, ranging from 0.58 to 0.68 for different fingers. The high value for the total count results from the multiple measurement which eliminates all but one-tenth of the environmental component V_{Es} affecting each finger separately, as explained in Chapter 8.

The effects of common environment present serious difficulties in the interpretation of the correlations between relatives in man, especially for characters influenced by cultural transmission. These difficulties in arriving at a meaningful estimate of the heritability cannot be discussed here. It must suffice to say that they may be at least partly overcome by utilizing correlations of several different sorts of relatives and having an index that quantifies the environment to which each family is subject (see Morton, 1974; Rao, Morton, and Yee, 1974, 1976). Methods of dealing with cultural transmission and estimating its effect are described by Eaves (1976) and by Cloninger, Rice, and Reich (1979). These last authors conclude, for example, that the heritability of IQ scores in their data was 33 per cent but the 'total transmissible variance' was 69 per cent. Rather than trying to estimate genetic parameters such as the heritability, it is perhaps more important to test whether the parameters are non-zero. This is done by 'model-fitting'. A 'model' is simply a series of expectations for correlations between relatives based on the hypothesis to be tested. The hypothesis might be, for example, that there is variation due to common environment but no genetic variance. If the data give a significantly bad fit to the expectations of the model, the hypothesis is disproved. The application of these methods to psychological characters in man is reviewed by Eaves et al. (1978).

Assortative mating

Assortative mating means mating 'like with like' and is seen in a correlation between the phenotypic values of mated individuals. Mating in human populations is assortative with respect to many metric characters, such as stature and IQ scores, though not necessarily by deliberate choice of mates. The questions to be considered in this section are how assortative mating affects the estimation of heritability, and whether the use of assortative mating as a deliberate breeding policy has any advantages in this respect. The genetic consequences of assortative mating are rather complex and only the conclusions can be given here, with no more than brief indications of how they are arrived at. Full explanations are given by Crow and Kimura (1970).

The degree of assortative mating is expressed as the correlation r between the phenotypic values of the mated individuals, and this is what can be observed. The genetic consequences, however, depend on the correlation m between the breeding values of the mates. To deduce the connection between m and r it is necessary to know what governs the choice of mates – whether the primary cause of the resemblance is phenotypic, genetic, or environmental. Primary phenotypic resemblance means that the mates are chosen on the basis of their phenotypic values of the character under consideration. This is how assortative mating would be applied in a breeding programme. The relationship between the two correlations can then be shown to be $m = rh^2$, where h^2 is the heritability of the character by which the mates are chosen. (The derivation of this relationship will be explained later.) The consequences to be described are restricted to primary phenotypic resemblance as a cause of assortative mating. Assortative mating in man, however, probably seldom arises purely in this way and caution is needed in applying the results to human data, particularly in assuming the relationship $m = rh^2$ to be applicable.

Primary genetic or primary environmental resemblance occurs if matings take place within groups that are differentiated from each other genetically or environmentally. This is probably how much of the assortative mating in man arises. The observed phenotypic correlation r is then a 'secondary' correlation resulting from the 'primary' correlation of breeding values or of environmental deviations. The primary correlations cannot be deduced from r unless one of them can be estimated by other means, and the genetic consequences of the assortative mating cannot be deduced without a knowledge of m. If the primary correlation is wholly environmental ($m = 0$), there will be no genetic consequences of the assortative mating (except that the increased variance of mid-parent phenotypic values will reduce the regression of offspring on mid-parent.) Environmental correlation may be the basis of the assortative mating for IQ in man. An analysis of family data on IQ scores (Rao, Morton, and Yee, 1976) showed that the phenotypic correlation between husband and wife of $r = 0.5$ could be largely, perhaps wholly, attributed to people choosing a spouse from those with a family background similar to their own.

Returning to assortative mating by phenotypic value, the genetic consequences are, in summary, as follows. The resulting correlation m between breeding values causes an increase of the additive variance, and consequently of the heritability. The correlations between relatives, however, are increased by more than would result from the increased heritability alone. There is therefore a possible ambiguity in the meaning of heritability under assortative mating. It may be thought of as the determinant of the resemblance between relatives, as expressed in equation [10.5], or as the ratio of variance components, V_A/V_P, and the two are not the same under assortative mating. The definition as V_A/V_P will be retained here. The questions with which we shall be mainly concerned are: by how much is the heritability increased, and how is the heritability (defined as V_A/V_P) to be estimated from the resemblance between relatives?

Other aspects of assortative mating that must be noted are the following. (1) The full effects are not immediate; it takes some generations following random mating to reach an equilibrium state. (2) The effects are dependent on the number of loci influencing the character: it will be assumed that the number is large. (3) The effect on the dominance variance is small and may be neglected. Attention will be restricted to pair-matings producing full-sib families in the progeny. Linkage will be disregarded.

The consequences of assortative mating can be worked out by consideration of the covariances of mated pairs, of which three are needed. These are given in Table 10.6 in terms of the two correlations, r (between phenotypic values) and m

(between breeding values), and the variance components in the generation to which the mated pairs belong. The relationship $m = rh^2$, stated earlier, can now be derived as follows. $\text{cov}(A_1 A_2) = \text{cov}(h^2 P_1, h^2 P_2) = h^4 \text{cov}(P_1 P_2) = h^4 r V_P = rh^2 V_A$; by (2) of Table 10.6, $\text{cov}(A_1 A_2) = m V_A$, so $m = rh^2$. It is important to note that h^2 here is the heritability of the character measured at the age at which the choice of mates takes place. The variance in the progeny is obtained as follows. The covariance of breeding values of the parents increases the variance of mid-parent breeding values and so increases the variance between family-means. The variance within families is due to segregation and is not affected provided the number of segregating loci is not small. Adding together the between-family and the within-family components gives the total variance in the offspring generation. Equations relating the additive and phenotypic variances to those in the random breeding base population are given in Table 10.6. In each case the first equation, (4) and (6), refers to the offspring of one generation of assortative mating, and the second, (5) and (7), to a population that has reached equilibrium, when the

Table 10.6. Assortative mating. Approximate expressions for variances and covariances. (For meanings of symbols, see notes below.)

Covariances of mates:

phenotypic values, $\text{cov}(P_1 P_2) = r V_P$ (1)

breeding values, $\text{cov}(A_1 A_2) = m V_A = mh^2 V_P$ (2)

breeding value of one with environmental deviation of the other,

$$\text{cov}(A_1 E_2) = (r - m) V_A = (r - m)h^2 V_P \qquad (3)$$

Variances:

	Additive	Phenotypic	Heritability
1 generation	$V_{A1} = V_{A0}(1 + \tfrac{1}{2}m)$ (4)	$V_{P1} = V_{P0}(1 + \tfrac{1}{2}mh^2)$ (6)	$h_1^2 = h_0^2 \left[\dfrac{1 + \tfrac{1}{2}m}{1 + \tfrac{1}{2}mh^2} \right]$ (8)
equilibrium	$V_{A0} = V_A(1 - m)$ (5)	$V_{P0} = V_P(1 - mh^2)$ (7)	$h_0^2 = h^2 \left[\dfrac{1 - m}{1 - mh^2} \right]$ (9)

Relatives:

	Covariance		Regression (b) or correlation (t)	
Offspring, mid-parent	$\tfrac{1}{2}V_A(1 + r)$	(10)	$b = h^2$	(13)
Offspring, one parent	$\tfrac{1}{2}V_A(1 + r)$	(11)	$b = \tfrac{1}{2}h^2(1 + r)$	(14)
Full sibs	$\tfrac{1}{2}V_A(1 + m)$	(12)	$t = \tfrac{1}{2}h^2(1 + m)$	(15)

Notes:

r = correlation between phenotypic values of mates.

m = correlation between breeding values of mates. When choice of mates is purely by phenotypic values, $m = rh^2$.

h^2 = heritability, defined as V_A/V_P, at the age of mating.

Covariances of mates: subscripts 1 and 2 refer to the two mated individuals; E includes non-additive genetic deviations; V_A and V_P refer to the generation of the mated pairs.

Variances: subscript 0 refers to the random breeding base, 1 refers to the offspring of 1 generation of assortative mating.

Relatives: dominance, V_D, and common environment, V_{Ec}, are omitted; V_A and h^2 refer to the parental generation with correlations r and m.

variances remain constant. Dividing the additive by the phenotypic variance gives the heritability in equations (8) and (9).

With these equations we can answer two questions about the heritability. First, by how much does one generation of assortative mating increase the heritability? Equation (8) with substitution of $m = rh^2$ shows that it increases it by a factor of $(1 + \frac{1}{2}h_0^2 r)/(1 + \frac{1}{2}h_0^4 r)$, where h_0^2 is the original heritability. The increase may be useful in improving the accuracy with which individuals' breeding values can be predicted from their phenotypic values. The increase of the heritability, however, is never very great – at most about 10 per cent. For an experiment with *Drosophila* on the use of assortative mating in this way, see McBride and Robertson (1963). The second question is this: if we have a population in equilibrium and estimate the heritability in it, what would the heritability be if the population were mating at random? Equation (9) with substitution of $m = rh^2$ shows that if, for example, the heritability were 0.50 under assortative mating of $r = 0.5$, the random-mating heritability would be 0.43. This question is relevant to human populations if comparisons are to be made with other species, but equation (9) can be applied to human data only if m is known or guessed, because $m = rh^2$ is probably seldom true for the reasons already given.

A final question to consider is the estimation of the heritability, defined as V_A/V_P, in a population with assortative mating. The covariances of relatives can be worked out in the manner described in Chapter 9, taking account of the parental covariances given in (1), (2), and (3) of Table 10.6. The covariance and regression or correlation are given for three relationships in Table 10.6. (For the correlations of other relatives, see Nagylaki, 1978). These expressions apply to any generation, V_A and h^2 being the values in the parental generation. V_A is increased over its random breeding value, as shown by equation (5), and the covariances are increased by a further factor of $(1 + r)$ for offspring and parents and by $(1 + m)$ for full sibs. As before, the offspring–parent covariance is the same for mid-parent values as for single parents. The variance of mid-parent values, however, is increased by the same amount as the covariance; so the regression of offspring on mid-parent values is equal to the heritability, as in a random-breeding population. This conclusion has important practical consequences for the estimation of the heritability. Assortative mating among the parents does not affect the regression of offspring on mid-parent values and so the regression provides a valid estimate of the heritability in the population from which the parents came. Assortative mating, however, has the advantage of increasing the precision of the estimate, the standard error being reduced because the variance of mid-parent values is increased, as will be explained in the next section (for details see Reeve, 1961).

The regression of offspring on single parents and the full-sib correlation are both affected by assortative mating. The variance of single parents is simply the phenotypic variance, but because of the correlation with the unmeasured parent the regression of offspring on single parents is increased by the factor $(1 + r)$, and with perfect assortative mating $(r = 1)$ it would be the same as the regression on mid-parent values. The correlation of full sibs in equation (15) of Table 10.6 omits dominance and common environment. If these were assumed to be negligible, the heritability could be estimated by substituting $m = rh^2$. The equation is then quadratic with the solution $h^2 = [-1 + \sqrt{(1 + 8rt)}]/2r$.

Precision of estimates and design of experiments

The precision of an estimate of heritability, indicated by its standard error, is easily obtained from the standard error of the regression or correlation from which the heritability is estimated. Standard errors of heritability estimates are uncomfortably large unless the regression or correlation is based on very large numbers, so it is important to do everything possible to minimize the standard error. In planning an experiment to estimate a heritability, one wants to know how many observations are needed to achieve a given degree of precision; and to achieve the greatest possible precision, within the limitations imposed by the available facilities, one needs to know what is the best method and the best design of the experiment. These are the problems to be considered now. The choice of method is between regression of offspring on one or on both parents, and sib-correlations. The choice of method, however, is usually determined more by practical considerations and by freedom from bias, than by precision. We shall therefore not give much attention to the comparison of methods. The question of design concerns the number of individuals per family. The total number of individuals that can be measured is limited by space, labour, or cost. Increasing the number of individuals per family therefore reduces the number of families. The problem is to find the best compromise between large families and many families that will minimize the sampling variance of the regression or correlation.

In assessing the relative efficiencies of different methods and designs, we have to compare experiments made on the same scale; that is to say, with the same total expenditure in labour or cost. We must therefore decide first what are the circumstances that limit the scale of the experiment. If the labour of measurement is the limiting factor, as for example in experiments with *Drosophila*, then the limitation is in the total number of individuals measured, including the parents if they are measured. If, on the other hand, breeding and rearing space is the limiting factor, as it generally is with larger animals, the limitation may be either in the number of families or in the total number of offspring that can be produced for measurement, and measurements of the parents may be included without additional cost.

We cannot take account here of all the ways in which the scale of an experiment may be limited. For the sake of illustration, the limitation will be taken to be the number of individuals that can be measured in one generation, implying that equal numbers of parents and offspring can be measured. After finding the optimal design for offspring–parent regressions, we shall deal with assortative mating and selection of parents and with the weighting of families when the number of offspring per family is not limited. Then we shall consider the optimal design for estimation from sib-correlations. The principles of finding the optimal design are described by Latter and Robertson (1960) for offspring–parent regressions and by A. Robertson (1959a) for sib-correlations.

Offspring–parent regression
Consider first estimates based on the regression of offspring on parents. Let X be the independent variate, which may be either the value of a single parent or the mid-parent value. Let Y be the dependent variate, which may be either a single offspring of each parent or the mean of n offspring. Let σ_X^2 and σ_Y^2 be the variances of X and Y respectively; let b be the regression of Y on X, and N the number of

paired observations of X and Y, which is equivalent to the number of families in the experiment. By rearrangement of the standard formula, the sampling variance of a regression coefficient can be expressed as

$$\sigma_b^2 = \frac{1}{N-2}\left[\frac{\sigma_Y^2}{\sigma_X^2} - b^2\right]$$

For use as a guide to design, this formula is more convenient if put in a simplified and approximate form. The regression coefficient is usually small enough that b^2 can be ignored; and we may suppose that N is fairly large, so that the variance of the estimate may be put, approximately, as

$$\sigma_b^2 = \frac{1}{N}\frac{\sigma_Y^2}{\sigma_X^2} \qquad \text{(approx.)} \qquad \dots [10.6]$$

Equation [10.6] can be expressed in terms of numbers in the following way. Let n be the number of offspring measured per family and k the number of parents, i.e., 1 or 2. Then, provided the parents are not mated assortatively, the variance of single parents, or of mid-parent values, is

$$\sigma_X^2 = \frac{1}{k}V_P$$

The variance of the offspring values, σ_Y^2, is the variance of the observed means of families of n individuals. This depends on the phenotypic correlation t between members of families, in a manner that will be explained in Chapter 13 (see Table 13.3), where it will be shown that

$$\sigma_Y^2 = \frac{1 + (n-1)t}{n}V_P \qquad \dots [10.7]$$

By substitution for σ_X^2 and σ_Y^2 in equation [10.6], the sampling variance of the regression becomes

$$\sigma_b^2 = \frac{k[1 + (n-1)t]}{nN} \qquad \text{(approx.)} \qquad \dots [10.8]$$

This approximate expression for the sampling variance allows one to compare the methods – one or both parents measured – and to decide how many offspring should be measured per family.

One parent. Consider first the measurement of only one parent. The denominator, nN, of equation [10.8] is the total number of offspring measured. If this is what limits the scale of the experiment, then nN is fixed and the sampling variance is minimal when $n = 1$, i.e., $(n-1)t = 0$. Thus the most efficient design under these circumstances is to have as many families as possible and to measure only one offspring per family. The standard error of the estimate of the heritability will then be as follows:

$$\text{s.e.}(h^2) = 2\sigma_b = 2/\sqrt{N} \qquad \text{(approx.)} \qquad \dots [10.9]$$

To achieve a standard error of 0.1 it is necessary to measure 400 parents and 400 offspring. This illustrates the fact that large numbers are needed to give estimates of even very modest precision. If only 100 families could be measured, the standard error would be about 0.2 and no estimates under about 0.4 would be significantly different from zero. If the number of offspring per family can be increased without reducing the number of families, this will increase the precision. The increase of precision, however, depends on the sib-correlation t, because additional offspring give more information about the family-mean when the correlation is low. The advantage gained can be worked out from equation [10.8] if the value of t is known.

Both parents.   Now consider the measurement of both parents for the regression on mid-parent values. If only one offspring was measured per family, substitution of $k = 2$, $n = 1$ in equation [10.8] gives

$$\text{s.e.}(h^2) = \sigma_b = \sqrt{(2/N)} \quad \text{(approx.)} \qquad \qquad \ldots [10.10]$$

However, if both parents are measured, the same facilities will allow two offspring per family to be measured. Substituting $k = 2$, $n = 2$ into equation [10.8] gives the standard error of the estimate as $\sqrt{[(1 + t)/N]}$, where t is the full-sib correlation. Comparison will show that, under most circumstances, regression on mid-parent gives better precision than regression on one parent.

Assortative mating.   Mating the parents assortatively increases the precision. Both parents must, of course, be measured, so only the regression on mid-parent values need be considered. The effect comes from the increase of the variance of mid-parent values. The variance of offspring is also increased but not by much, and this will be neglected for the sake of simplicity. The variance of mid-parent values under assortative mating is $\frac{1}{2}V_P(1 + r)$, where r is the correlation between mates. Substituting this for σ_X^2 in equation [10.6], and taking σ_Y^2 from equation [10.7] as before, shows that the sampling variance of the regression with assortative mating is approximately $1/(1 + r)$ times the sampling variance with random mating. Thus the precision, in terms of standard errors, is increased by a factor of $\sqrt{(1 + r)}$, or by $\sqrt{2}$ if assortative mating is complete, i.e., if $r = 1$.

Weighting families of unequal size.   It is often possible to measure as many offspring as there are in each family without reducing the number of families. The number of offspring per family then varies among families and this introduces the problem of how to weight the families according to the number of offspring. The appropriate weighting depends on the phenotypic correlation t between the offspring in the families. The principle of the weighting is that families of size n are weighted in proportion to the reciprocal of the variance of the regression that would be obtained if all families were of size n. The weighting is described by Kempthorne and Tandon (1953). The following procedure (Falconer, 1963) is a modification which adjusts the weights so that families of size $n = 1$ always have a weight of 1. First, the intraclass correlation t must be calculated from an analysis of variance between and within families of offspring. Second, the regression coefficient to be estimated must be guessed at, or estimated approximately from unweighted means of families. The weight w_n to be given to the mean of n offspring is then

$$w_n = (n + nT)/(1 + nT) \qquad \qquad \dots [10.11]$$

where $T = (t - b^2)/(1 - t)$ for regression on single parents, and $T = (t - \frac{1}{2}b^2)/(1 - t)$ for regression on mid-parent values. The weighting does not have much effect on the precision unless n varies substantially. Bohren, McKean, and Yamada (1961) examine its merits.

Sib analyses

Now let us consider estimates obtained from the intraclass correlation of full-sib or half-sib families. We shall at first suppose for simplicity that half-sib families are not subdivided into full-sib families, i.e., that only one offspring from each dam is measured in paternal half-sib families. Let N be the number of families, and n the number of individuals per family, so that the total number of individuals measured is $T = nN$. Let the intraclass correlation be t. The sampling variance of the intraclass correlation is then

$$\sigma_t^2 = \frac{2[1 + (n - 1)t]^2(1 - t)^2}{n(n - 1)(N - 1)} \qquad \qquad \dots [10.12]$$

When the value of $T = nN$ is limited by the size of the experiment, it can be shown that the sampling variance of the intraclass correlation is minimal when $n = 1/t$, approximately. Thus the optimal family size depends on the correlation and therefore on the heritability. Assuming that variance due to dominance and common environment are negligible, with full sibs $t = h^2/2$, and with half sibs $t = h^2/4$. So the family sizes giving the most efficient design are $n = 2/h^2$ for full-sib families, and $n = 4/h^2$ for half-sib families. In the case of half sibs we are assuming that only one offspring per dam is measured, so n is the optimal number of dams per sire. Since prior knowledge of the heritability will be at the best only approximate, the optimal family size cannot be exactly determined before-hand. The loss of efficiency, however, is much greater if the family size is below the optimum than if it is above. It is therefore better to err on the side of having too large families. A. Robertson (1959a) shows that, in the absence of prior knowledge of the heritability, half-sib analyses should generally be designed with families of between 20 and 30.

The sampling variance of the correlation when the experiment has the optimal design is obtained by substituting $n = 1/t$ in equation [10.12]. Making some approximations, this leads to

$$\sigma_t^2 = 8t/T \qquad \text{(approx.)} \qquad \qquad \dots [10.13]$$

To get the sampling variance of the heritability, the variance of the full-sib correlation must be multiplied by 4, and the variance of the half-sib correlation by 16. Then, by substituting $t = h^2/2$ in equation [10.13], the sampling variance of the heritability estimated from full sibs becomes

$$\sigma_{h^2}^2 = 4\sigma_t^2 = 16h^2/T \quad \text{(approx.)} \qquad \qquad \dots [10.14]$$

And, by substituting $t = h^2/4$ in equation [10.13], the sampling variance of the estimate from half sibs becomes

$$\sigma_{h^2}^2 = 16\sigma_t^2 = 32h^2/T \quad \text{(approx.)} \qquad \ldots [10.15]$$

Thus, other things being equal, an estimate from full-sib families is twice as precise, in terms of their variances, as one from half-sib families,

It is sometimes desirable to design an experiment for estimating the heritability both from offspring–parent regression and from sib-correlation. Hill and Nicholas (1974) show that the optimal design does not differ much from what would be best for either method alone.

Selection of parents

In experimental populations and in farm animals the parents used are often a selected group. They may be selected on the basis of the character whose heritability is being estimated, or on the basis of some other character correlated with it. The selection causes the variance between parents to be reduced and consequently the covariance of sibs to be reduced. As a result, the heritability estimated from intraclass correlations is biased downwards, and can be as much as 50 per cent below its true value (see Ponzoni and James, 1978). If the selection of parents is based on the character whose heritability is being estimated, it does not affect the regression of offspring on parents, either single parents or mid-parent values, but it reduces the precision because it reduces the variance of the parents, σ_X^2, in equation [10.6] (see A. Robertson, 1977a). Selection can, however, improve the precision if two groups of parents are selected, one with high values and one with low values, and offspring are reared only from these selected groups. The gain in precision comes from devoting all the available facilities to the more extreme families, which give the most information about the regression. When equal numbers of offspring and selected parents are measured, the optimal proportion of parents to select in each group is about 5 per cent. For details see Hill and Thompson (1977).

11 SELECTION:
I. The response and its prediction

Up to this point the treatment of metric characters has been mainly concerned with the description of the genetic properties of a population as it exists under random mating, with no influences tending to change its properties; now we have to consider the changes brought about by the action of a breeder or experimenter. There are two ways in which the action of the breeder can change the genetic properties of the population; the first by the choice of individuals to be used as parents, which constitutes selection, and the second by control of the way in which the parents are mated, which embraces inbreeding and cross breeding. Selection in one form or another is the means whereby all improvement of domesticated animals and plants has been made. In this chapter, therefore, we start consideration of the most important application of quantitative genetics. Selection means breeding from the 'best' individuals, whatever 'best' may be. The ways in which the theory of quantitative genetics can help in this are, first, by showing how to choose individuals with the best breeding values and, second, by predicting the outcome so that different breeding schemes can be compared. The simplest form of selection is to choose individuals on the basis of their own phenotypic values. This is the form of selection to be considered in this chapter. Chapter 13 will show how information from relatives can help to identify individuals with the best breeding values. Experimental selection in laboratory animals provides a means of studying the genetics of metric characters. This aspect of selection will be dealt with in the next chapter.

In any practical selection programme the number of parents used is more or less restricted, with the result that some inbreeding inevitably takes place, and its effects are superimposed on those of the selection. Any inbreeding effects that there may be will at first be ignored, but they will have to be taken into consideration later.

The basic effect of selection is to change the array of gene frequencies in the manner described in Chapter 2. The changes of gene frequency themselves, however, are now almost completely hidden from us because we cannot deal with the individual loci concerned with a metric character. The effects of selection that can be observed are therefore restricted mainly to changes of the population mean. Let us, however, consider the underlying changes of gene frequencies a little further in general terms.

To describe the change of the genetic properties from one generation to the next we have to compare successive generations at the same point in the life-cycle of the individuals, and this point is fixed by the age at which the character under study is measured. Most often the character is measured at about the age of sexual maturity or on the young adult individuals. The selection of parents is made after the measurements, and the gene frequencies among these selected individuals are different from what they were in the whole population before selection. If there are no differences of fertility among the selected individuals or of viability among their progeny, then the gene frequencies are the same in the offspring generation as in the selected parents. Thus artificial selection – that is, selection resulting from the action of the breeder in the choice of parents – produces its change of gene frequency by separating the adult individuals of the parent generation into two groups, the selected and the discarded, that differ in gene frequencies. Natural selection, operating through differences of fertility among the parent individuals, or of viability among their progeny, may cause further changes of gene frequency between the parent individuals and the individuals on which measurements are made in the offspring generation. Thus there are three stages at which a change of gene frequency may result from selection: the first through artificial selection among the adults of the parent generation; the second through natural differences of fertility, also among the adults of the parent generation; and the third through natural differences of viability among the individuals of the offspring generation. Though natural differences of fertility and viability are always present, they are not necessarily always relevant, because they are not necessarily connected with the genes concerned with the metric character.

Response to selection

The change produced by selection that chiefly interests us is the change of the population mean. This is the *response* to selection, which will be symbolized by R; it is the difference of mean phenotypic value between the offspring of the selected parents and the whole of the parental generation before selection. The measure of the selection applied is the average superiority of the selected parents, which is called the *selection differential*, and will be symbolized by S. It is the mean phenotypic value of the individuals selected as parents expressed as a deviation from the population mean, that is from the mean phenotypic value of all the individuals in the parental generation before selection was made. To deduce the connection between response and selection differential, let us imagine two successive generations of a population mating at random, as represented diagrammatically in Fig. 11.1. Each point represents a pair of parents and their progeny, and is positioned according to the mid-parent value measured along the horizontal axis and the mean value of the progeny measured along the vertical axis. The origin represents the population mean, which is assumed to be the same in both generations. The sloping line is the regression line of offspring on mid-parent. (A diagram of this sort, plotted from actual data, was given in Fig. 10.1.) Now let us regard a group of individuals in the parental generation as having been selected – say those with the highest values. These pairs of parents and their offspring are indicated by solid dots in the figure. The parents have been selected on the basis of their own phenotypic values, without regard to the values of their

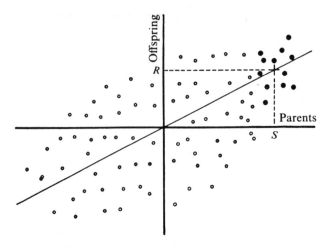

Fig. 11.1. Diagrammatic representation of the mean values of progeny plotted against the mid-parent values, to illustrate the response to selection, as explained in the text.

progeny or of any other relatives. Let S be the mean phenotypic value of these selected parents, expressed as a deviation from the population mean. And similarly let R be the mean deviation of their offspring from the population mean. Then S is the selection differential and R is the response. The point marked by the cross represents the mean value of the selected parents and of their progeny, and its expected position is on the regression line as shown. Thus the ratio R/S is equal to the slope of the regression line. The connection between the response and selection differential is therefore given by

$$R = b_{\overline{OP}}S \qquad \qquad \ldots [11.1]$$

We saw in the last chapter that the regression of offspring on mid-parent is equal to the heritability, provided there is no non-genetic cause of resemblance between offspring and parents. To this we must add the further condition that there should be no natural selection: that is to say, that fertility and viability are not correlated with the phenotypic value of the character under study. Provided these conditions hold, therefore, the ratio of response to selection differential is equal to the heritability, and the response is given by

$$R = h^2 S \qquad \qquad \ldots [11.2]$$

The connection between the response and the selection differential, expressed in equation [11.2], follows directly from the meaning of the heritability. We noted in the last chapter (equation [10.2]) that the heritability is equivalent to the regression of an individual's breeding value on its phenotypic value. The deviation of the progeny from the population mean is, by definition, the breeding value of the parents, and so the response is equivalent to the breeding value of the parents. Thus it follows that the expected value of the progeny is given by $R = h^2 S$.

There is one point at which the situation envisaged in deducing the equations of response does not coincide with what is actually done in selection. We supposed the individuals of the parent generation to have mated at random and the selection to have been applied subsequently. In practice, however, the selection is usually made before mating, on the basis of the individuals' values and not the mid-parent values. The effect of this is that the individuals, when regarded as part of the whole parental population, have been mated assortatively. Assortative mating, however, has very little effect on the mid-parent regression, as we noted in the last chapter, and this feature of selection procedure can therefore be disregarded.

Prediction of response

The chief use of these equations of response is for predicting the response to selection. Let us consider a little further the nature of the prediction that can be made. First, it is clear that equation [11.1] is not a prediction but simply a description, because the regression of offspring on parent cannot be measured until the offspring generation has been reared. The equation $R = h^2 S$, however, provides a means of prediction from knowledge of the heritability obtained from previous generations. The heritability for use in the prediction can be estimated by any method, such as a sib-correlation, and does not have to be estimated from the offspring–parent regression. The selection differential S cannot be known till after the parents have been selected, but its expected value can be predicted, as will be explained in the next section. The following example illustrates the calculation of the selection differential and response, and the prediction of the response by equation [11.2].

Example 11.1 The data in the table come from the experiment of Clayton, Morris, and Robertson (1957) on selection for abdominal bristle number in *Drosophila melanogaster*. The heritability of bristle number was first estimated in the base population before selection and found to be 0.52, as stated in Example 10.1. The parents selected for high bristle number had a mean superiority of $S = 40.6 - 35.3 = 5.3$ bristles. The predicted response, by equation [11.2], is $0.52 \times 5.3 = 2.8$. The observed response was $37.9 - 35.3 = 2.6$ bristles.

Generation	Mean of all measured	Mean of those selected	Selection differential	Response Exp.	Obs.
Parents	35.3	40.6	5.3	2.8	—
Offspring	37.9	—	—	—	2.6

The prediction of response is valid, in principle, for only one generation of selection. The response depends on the heritability of the character in the generation from which the parents are selected, so responses in later generations cannot, strictly speaking, be predicted without redetermining the heritability in each generation. There are two reasons why the heritability is expected to change. First, if there is a response the gene frequencies must change, and the heritability depends on the gene frequencies. This change is not likely to be apparent for some considerable time because gene frequency changes are small unless only a few loci are involved. Second, the selection of parents reduces the variance and the heritability. This takes place in the early generations. It will be explained briefly later and will be ignored meantime. These expected changes in the heritability are not large, however, and experiments have shown that the response is usually maintained with little change over several generations – up to five, ten, or even more. This will be seen in the graphs of responses to selection given later in this chapter and in the next.

Selection differential and intensity of selection

The selection differential can be predicted in advance provided that two conditions hold: the phenotypic values of the character being selected are normally distributed, and selection is by *truncation*. Truncation selection means that individuals are chosen strictly in order of merit as judged by their phenotypic values, no individual being selected that is less good than any of those rejected. Under these conditions the selection differential depends only on the proportion of the population included among the selected group, and the phenotypic standard deviation of the character. The dependence of the selection differential on these two factors is illustrated diagrammatically in Fig. 11.2. The graphs show the distribution of phenotypic values, which is assumed to be normal. The individuals with the highest values are supposed to be selected, so that the distribution is sharply divided at a point of truncation, all individuals above this value being selected and all below rejected. The arrow in each figure marks the mean value of the selected group, and *S* is the selection differential. In graph (*a*) half the population is selected, and the selection differential is rather small: in graph (*b*) only 20 per cent of the population is selected, and the selection differential is much larger. In graph (*c*) 20 per cent is again selected, but the character represented is less variable and the selection differential is consequently smaller. The standard deviation in (*c*) is half as great as in (*b*) and the selection differential is also half as great.

The standard deviation, which measures the variability, is a property of the character and the population, and it sets the units in which the response is expressed, i.e., so many pounds, millimetres, bristles, etc. The response to selection may be generalized if the selection differential is expressed in terms of the phenotypic standard deviation, σ_P. This standardized selection differential S/σ_P is called the *intensity of selection*, symbolized by *i*. Then the selection

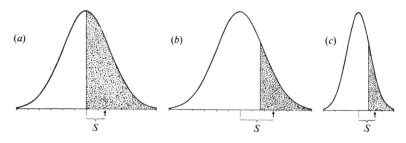

Fig. 11.2. Diagrams to show how the selection differential, *S*, depends on the proportion of the population selected, and on the variability, of a normally distributed character. All the individuals in the stippled areas, beyond the points of truncation, are selected. The axes are marked in hypothetical units of measurement.

  (*a*) 50 per cent selected; standard deviation 2 units: $S = 1.6$ units
  (*b*) 20 per cent selected; standard deviation 2 units: $S = 2.8$ units
  (*c*) 20 per cent selected; standard deviation 1 unit : $S = 1.4$ units

differential is

$$S = i\sigma_P$$

and the expected response in equation [11.2] becomes

$$R = ih^2 \sigma_P \qquad \dots [11.3]$$

By noting that $h = \sigma_A/\sigma_P$, where σ_A is the standard deviation of breeding values (square-root of the additive genetic variance), we may write this equation in the form

$$R = ih\sigma_A \qquad \dots [11.4]$$

which is sometimes used in comparisons of different methods of selection.

The intensity of selection, i, depends only on the proportion of the population included in the selected group and, provided the distribution of phenotypic values is normal, it can be determined from tables of the properties of the normal distribution. If p is the proportion selected, i.e., the proportion of the population falling beyond the point of truncation, and z is the height of the ordinate at the point of truncation, then it follows from the mathematical properties of the

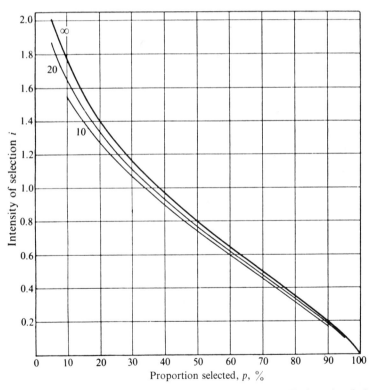

Fig. 11.3. Intensity of selection in relation to proportion selected. The intensity of selection is the mean deviation of the selected individuals, in units of phenotypic standard deviations. The upper graph refers to selection out of a large total number of individuals measured: the lower two graphs refer to selection out of totals of 20 and 10 individuals respectively. A normal distribution is assumed.

normal distribution that

$$\frac{S}{\sigma_P} = i = \frac{z}{p} \qquad \ldots [11.5]$$

Thus, given only the proportion selected, p, we can find out by how many standard deviations the mean of the selected individuals will exceed the mean of the population: that is to say, the intensity of selection, i. The graphs in Fig. 11.3 show the relationship between i and p. Values of i for given values of p are tabulated in Appendix Table A. The relationship between i and p given in equation [11.5] applies, strictly speaking, only to a large sample: that is to say, when a large number of individuals have been measured, among which the selection is to be made. When selection is made out of a small number of measured individuals, the mean deviation of the selected group is a little less. The intensity of selection can be found from tables of deviations of ranked data (Table XX of Fisher and Yates, 1963). The two lower curves in Fig. 11.3 show the intensity of selection when selection is made from samples of 20 and of 10 measured individuals. Appendix Table B gives some values of i when selection is made from small numbers.

> **Example 11.2** A comparison of the expected and observed responses under different intensities of selection was made by Clayton, Morris, and Robertson (1957), studying abdominal bristle number in *Drosophila*. The heritability was first determined by three methods which yielded a combined estimate of 0.52 (see Example 10.1). The standard deviation of bristle number (average of the two sexes) was 3.35. Selection at four different intensities was carried on for five generations, both upward and downward (i.e., both for increased and for decreased bristle number). In each case 20 males and 20 females were selected as parents, the intensity being varied by the number out of which these were selected, as shown in the first column of the table. The intensities of selection corresponding to these proportions selected are given in the second column of the table. The expected responses are then found from equation [11.3]. Under the most intense selection, for example, it is $R = 1.4 \times 3.35 \times 0.52 = 2.44$. The observed responses are given in the last two columns of the table. Although they do not agree very precisely with expectation, they show how the change made by selection falls off as the intensity of selection is reduced.

		Mean response per generation		
			Observed	
Proportion selected, p	Intensity of selection, i	Expected	Up	Down
20/100 = 0.20	1.40	2.44	2.62	1.48
20/75 = 0.267	1.23	2.14	2.20	1.26
20/50 = 0.40	0.97	1.65	1.46	0.79
20/25 = 0.80	0.34	0.59	0.28	-0.08

The selection differential, S in equation [11.2], and the intensity of selection, i in equation [11.3], refer to the mean superiority of all the parents used. Males and females may differ in the amount of selection that can be applied to them. Some characters, for example, can be measured only on one sex, so that no selection can be applied to the other sex. If the selection applied to males and females differs, the values of S or i to be used are the unweighted means for the two sexes, i.e.,

$$S = \tfrac{1}{2}(S_m + S_f) \qquad \qquad \text{... [11.6a]}$$

or

$$i = \tfrac{1}{2}(i_m + i_f) \qquad \qquad \text{... [11.6b]}$$

Thus if only females, for example, can be selected, $S = \tfrac{1}{2}S_f$, and the expected response is $R = \tfrac{1}{2}h^2 S_f$. This can be related to equation [11.1] by noting that $\tfrac{1}{2}h^2$ is the regression of offspring on single parents. The sexes may also differ in the numbers used as parents. The value of S or i is then again as given in equations [11.6] because half the genes in the offspring come from each sex of the parents regardless of the numbers.

Improvement of response

The ways in which the breeder might improve the rate of response can be seen from the equation $R = ih^2\sigma_P$. The phenotypic standard deviation, σ_P, merely specifies the units of measurement. The heritability can be increased by reducing the environmental variation through attention to the techniques of rearing and management, by multiple measurements when these are possible, and to a small extent by assortative matings as explained in the last chapter. Increasing the intensity of selection seems at first sight to be a straightforward way of improving the response, but there are two factors that limit what the breeder can do in this way. First, the reproductive rate of the organism limits the intensity of selection because the proportion selected for breeding can never be less than the proportion needed for replacement. That is to say, two individuals are needed on average to replace each pair of parents. So the more prolific the organism the more intense the selection that can be applied. If males mate with more than one female, males have more offspring than females, and selection can be more intense on males than on females. Suppose, for example, that each male mates with 10 females, and the females have on average 5 daughters each. To allow for replacement of the females the proportion of females selected cannot be less than 1/5, but males have on average 50 sons, allowing selection of 1/50 to replace the males. The upper limits of the intensity of selection in this case would be $i_f = 1.40$ for females and $i_m = 2.42$ for males, giving a net intensity of $i = 1.91$ by equation [11.6b]. The second factor that limits the intensity of selection is the consequence of population size and inbreeding. Inbreeding almost always reduces reproductive fitness and characters related to it, as will be explained in Chapter 14. So the number of parents used must be large enough to keep this inbreeding depression to an acceptable level when balanced against the gain from selection. In experimental work, for example, one might decide to use not fewer than 10 or 20 pairs of parents. When the number of parents to be used is fixed in this way, the intensity of selection can be increased only by measuring more individuals out of which to select the parents.

Generation interval. The number of offspring available for selection depends not only on the parents' reproductive rate but, in many organisms, also on how long the breeder is willing to wait before he makes the selection. The progress per unit of time is usually more important than progress per generation which has been dealt with so far. The interval of time between generations is therefore an important factor in reckoning the response to selection. By waiting until more offspring have been reared before he makes the selection, the breeder can increase the intensity of selection and the response per generation, but in doing so he inevitably increases the generation interval and may thereby reduce the response per unit of time. There is thus a conflict of interest between intensity of selection and generation interval, and the best compromise must be found between the two. Increasing the number of offspring will pay up to a certain point, and beyond this point it will not. The optimal number of offspring cannot be stated in general terms, and each case must be worked out according to its special circumstances.

In reckoning the generation interval under any scheme of selection, distinction must be made between discrete and overlapping generations. When generations are discrete the offspring are kept till the last-born are mature; selection is then made and the selected individuals are all mated at more or less the same time. The generation interval is the interval between the matings made in successive generations. When generations overlap, replacement of the parents by selected offspring is a more or less continuous process, the selected offspring being mated as soon as they are mature. The generation interval can be calculated as the average age of the parents at the birth of their selected offspring. The problem is to find the optimal age for discarding the parents. Example 11.3 below illustrates the calculation of the optimal procedure both when generations are discrete and when they are overlapping. When fewer male parents are used than female, the generation intervals of males and females, L_m and L_f, must be distinguished as well as the intensities of selection, i_m and i_f. What has to be maximized is the ratio $(i_m + i_f)/(L_m + L_f)$. The solution, which can be found graphically, is explained by Ollivier (1974), where the solutions for the main farm animals are given. Sometimes it is necessary to distinguish four intensities of selection and generation intervals, according to whether the male and female parents are used to breed sons or daughters. The overall ratio i/L is then calculated as $\sum i / \sum L$, where $\sum i$ is the sum of the four intensities and $\sum L$ is the sum of the four generation intervals (Rendel and Robertson, 1950).

> **Example 11.3** Let us suppose that selection is to be applied to some character in mice, and that speed of progress per unit of time is the aim. The question is: how many litters should be raised? To find the number of litters that will give the maximum speed of progress, we have to find the intensity of selection and the generation interval. The ratio of the two will then give the relative speed. The actual speed could be obtained by multiplying by the heritability and the standard deviation, but these factors will be assumed to be independent of the number of litters raised. A comparison of the expected rates of progress per week is made in the table. The comparison is made for two different average sizes of litter, meaning the number of young reared per litter. It is assumed that the character to be selected can be measured before sexual maturity, and that first litters are born when the parents are 9 weeks old, subsequent litters following at intervals of 4 weeks. It is assumed also that the population is large enough to be treated as a large sample in reckoning the intensity of selection, and that equal numbers of males and females are selected.
>
> The generation interval depends on the procedure for selection and mating. Two different procedures are considered. First, selection is deferred till all the litters have been measured. All the selected mice are then mated at the same time, and generations

are discrete. The generation interval, tabulated under L, is the age of the parents at the birth of the last litter to be raised. This is a realistic procedure for laboratory experiments. Second, the mice required are selected equally from all the litters raised. For example, if two litters are raised, one per litter is selected from first litters and one per litter from second litters, making a total of two per family. The intensity of selection is the same as by the first procedure provided the total numbers are large. The selected mice are mated as soon as they are mature, and generations are therefore overlapping. The generation interval, tabulated under $\bar{L}$, is the parents' mean age at the birth of their litters. This procedure is more realistic for a practical breeding programme.

The optimal number of litters is shown by the maximal value of the ratio i/L or $i/\bar{L}$. It differs according to the litter size and the procedure. With the first procedure, if 6 young are raised per litter the maximum rate of response is attained by rearing only one litter; if 4 young are reared it is worth while to wait for second litters but not for third litters. With the second procedure, if 6 young are reared the parents should be discarded after their second litters; if 4 young are reared, the rate of progress is the same when parents are discarded after their second or their third litters. These conclusions could hardly have been guessed at without the computations shown in the table.

			$n = 6$				$n = 4$			
N	L	$\bar{L}$	p	i	i/L	$i/\bar{L}$	p	i	i/L	$i/\bar{L}$
1	9	9	0.333	1.10	0.122	0.122	0.500	0.80	0.089	0.089
2	13	11	0.167	1.50	0.115	0.136	0.250	1.27	0.098	0.115
3	17	13	0.111	1.71	0.101	0.132	0.167	1.50	0.088	0.115
4	21	15	0.083	1.85	0.088	0.123	0.125	1.65	0.079	0.110

n = number of young reared per litter
N = number of litters raised
L = generation interval, in weeks, to last litter
$\bar{L}$ = mean generation interval, all litters (See text).
p = proportion selected.
i = intensity of selection.

Effect of selection on variance

It was stated earlier that selection of parents reduces the variance, and the consequences of this must now be explained. The consequences are rather complicated and can be only briefly described. Details will be found in Bulmer (1971, 1976) and Robertson (1977a). A group of selected parents represents one tail of the phenotypic distribution, and in consequence their phenotypic variance must be less than that of the whole population from which they are selected. Let V_P be the variance in the whole population, V_P' the variance in the selected parents, and k the factor by which the variance is reduced, so that

$$V_P' = V_P(1 - k)$$

The factor k depends on the intensity of selection and is given by

$$k = i(i - x)$$

where i is the intensity of selection and x is the deviation of the point of truncation from the population mean in standard deviation units. (Values of x and i are given in Appendix Table A.) The factor by which the additive genetic variance in the parents, V_A', is reduced depends on the heritability and is $h^2 k$, so that

$$V_A' = V_A(1 - h^2 k)$$

The selection of parents affects the genetic variance in the progeny by generating gametic phase disequilibrium. This was mentioned in Chapter 8 as the consequence of one form of non-random mating. The disequilibrium generated causes the covariance term in equation [8.10] to be negative, and this is how the genetic variance is reduced. With unlinked loci the disequilibrium is halved in the next generation, so the total additive genetic variance in the progeny is $(1 - \frac{1}{2}h^2 k)V_A$. The reduced genetic variance in the parents appears in the progeny as reduced variance between families; the variance within families is not affected because it is restored by segregation. The additive variance in the progeny is therefore distributed between and within full-sib families as follows, where V_A is the additive variance in the parents before selection.

Total	$(1 - \frac{1}{2}h^2 k)V_A$
Between families	$(\frac{1}{2} - \frac{1}{2}h^2 k)V_A$
Within families	$\frac{1}{2}V_A$

The environmental variance among the progeny is not affected until they themselves are selected. The reduced additive variance therefore results in a reduced heritability among the progeny. In the second generation of selection the genetic variance is further reduced, but not by so much, and after a few generations there is no further change.

To give some idea of how much the reduced variance affects the response to selection, Table 11.1 shows a specific example. The hypothetical character has a phenotypic variance of 100 units and a heritability of 50 per cent. The proportion selected is 20 per cent. All loci are assumed to be unlinked, and gene frequency changes are neglected in respect of their effect on the variance. $V_{(diseq.)}$ is the variance due to disequilibrium. The heritability is reduced to 43 per cent and most of this reduction is in the first generation. The response to selection in the first generation is 7.0 units, which is the prediction from $R = ih^2\sigma_p$ with $h^2 = 0.5$.

Table 11.1 Illustration of the way in which the reduced variance resulting from selection of parents affects the response. $V_{(diseq.)}$ is the contribution to the variance made by gametic phase disequilibrium. (Reprinted from Bulmer (1971) by permission of The University of Chicago Press, © 1971 by the University of Chicago.)

	Generation				
	0	1	2	3	4 and after
$V_{(diseq.)}$	0	− 9.8	− 11.9	− 12.4	− 12.5
V_P	100	90.2	88.1	87.6	87.5
V_A	50	40.2	38.1	37.6	37.5
h^2	0.5	0.446	0.432	0.429	0.428
Response	—	7.0	5.9	5.7	5.6

The response, however, falls to 5.6 units as a result of the reduced heritability and reduced phenotypic standard deviation.

Most of the effect of the reduced variance on the response is in the second generation of selection, and the response does not change much after that. The constancy of response usually found in experiments is therefore not in conflict with the effect of selection on variance, at least after the first generation. But the response observed after the first generation is expected to be somewhat less than would be predicted from the heritability in the base population. The discrepancy will be small when the heritability is low, and greater when it is higher.

Measurement of response

When one or more generations of selection have been made, the measurement of the response actually obtained introduces several problems. These are matters of procedure rather than of principle and will be only briefly discussed.

Variability of generation means

The first problem arises from the variability of generation means. Inspection of any graph of selection shows that the generation means do not progress in a simple regular fashion, but fluctuate erratically and more or less violently. The consequence of this variation between generation means is that the response can seldom be measured with any pretence of accuracy until several generations of selection have been made. The best measure of the average response per generation is then obtained from the slope of a regression line fitted to the generation means, as illustrated in the following example, the assumption being made that the true response is constant over the period.

Example 11.4 Figure 11.4 shows the results of 11 generations of two-way selection for body weight in mice (Falconer, 1953). On the left the 'up' and 'down' lines are shown separately, and on the right the divergence between the two is shown. Linear regression lines are fitted to the observed generation means. (The first generation of selection is disregarded because the method of selection was different.) The estimates of the average response per generation, with their standard errors, are as follows:

Up	0.27 ± 0.050
Down	0.62 ± 0.046
Divergence	0.88 ± 0.036

The difference between the upward and downward responses will be discussed in the next chapter.

The causes of variation of the generation means will be considered more fully in the next chapter. Here we simply note what the causes are, and consider what can be done to reduce this variation in the response from generation to generation. The causes of the variation are: random genetic drift, sampling errors in estimating the generation means, differences in the selection differential, and environmental factors. Variation due to random drift and sampling errors can be reduced only by increasing the numbers selected and measured. Differences in the selection differential can be allowed for in a way to be explained in the next sections. Environmental differences between generations can arise from many causes, climatic, nutritional, and general management. The obvious way of

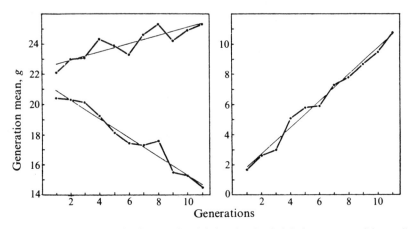

Fig. 11.4. Two-way selection for 6-week weight in mice. On the left the responses of the two lines are shown separately. On the right the 'divergence' is shown, i.e., the difference between the upward- and downward-selected lines. See Example 11.4. (*Based on Falconer, 1953.*)

eliminating environmental fluctuations from an assessment of the rate of response is by keeping an unselected control population. On the assumption that environmental differences affect the selected and control populations alike, the difference between them estimates the genetic improvement made by selection. The use of a control, however, does not always improve the precision with which the response is estimated. Both populations are subject to random drift and to sampling errors, and the difference between the two is subject to variance from these causes that is the sum of the variances of the two lines. Furthermore, the scale of an experiment is usually limited by the facilities available, so that the use of a control necessitates a reduced population size of the selected line. If the selected line and the control both have half the population size of a single selected line, then the use of a control quadruples the sampling variance of the response measured as deviations from control, and so doubles the standard error. This loss of accuracy may counterbalance the gain from eliminating environmental differences. The relative accuracy of the response measured by use of a control can be improved if the 'control' is not an unselected population but is selected in the opposite direction. This is known as 'two-way', or 'divergent', selection and is illustrated in Fig. 11.4. Each selected line acts as a 'control' for the other and the response is measured as the divergence between the two lines. The elimination of some of the variation from generation to generation by this means is clearly seen in Fig. 11.4. In the absence of environmental differences between generations, the precision is the same, relative to the magnitude of the response, as that of a single line occupying the same total facilities. The reason for this is that the standard error of the difference between the lines is doubled, as explained above, but the response is also approximately doubled because both lines are selected. An unselected control, however, is preferable to two-way selection if, for practical reasons, one is interested only in the change in one direction, because the response is not always equal in the two directions, a point that will be discussed in the next chapter.

Random changes of environment reduce the precision with which the response is estimated, but they do not bias the estimate. A more serious difficulty arises from environmental trends, i.e., progressive changes with time, because what looks like a response to selection may really be due to an environmental trend. This makes it difficult to assess the effectiveness of selection in the improvement of domesticated animals and plants, because without a control there is no sure way of deciding how much of the improvement is due to selection and how much to improved management. However, when generations overlap, as with farm animals, it is possible to assess the genetic improvement without an unselected control by making comparisons between contemporary individuals belonging to different generations (see Smith, 1962). The theoretical considerations involved in the use of controls for measuring responses are described by Hill (1972a) and the experimental evidence about the usefulness of controls is reviewed by Hill (1972b).

Weighting the selection differential

In experimental selection the selection differential as well as the response has to be measured because it is the relationship between the two, and not the response alone, that is of interest from the genetic point of view. We have to distinguish between the expected and the effective selection differential, because in practice the individual parents do not contribute equally to the offspring generation. Differences of fertility are always present, so that some parents contribute more offspring than others. To obtain a measure of the selection differential that is relevant to the response observed in the mean of the offspring generations, we therefore have to weight the deviations of the parents according to the number of their offspring that are measured. The expected selection differential is the simple mean phenotypic deviation of the parents as defined at the beginning of this chapter; the effective selection differential is the weighted mean deviation of the parents, the weight given to each parent, or pair of parents, being their proportionate contribution to the individuals that are measured in the next generation.

The weighting of the selection differential takes account of a good part of the effects of natural selection. If the differences of fertility are related to the parents' phenotypic values for the character being selected, then this natural selection will either help or hinder the artificial selection. If, for example, the more extreme phenotypes are less fertile, or more frequently sterile, then natural selection is working against artificial selection. By weighting the selection differential we measure the joint effects of natural and artificial selection together. A comparison of the effective (i.e. weighted) with the expected selection differential may thus be used to discover whether natural selection is operative.

Example 11.5 The experiment with mice, shown in Fig. 11.4, was carried through 30 generations in the upward direction and 24 generations in the downward direction (see Falconer, 1955). Comparisons are made in the table between the effective (weighted) and the expected (unweighted) selection differentials in the two lines. The period of selection is divided into two parts and the comparisons are made separately in each. Throughout the whole of the upward selection there was virtually no difference between the effective and expected selection differentials, and we can conclude that natural selection was unimportant as a factor influencing the response. The situation in the downward selected line, however, is different, the effective

		Selection differential per generation (g.)		
Direction of selection	Generation numbers	Expected	Effective	$\dfrac{Effective}{Expected}$
Upwards	1–22	1.39	1.36	0.98
	23–30	1.08	1.09	1.01
Downwards	1–18	1.03	0.96	0.93
	19–24	0.82	0.70	0.86

selection differential being less than the expected, especially in the second part. From this we can conclude that natural selection was operating in favour of large size, thus hindering the artificial selection and reducing the response obtained, particularly in the latter part of the experiment. The main way in which natural selection acted in the small line was through fertility. The smaller mice tended to have fewer young in their litters and to be more often sterile, with the result that the smallest of the selected parents were represented by fewer measured offspring. There was also another way in which the effective selection differential was reduced which was formally equivalent to natural selection. Smaller mice tend to be more reactive and 'jumpy' than larger mice, and very small mice often escape and are lost during the changing of cages. Those lost in this way before they reproduced proved to be the smallest of the selected mice.

The weighting of the selection differential does not take account of the whole effect of natural selection, because it makes no allowance for any differences of viability among the offspring that may be related to their phenotypic values.

Realized heritability

The response per generation, such as is illustrated in Fig. 11.4, describes what happened, but it takes no account of the amount of selection applied. A means is therefore needed of showing how the response is related to the selection differential. This is done by expressing the response as a proportion of the selection differential, i.e., the ratio R/S, in the following way. The generation means are plotted against the cumulated selection differential. That is to say, the selection differentials, appropriately weighted, are summed over successive generations so as to give the total selection applied up to the generation in question. Responses plotted in this way are illustrated in Fig. 11.5. The average value of the ratio R/S is then given by the slope of the regression line fitted to the points.

The response to selection can be used as a means of estimating the heritability in the base population, because the expected value of the ratio R/S is the heritability, rearrangement of equation [11.2] giving

$$h^2 = \frac{R}{S} \qquad \qquad \dots [11.7]$$

The heritability estimated in this way is called the *realized heritability* because it is primarily a description of the response and may, for several reasons, not provide a valid estimate of the heritability in the base population. If there are systematic changes due to environmental trends or inbreeding depression, these will be

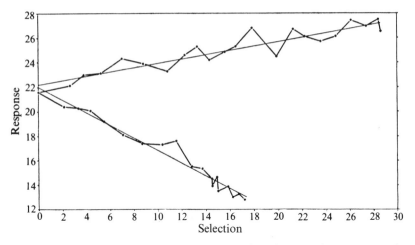

Fig. 11.5. Two-way selection for 6-week weight in mice. The generation means are plotted against the cumulated selection differentials, as explained in the text. The slopes of the regression lines fitted to the points measure the realized heritabilities, which were 0.175 for upward selection and 0.518 for downward selection. (*After Falconer, 1954.*)

included in the response unless they are removed by comparison with a control line. Changes due to random drift are also confounded with the response. The effects of random drift, which will be discussed in the next chapter, can be assessed only by replication of the selection. Figure 11.5 provides a clear example of realized heritabilities being disturbed by other factors. Selection in the two directions yielded very different realized heritabilities; each is a valid description of the response, but they cannot both be valid estimates of the heritability in the base population. The reasons for responses being different in the two directions will also be discussed in the next chapter.

Change of gene frequency under artificial selection

It was pointed out at the beginning of this chapter that the change of the population mean resulting from selection is brought about through changes of the gene frequencies at the loci that influence the character selected. But since the effects of the loci cannot be individually identified, the changes of gene frequency cannot be followed in practice. It is possible, however, to deduce an approximate expression connecting the intensity of selection, i, with the coefficient of selection, s, acting on individual loci. The approximate change of gene frequency can then be found by substituting the appropriate value of s in the formulae given in Chapter 2.

The effect of selection for a metric character on one of the loci concerned may be pictured in the manner illustrated in Fig. 11.6. This refers to a locus with two alleles and shows only the two homozygous genotypes, A_1A_1 and A_2A_2. The position of the heterozygous genotype will be considered later. The two genotypes are shown as having equal frequencies, but this is not necessary to the argument. The homozygous genotypes differ in their mean phenotypic values by $2a$ units of measurement in the notation of earlier chapters (see Fig. 7.1). The

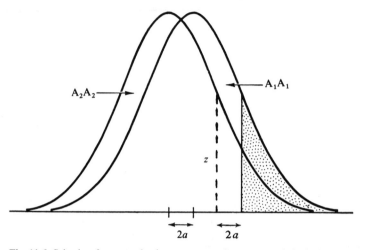

Fig. 11.6. Selection for a metric character operating on one of the loci concerned.

distribution of phenotypic values within each genotype is depicted, this variation arising from other loci as well as from environmental causes. It is assumed that both distributions are normal and that the variance within each genotype is the same. The solid vertical line marks the point of truncation, to the right of which all individuals of both genotypes are selected.

The coefficient of selection is deduced by finding the relative fitness of the two genotypes, i.e., the relative proportions that survive through being selected. Let p be the proportion of A_1A_1 that survive, shown by stippling in the figure. Now imagine the truncation point is moved down by $2a$ units, as shown by the broken line, so that the same proportion p is cut off from the A_2A_2 genotype by this new truncation. The proportion of the A_2A_2 genotype that is actually selected is p minus the proportion lying between the two truncation lines. Provided the separation between the two genotypes, i.e., $2a$, is small relative to the standard deviation, the area of the A_2A_2 curve between the two truncation lines is $2az$, where z is the height of the ordinate. The proportion of the A_2A_2 genotype corresponding to this area is $2az/\sigma_P$, where σ_P is the phenotypic standard deviation. From equation [11.5], $z = ip$, and so the proportion of A_2A_2 actually selected is $p - 2aip/\sigma_P = p(1 - 2ai/\sigma_P)$. In Chapter 2 the coefficient of selection referred to the reduced fitness of the genotype selected against, which is here A_2A_2. The coefficient of selection against A_2A_2 is obtained from its relative fitness as

$$1 - s = \frac{\text{fitness of } A_2A_2}{\text{fitness of } A_1A_1} = \frac{p(1 - 2ai/\sigma_P)}{p}$$

from which the coefficient of selection against A_2A_2 becomes

$$s = i\frac{2a}{\sigma_P} \qquad \text{(approx.)} \qquad\qquad \dots [11.8]$$

This relationship is valid for any degree of dominance, though only two genotypes have been considered in its derivation. If A_1 is completely dominant, the heterozygotes are included with the A_1A_1 genotype, and if A_1 is completely recessive they are included with A_2A_2. In either case the change of gene frequency is given approximately by equation [2.8], which is $\Delta q = -sq^2(1-q)$, where q is the frequency of A_2. If there is no dominance, the coefficient of selection acting on the heterozygote is defined in Chapter 2 as $\frac{1}{2}s$. The difference in mean between A_1A_2 and A_1A_1 is a, and the relationship equivalent to equation [11.8] is $\frac{1}{2}s = ia/\sigma_P$, which is the same as equation [11.8]. The change of gene frequency can therefore be obtained by substituting s from equation [11.8] into equation [2.7], which is $\Delta q = -\frac{1}{2}sq(1-q)$. The conditions under which equation [11.8] provides a reasonable approximation for s are examined by Latter (1965), and a more general derivation is given by Kimura and Crow (1978).

Equation [11.8] shows how the two ways of expressing the 'strength' of selection – by the intensity and by the coefficient of selection – are related to each other. The change of mean of a quantitative character can be derived in the two corresponding ways, and the equivalence of the two derivations provides a check on equation [11.8]. The change of mean can be derived from s and Δq as follows. Consider a single locus with no dominance. From equation [7.2], the mean with $d=0$ is $M = a(p-q) = a(1-2q)$, where q is the initial gene frequency. The change of mean is $\Delta M = a(1-2q_1) - a(1-2q)$, where q_1 is the new gene frequency. This reduces to $\Delta M = -2a\Delta q$. Substituting for Δq from equation [2.7] gives $\Delta M = asq(1-q)$. Equation [11.8] allows us to put s into terms of i and this gives $\Delta M = 2ia^2q(1-q)/\sigma_P$. Now, when $d=0$, $2a^2q(1-q) = V_A$, by equation [8.5], V_A being the additive variance due to the locus under consideration. Summing over all loci gives the change of mean in terms of the total additive variance as $\Delta M = iV_A/\sigma_P = (iV_A/V_P)\sigma_P = ih^2\sigma_P$, and this is the response to selection given by equation [11.3]. Thus equation [11.8] provides a connection between the change of mean derived from s and Δq, and the response derived from i and h^2. The same equivalence can be demonstrated for a dominant gene if the approximation of neglecting $(\Delta q)^2$ is made.

The quantity $2a/\sigma_P$ in equation [11.8] is the difference of value between the two homozygotes, expressed in terms of the phenotypic standard deviation. This quantity will be referred to as the *standardized effect* of the locus. Equation [11.8] will be used in the next chapter to draw some tentative conclusions about the standardized effects of loci giving rise to selection responses.

I2 SELECTION:
II. The results of experiments

In the last chapter we saw that the theoretical deductions about the effects of artificial selection are limited to the change of the population mean, and strictly speaking over only one generation. By changing the gene frequencies, selection changes the genetic properties of the population upon which the effects of further selection depend. And, because the effects of the individual loci are unknown, the changes of gene frequency cannot be predicted, and so the response to selection can be predicted only for as long as the genetic properties remain substantially unchanged. Thus there are many consequences of selection that can be discovered only by experiment. The object of this chapter is to describe briefly what seem to be the most general conclusions about these consequences that have emerged from experimental studies of selection. The most important questions to be answered by experiment concern the long-term effects of selection. For how long does the response continue? By how much can the population mean ultimately be changed? What is the genetic nature of the limit to further progress? Before dealing with the long-term effects, however, there are two questions to be considered concerning the earlier generations during which the rate of response remains more or less constant. These are the repeatability of responses and asymmetry of responses to selection in opposite directions.

Short-term results

Repeatability of response
The variability of the responses from one generation to the next was commented on in the last chapter, and four causes were given: random drift due to the restricted number of parents, sampling error in estimating the generation mean, variation of selection differentials, and environmental factors. The question to be considered now is the variability of the overall response: if the experiment were repeated, how closely would the results agree? An answer to this question is needed before a standard error can be attached to the realized heritability as an estimate of the heritability in the base population, or to allow comparisons to be made between different experiments. We are concerned here only with the period of selection during which the response remains constant and a linear regression can be fitted to the generation means. The standard error of the slope of the fitted regression line can, of course, be calculated. But this does not tell us how much variation in slope there would be between replicates, because it

does not take account of the variation between replicates arising from random drift. Random drift causes changes of gene frequencies which are reflected in changes of the generation mean. In consequence, replicate lines selected independently come to drift apart in the manner that was illustrated in Fig. 3.2. The changes due to drift are cumulative, any change in one generation being carried on as the starting point for the change in the next generation. Because of the cumulative nature of the changes due to drift, the deviations from regression of any one line do not include all the variation due to drift. For this reason the standard error of the fitted regression underestimates the variation between replicates. This point is illustrated in Fig. 12.1. In the experiment illustrated there were six replicates. Fig. 12.1(a) shows the response of all the replicates together, treated as if they were a single large population. Fig. 12.1(b) shows the responses in three of the six replicates. Regression lines were fitted to each of the six replicates and their slopes are given in Table 12.1. The realized heritabilities can be estimated either from the single regression lines in Fig. 12.1(a) or from the means of the six replicates in Table 12.1; there is little difference between them. The standard errors, however, differ greatly. The standard errors of the single

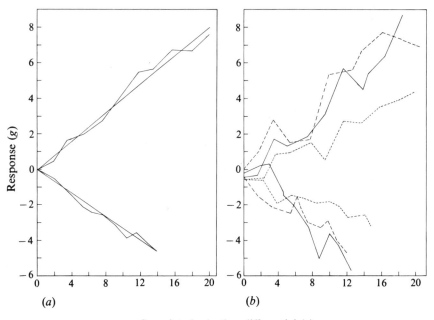

Cumulated selection differential (g)

Fig. 12.1. Ten generations of two-way selection for six-week weight in mice (Falconer, 1973). Generation means, measured as deviations from controls, are plotted against cumulated selection differentials (*original data*).

    (a) The whole 'population' consisting of six replicates selected in each generation. The fitted regression lines have slopes ± s.e. of: Upward, $b = 0.398 \pm 0.020$; Downward, $b = 0.328 \pm 0.014$.

    (b) Three of the six replicates selected in each direction. Each replicate was bred from 8 pairs of parents with minimal inbreeding, giving $N_e = 31$ (by equation [4.9]). The realized heritabilities of these and the other replicates are given in Table 12.1.

Table 12.1 Regression coefficients estimating realized heritability in replicate selection lines. The replicates are listed in order of the magnitude of their responses; those shown in Fig. 12.1 (*b*) are marked*.

	Upward selection	Downward selection
	0.457*	0.501*
	0.448	0.376
	0.438	0.365
	0.390	0.301
	0.385*	0.288*
	0.251*	0.159*
Mean	0.395 ± 0.031	0.331 ± 0.046
Difference	0.064 ± 0.055	

regression lines are 0.020 upwards and 0.014 downwards; the standard errors calculated from the actual variation between replicates are 0.031 upwards and 0.046 downwards.

The theory of the sampling variance of realized heritabilities has been developed by Hill (1971, 1972c,d, 1977) and formulae are given from which it is possible to calculate the approximate standard error when there is only one selection line. The formulae and their application, however, are too complicated to be explained here. The conclusion of general importance is that the standard error of the regression coefficient seriously underestimates the standard error of the realized heritability. For example, the true standard error can be between 3 and 5 times the standard error of the regression coefficient under a wide range of circumstances. A rough idea of the amount of variation expected from random drift can be obtained from the theory explained in Chapter 15 (see Table 15.1). The expected variance between replicate lines at any generation is approximately $2Fh^2V_P$, where F is the inbreeding coefficient, h^2 is the heritability, and V_P the phenotypic variance in the base population. This expectation refers to lines maintained without selection but is at least roughly true also of selected lines (Hill, 1977). Hill's papers cited above, and Bohren (1974), should be consulted for details of the efficiency and optimal design of experiments for estimating realized heritabilities.

Asymmetry of response

The experiment illustrated in Figs. 11.4 and 11.5 showed different rates of response to selection in opposite directions. Selection for increased body weight was only one-third as effective as selection for decreased body weight, when compared by the realized heritabilities. Asymmetrical responses have been found in many two-way selection experiments, and indeed most experiments show asymmetry in some degree. Asymmetry of response has important practical consequences for the following reason. The prediction of a response is made from the heritability estimated in the base population. This can be presumed to predict the mean of the responses in the two directions, and if there is asymmetry the response in one direction will fall short of expectation. Thus if a character of economic importance is selected in one direction only, the response may disappoint the breeder by being less than was expected. It would be useful to be able to predict when asymmetry is likely to occur, and particularly its direction, but this can be done only to a very limited extent. The reason is that there are several possible causes of asymmetrical responses, and only a few of these can be revealed by observations made before selection has been applied. The main causes that may generate asymmetrical responses are as follows.

1. *Random drift.* If there is only one selection line in each direction, asymmetry of response can easily result from random drift, as explained in the preceding section. In any particular case, therefore, the first question must be whether the asymmetry is real in the sense that the realized heritabilities in the two directions are significantly different. Without replication of the selection lines it is not easy to prove the reality of the asymmetry. In Fig. 11.5 there was no replication and the reality of the asymmetry is therefore doubtful. In Fig. 12.1(a) there was replication and the asymmetry was proved to be no more than was expected by chance. Asymmetry due to random drift cannot be predicted.

2. *Selection differential.* The selection differential may differ between the upward and downward selected lines, for several reasons. (i) Natural selection may aid artificial selection in one direction or hinder it in the other. (ii) The fertility may change so that a higher intensity of selection is achieved in one direction than in the other. (iii) The variance may change as a result of the change of mean: the selection differential will increase as the variance increases and decrease as it decreases. This is a 'scale-effect', to be discussed more fully in Chapter 17. Differences of the selection differential influence the response per generation, and the agreement between observed and predicted responses, but they affect the realized heritability only a little (Falconer, 1954). Therefore asymmetry of realized heritabilities cannot be attributed to any cause operating through the selection differential.

3. *Inbreeding depression.* Most experiments on selection are made with populations not very large in size, and there is usually therefore an appreciable amount of inbreeding during the progress of the selection. If the character selected is one subject to inbreeding depression, there will be a tendency for the mean to decline through inbreeding. This will reduce the rate of response in the upward direction and increase it in the downward direction, thus giving rise to asymmetry. An unselected control population, subject to the same inbreeding depression, will reveal how much asymmetry can be attributed to this cause. Prior knowledge of the rate of inbreeding depression would allow the asymmetry of response to be predicted.

4. *Maternal effects.* Characters complicated by a maternal effect may show an asymmetry of response associated with the maternal component of the character. The asymmetry in the experiment of Fig. 11.5 was of this sort. The character selected – 6 week weight of mice – may be divided into two components, weaning weight and post-weaning growth, the former being maternally determined. The weaning weights increased hardly at all in the large line but decreased very much in the small lines. Thus the asymmetry of response was all in

the maternal component of the character. To attribute asymmetry of response to a maternal effect, however, only transfers the problem from the character selected to another and does not explain the asymmetry.

5. *Genetic asymmetry.* The additive genetic variance and the heritability depend on the gene frequencies. Additive genes contribute maximally to the heritability when the gene frequency is 0·5, and recessive genes when the recessive allele has a frequency is 0·75 (see Fig. 8.1). These will be called the 'symmetrical' gene frequencies. If all the genes affecting the character were at these symmetrical frequencies in the initial population, the realized heritabilities would gradually diminish as the gene frequencies became changed, but the diminution would be roughly equal in lines selected in opposite directions and there would be no asymmetry. Suppose, however, that the population starts with gene frequencies above or below these values. In one line the frequencies will then move away from the symmetrical values and the heritability will diminish. But in the line selected in the opposite direction the gene frequencies will move toward the symmetrical values and the heritability will increase. Thus asymmetry will develop as the gene frequencies become different in the up- and down-lines. The response observed depends on the combined effects of all the loci, and asymmetry is to be expected if the 'average' gene frequencies are different from the symmetrical values of 0.5 for additive genes and 0.75 for recessive genes, the 'averages' being weighted for the gene effects. Asymmetry of response from this cause, however, would not be expected to appear immediately in the first few generations because it depends on differentiation in gene frequencies. Furthermore, it would be associated with non-linear responses, because it depends on the response decelerating in one line and accelerating in the other. So it could not readily explain asymmetry of responses that are not detectably non-linear. Genetic asymmetry can be looked at in a different way as the relation of the starting point to the selection limits. The theoretical limits to selection are when all favourable alleles have been fixed. Asymmetry of response will result if the initial population is not mid-way between the two limits in phenotypic value, so that the selection response has further to go in one direction than in the other. If selection favours heterozygotes, the situation is a little different because the limit in one direction is not fixation but is the equilibrium gene frequency. Asymmetry will result if the initial population is not at the point, in respect of gene frequencies, where the additive variance is maximal.

6. *Genes with large effects.* Asymmetry of response that appears immediately in the first generation can result from genetic asymmetry of genes with large effects. The reasons why the asymmetry is immediate is that the first selection of parents produces a large change of gene frequency, equivalent to many generations of selection on genes with small effects. The asymmetry results from the initial gene frequencies not being at the symmetrical points. If the first response is asymmetrical it follows that the regression of offspring on mid-parent values in the base population will be non-linear (Robertson, 1977*b*). Asymmetrical responses of this sort should therefore be predictable.

7. *Scalar asymmetry.* The genetic and environmental variation may be skewed to different degrees or in opposite directions. The genetic variation will

then make up a larger proportion of the total at one end of the distribution than at the other. In consequence the offspring–parent regression in the base population will be non-linear and the response asymmetrical in the first generation. The situation envisaged is shown diagrammatically in Fig. 12.2(a), where the genetic and environmental variance are skewed in opposite directions. The difference in skewness may be a scale effect, as will be explained in Chapter 17, or it may be due to genotype–environment interaction in the following way. Individuals that experience a good environment may exhibit less genetic variation than those that experience a poor environment; this is illustrated in Fig. 12.2(b). Or, individuals with high genetic values may be more susceptible to environmental variation than individuals with low genetic values, as in Fig. 12.2(c). In either case, individuals with high values will exhibit a lower heritability than those with low values. The difference in skewness could equally well be the other way round from that shown in Fig. 12.2, in which case the upward heritability would be greater than the downward. This form of asymmetrical response should, again, be predictable from a non-linear

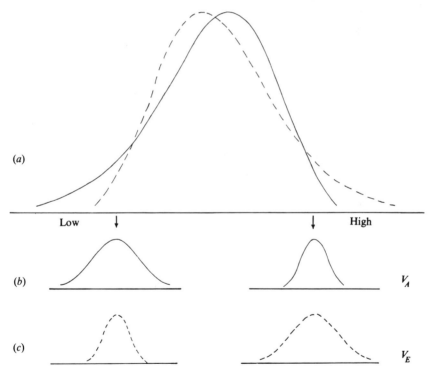

Fig. 12.2. Frequency distributions illustrating scalar asymmetry. (a) shows the additive genetic variation (solid line) with negative skewness, and the environmental variation (broken line) with positive skewness. (b) shows the additive genetic variation among individuals whose environmental values are at the points marked by the arrows. (c) shows the environmental variation among individuals whose breeding values are at the points marked by the arrows. In both situations upward selection will give a lower realized heritability than downward selection. The figure is drawn roughly to scale, and if the distributions in (b) and (c) represented selected individuals the upward heritability would be about 0.1 and the downward about 0.4.

offspring–parent regression in the base population. For details see Robertson (1977*b*).

8. *Indirect selection.* Sometimes the criterion of selection is not quite the same as the character measured for assessing the response. Then, if the measured character is not linearly related to the selection criterion, asymmetry of response may result. Baptist and Robertson (1976) selected *Drosophila* for body size by inducing them to crawl as far as they could through a series of slits of decreasing diameter. The criterion of selection was the number of slits traversed, and this responded symmetrically. The procedure, however, selected not only for body size but also for activity. Small flies are more active than large and this led to a non-linear relation between body size and slit score. In consequence, body size responded less to upward selection than to downward.

With all these possible causes it is not surprising that asymmetrical responses are often found. Nor is it surprising that the cause operating in a particular case is hard to identify. Some of the causes make prediction of asymmetry possible but others do not and, until asymmetrical responses are better understood, the prediction of rates of response from the heritability in the base population will remain somewhat unreliable. There is, however, one generalization that might be made, or at least suggested. It is that if the character selected is a component of natural fitness, asymmetry should be expected, with selection towards increased fitness giving a slower response than selection towards decreased fitness. The reasons are, first, that these characters usually show inbreeding depression, which is itself a cause of asymmetry, and, second, that if the character has been subject to natural selection the gene frequencies are likely to be above the symmetrical point, i.e., nearer the upper limit, thus giving rise to genetic asymmetry.

Long-term results

The response to selection cannot be expected to continue indefinitely. Sooner or later it is to be expected that all the favourable alleles originally segregating will be brought to fixation. As they approach fixation the genetic variance should decline and the rate of response diminish, till, when fixation is complete, the response should cease. When the response has ceased the population is said to be at the *selection limit*. The questions to be considered now are, first, what is the total response attainable and how long does it take to achieve; and second, what are the genetic properties at the limit – are all loci fixed, as expected from simple theory?

Total response and duration of response

We shall first look at the experimental evidence about the limits to selection, and then consider the theory. Figure 12.3 shows four examples of experiments in which the limits have been reached, or nearly so. The experimental evidence is not at all precise; nor is it altogether consistent. It is usually impossible to decide exactly at what point the limit is reached, because the limit is approached gradually, the response becoming progressively slower as the limit is approached. The number of generations required to reach the limit is therefore very imprecise.

The time-scale can, however, be more precisely expressed as the *half-life* of the response. This is the number of generations taken to go half way to the limit. Another problem is that the limit itself is often uncertain. Some experiments appear to reach a limit but, after some further generations, a renewed response takes place and a new limit is reached. Figures 12.3(a) and (d) show examples. Renewed responses like these are most probably due to mutation producing new favourable alleles. They could be due to recombination between loci whose favourable alleles were originally in close repulsion linkage, but this is less likely because the two alleles would not both remain long in the population unless they had almost equal effects or unless they were maintained by balanced lethals.

Bearing these uncertainties in mind, we may examine the results of some fairly typical experiments summarized in Table 12.2. Only two-way experiments are considered because of the complications of asymmetry, and the total response is taken as the sum of the responses in the two directions. This is the difference between the upper and lower limits and may be called the *range*, R. In Table 12.2 the range is expressed in three ways: as the ratio of the means at the two limits (H/L), in units of additive genetic standard deviations (R/σ_A), and in units of phenotypic standard deviations (R/σ_P). The experiments differ considerably and it is clear that precise conclusions cannot be drawn. In all but the mouse litter-size experiment the selection took the populations far beyond the range of variation in the base population, to between 4 and 10 phenotypic standard deviations in each direction, and between 8 and 14 additive genetic standard deviations. There is more consistency in the duration of the responses. All the experiments took roughly 20 to 30 generations to reach the limits, and the half-lives ranged from 7 to 12 generations. Other experiments, however, have given responses over much

Table 12.2 Limits to selection in four experiments. Explanation in text.

		Experiment			
		(1)	(2)	(3)	(4)
Observed ratio of means	H/L	8.0	1.3	2.5	1.6
Observed range	R/σ_A	28	20	16	3.6
	R/σ_P	20	12	8	1.7
Theoretical maximum	$R_{(max)}/\sigma_P$	82	52	22	8
Effective population size	N_e	28	28	15	32
Duration (generations):	Total (approx.)	30	20	25	20
	Half-life	10	7	9	12
	Half-life/N_e	0.4	0.3	0.6	0.4
Observed/Maximum:	Range	0.24	0.22	0.36	0.21
	Half-life	0.29	0.19	0.43	0.36
$n = R^2/8\sigma_A^2$		98	50	32	2
Standardized effect:	$2a/\sigma_P$	0.21	0.23	0.24	0.95

Experiments and sources
(1) *Drosophila*, abdominal bristles: Clayton and Robertson (1957).
(2) *Drosophila*, thorax length (Fig. 12.3c): F.W. Robertson (1955).
(3) Mouse, 6-week weight (Fig. 11.5): Falconer (1955), Roberts (1966a).
(4) Mouse, litter size: Falconer (1965b, 1971, 1977).
Population size in the *Drosophila* experiments is taken to be $N_e = 0.7N$ (see Ch. 4).

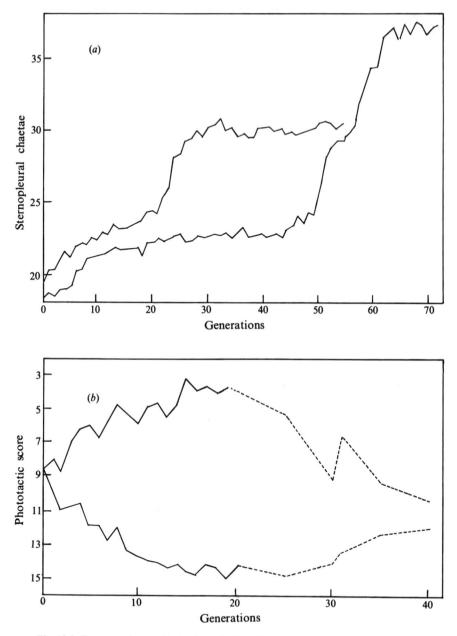

Fig. 12.3. Four experiments illustrating selection limits.

    (*a*) *Drosophilia melanogaster*, sternopleural bristles (chaetae). Two lines derived from different combinations of three stocks and selected independently for increased chaeta number. (*Adapted from Thoday, Gibson, and Spickett, 1964.*)

    (*b*) *Drosophila pseudobscura*, phototoxis. The upper graph is the line selected for avoidance of light. The phototactic score is the number of times the fly moved toward the light out of a total of 15 light–dark choices. (*After Dobzhansky and Spassky, 1969.*)

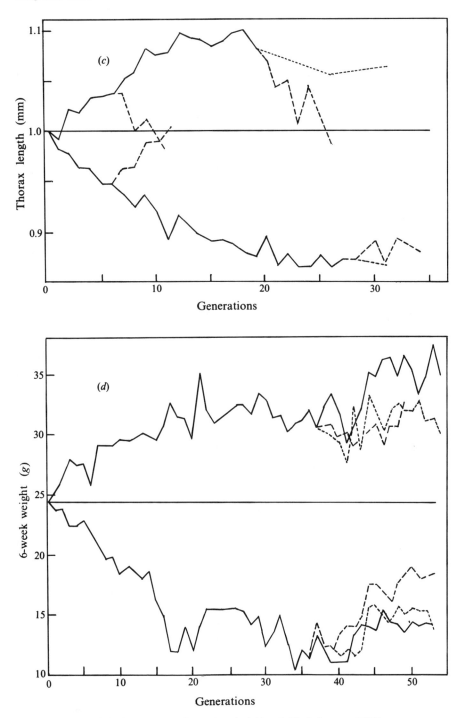

(c) *Drosophila melanogaster*, thorax length. (*After F. W. Robertson, 1955.*)

(d) Mouse, six-week body weight. (*Adapted from Roberts, 1966b.*)

Dashed lines are responses to selection in the reverse direction; dotted lines are responses to natural selection, with artificial selection suspended.

(*All figures redrawn from the above sources with permission of the authors and publishers.*)

longer periods. For example, pupa weight in *Tribolium* continued responding to upward selection for at least 75 generations with a half-life of 30 generations (Enfield, 1977); and the percentage of both oil and protein in maize showed no sign of approaching limits after 76 generations of upward selection (Dudley, 1977). The reasons for these differences between experiments are not known, but may be differences in population size. (See note added on p. 206.)

The total responses in Table 12.2, though they may be impressive when reckoned in terms of the variation present in the original population, are not at all spectacular when compared with the achievements of the breeders of domestic animals. For example, after selection to the two limits, the large mice were 2.5 times the weight of the small mice. In contrast, the weights of the largest breeds of dogs are about 100 times the weights of the smallest breeds (Stockard, 1941). The reason for the disappointing results of experimental selection when viewed against the differences between the breeds of domestic animals is that experiments are carried out with closed populations, usually of not very large size. The limits are set by the gene content of the foundation individuals and, furthermore, there is little opportunity for new mutation to add to the genetic content. The breeder of domestic animals in contrast, by intermittent crossing, casts his net far wider in the search for genes favourable to his purpose and has had more time to utilize new mutations, some of which may have large effects.

Theory of limits.   The total response relative to the initial genetic variation, R/σ_A, depends primarily on the number of loci contributing to the variation. If, for example, there were only one additive locus with gene frequencies of 0.5, the most extreme genotypes would each appear in the base population with frequencies of $1/4$; or if there were two such loci, with frequencies of $1/16$. The selection limits, which are represented by the most extreme genotypes, would then be well within the range of variation found in the base population. With larger numbers of loci, the extreme genotypes are rarer in the base population and the selection limits are further removed, in σ_A units, from the original mean. The numbers of genes and their effects are inversely related because, with a given amount of genetic variation, if there are few genes they must have large effects and if there are many genes they must have small effects. Since neither the number of genes nor their effects are known it is not possible to predict the limits. It is possible, however, to predict a 'theoretical maximum' for the limit, so let us pursue the theoretical consideration of the limits a little further, to find out first how the limit depends on the number of loci and then to see what the theoretical maximum can tell us.

First suppose that the population being selected is bred from a very large number of parents, so that no random fixation occurs; and suppose also that there are no overdominant loci. Then the favourable alleles at all relevant loci will be made homozygous at the limits. In the notation of Chapter 7, the range is $\sum 2a$ units of measurement, i.e., the homozygote difference summed over all loci. How then does the range relate to the original additive genetic variance? In order to express this relationship in a simple way we have to make two assumptions that are certainly not true, but we can see later how the error may affect the conclusions. The first assumption is that all the loci have the same magnitude of effect on the character selected. The range is then $R = 2na$, where n is the number of loci, each having a homozygote difference of $2a$. The second assumption is that all the genes start at frequencies of 0.5. With these assumptions the original additive variance, by equation [8.7], is $\sigma_A^2 = \frac{1}{2}\sum a^2 = \frac{1}{2}na^2$. (The symbol σ_A^2 is used here rather than V_A because it simplifies the formulation when the standard deviation σ_A is involved.) Note that when gene frequencies are 0.5 the dominance deviation d does not appear in the formulation of σ_A^2, so no assumption about the degree of dominance needs be made. The relationship between the range and the additive variance is obtained by squaring the range, which gives

$$\frac{R^2}{\sigma_A^2} = \frac{4n^2a^2}{\frac{1}{2}na^2} = 8n \qquad \ldots [12.1]$$

This equation will be considered later as a possible way of estimating the number of loci.

The total response considered above depends on the effective population size being very large. In practice the number of parents used is seldom large enough for random drift to be ignored. Some inbreeding therefore occurs, which leads to random fixation at some loci. In other words, unfavourable alleles are fixed at some loci despite the selection against them, with the result that the total response is less than that indicated by equation [12.1]. The limit achieved then depends on the chance of fixation of the favourable allele at each locus, this chance being determined partly by the selection and partly by the inbreeding. The way in which the inbreeding and selection interact has been worked out by Robertson (1960), Hill and Robertson (1966), and Robertson (1970). The main conclusions are as follows.

The number of loci affect the issue through the coefficient of selection s acting on each locus. The larger the number of loci, the smaller are their effects and the smaller the coefficient of selection. The chance of fixation of a favourable allele depends on its initial gene frequency; the rarer it is the more likely it is to be lost. Given the initial gene frequency, the chance of fixation is a function of $N_e s$, which is the product of the effective population size and the coefficient of selection in favour of the allele. The coefficient of selection is equal to $i(2a/\sigma_P)$, where i is the intensity of selection (equation [11.8]). Therefore with a given initial frequency and a given gene-effect $(2a/\sigma_P)$ the chance of fixation of a favourable allele is a function of $N_e i$, the product of the effective population size and the intensity of selection. Thus the total response should be greater with larger population sizes and with more intense selection. In practice, increasing the intensity of selection usually necessitates a reduction in the number of parents, which is the population size. The best compromise is to select 50 per cent. This maximizes the total response, though the rate of progress will be less than it could be with more intense selection (Robertson, 1960; Jódar and López-Fanjul, 1977).

When the population size is not large, and there is consequently some inbreeding, it is still true that the larger the number of loci the larger the response will be in relation to the original variance. If the number of loci is very large their effects must be very small and most loci become fixed by random drift before selection can fix the favourable alleles. In spite of this, the greatest response would be attained if the genetic variance were caused by a very large number of loci (strictly speaking an infinite number), even though only a small proportion of them are fixed by the selection. This is the theoretical maximum response; it is the total response that would be attained if the genetic variance were generated by an infinite number of loci. With additive genes, the theoretical maximum response is shown to be (Robertson, 1960)

$$R_{(max)} = 2N_e \, ih^2\sigma_P \qquad \qquad \ldots [12.2]$$

All the terms in this expression can be estimated, so the maximum response can be predicted. Note that $ih^2\sigma_P$ is the predicted response in one generation (equation [11.3]) or the observed response over the first few generations. The maximum response to divergent selection is obtained simply by putting i as the sum of the intensities of selection in the two directions. With recessive genes, however, the maximum response may be much greater, particularly if favourable recessives have low initial frequencies. The theoretical maximum response has a half-life of $1.4N_e$ generations if all the genes are additive, or up to about $2N_e$ generations if the genes are recessive.

The theoretical maximum response is not something that can be achieved by optimal selection procedures. The response that can be achieved depends on the number of genes. The theoretical maximum is what would be achieved in the most favourable genetic situation, which is a very large number of genes – in fact an impossibly large number. As a prediction, therefore, the theoretical maximum does no more than set an upper limit to what can be expected. There is, however, some interest in comparing observed responses with their theoretical maxima because the comparison shows roughly how much the response has been limited by inbreeding. If the observed response is not much less than the maximum, this shows that random fixation has limited the response and a greater response would have been attained with a larger population. If, on the other hand, the observed response is much less than the maximum, we can conclude that selection has fixed favourable alleles at most loci and that the response has not been much reduced by random fixation. Ratios of observed to maximum responses are given in Table 12.2 in respect of the range and of the half-life. The ratios are between about 0.2 and 0.4, which suggests that inbreeding has not been an important limitation to the response achieved in these experiments.

The extent to which fixation has been produced by inbreeding rather than selection has a bearing on the differences between replicate selection lines at the limits. If there has been much fixation by inbreeding, different replicates will have different alleles fixed at many loci. The replicates will then reach different limits and, furthermore, if they are crossed, some genetic variance will be restored on which further selection could act. If there has not been much fixation by inbreeding, selection will have fixed the same alleles at most loci. All replicates will then have the same limit and no genetic variation will be generated by crossing them.

The theory outlined above is based on the supposition that when the limits are reached all loci have been fixed and no genetic variance remains. Studies of populations at the limits to be described in the next section have shown, however, that this supposition is often not true. Conclusions based on the theory can therefore only be regarded as tentative. A test of the theory of limits has been carried out with *Drosophila* (Jones, Frankham, and Barker, 1968). Selection was carried out for 50 generations for increased abdominal bristle number in a number of lines with different intensities of selection and different population sizes. Many of the lines were still responding after 50 generations, so the limits were not precisely determined. The results agreed with the theoretical expectations in that larger population sizes and more intense selection both gave larger total responses. There were, however, large differences between replicate lines treated in the same way, and the authors' general conclusion was that the irregular patterns of response 'made the long-term behaviour of any particular line more or less unpredictable'.

Number of loci (effective factors) and standardized effects

Since the total response depends primarily on the number of loci, it is tempting to try to use the observed response to estimate the number of loci that have contributed to it. There are, however, serious difficulties in interpreting any number so obtained, centring on what is meant by a locus in this context. There are two main difficulties. First, we do not know the form of the distribution of gene effects: do most genes have effects of more or less equal magnitude, or – which seems more likely – are there a few genes with large effects and increasing numbers with smaller and smaller effects? Then where do we stop counting a locus as one affecting the character? The second difficulty concerns linkage. What we count as a locus is a segment of chromosome that has not recombined in the course of the selection. In recognition of linkage, loci in this context are referred to as *effective factors*, and the number of effective factors is sometimes called the *segregation index*. Despite these difficulties it is worth while to consider briefly how the number of effective factors may be estimated.

The number of effective factors, n, can be estimated by equation [12.1] as $n = R^2/8\sigma_A^2$, where R^2 is the square of the difference between the upper and lower selection limits, and σ_A^2 is the additive variance in the base population. The estimate made by equation [12.1] is valid on three conditions: (1) all the favourable alleles have been fixed at both limits; (2) all the genes have equal effects; and (3) all the genes have initial frequencies of 0.5. Failure of conditions (1) or (2) leads to the estimate of n being too low. Failure of condition (3) leads to n being overestimated because σ_A^2 will then be less than it would be with gene frequencies of 0.5 as required. The requirement of condition (3) can be met by estimating σ_A^2 not in the base population but in the F_2 and subsequent generations of a cross between lines at the upper and lower limits. All genes by which the lines differ are then at frequencies of 0.5. But the problem of linkage is then magnified, since all relevant genes are in complete linkage disequilibrium in the F_1. With the reduction of disequilibrium by recombination, the genetic variance will decrease progressively in the generations following the F_2 and the number of effective factors will correspondingly increase.

The numbers estimated by equation [12.1] are given for the four experiments in Table 12.2. They range from 2 to 98. The estimate of 2 for litter size seems too low to be believed, which suggests that the assumptions made were seriously in error. A similar experiment, but selecting only upward, yielded an estimate of $n = 164$ on the assumption that all the genetic variation was due to recessive genes at initial gene frequencies of 0.25 (Eklund and Bradford, 1977). Analysis of the upward response of the experiment in Table 12.2 with the same assumptions gives $n = 25$. This large difference according to the assumptions made emphasizes the dubious value of these estimates of gene numbers.

From the number of effective factors it is possible to obtain the standardized effects of the genes, subject to the same conditions and qualifications. Rearrangement of the expression for the additive variance given earlier, $\sigma_A^2 = \frac{1}{2}na^2$, leads to the standardized effects of the genes as $2a/\sigma_P = 2h\sqrt{(2/n)}$, where h is the square-root of the heritability. The values obtained are close to 0.2 in three of the experiments in Table 12.2; the litter size experiment gives a value of 0.95 for $n = 2$, or 0.21 for $n = 25$.

The use of marker genes in *Drosophila* makes it possible to localize and identify some of the genes that have contributed to a selection response. The situation to which these studies point is not one of a large number of genes all with more or less equal effect. It seems, rather, that a small number of genes with large effects are responsible for most of the response, the remainder of the response being due to a larger number of loci with small effects. For example, five loci accounted for 87.5 per cent of the response of a line of *Drosophila* selected for high sternopleural bristle number (Spickett and Thoday, 1966). Differences found between replicate lines of *Drosophila* selected for abdominal bristle number also point to genes of large effect being involved in the responses (Frankham, Jones, and Barker, 1968).

Nature of the selection limit

The foregoing account of selection limits was based on the assumption that when the limit has been reached, all relevant loci have been made homozygous, either by the selection or by the accompanying inbreeding. The assumed fixation leads to a progressive loss of genetic variance till, at the limit, none remains. Experiments, however, have shown conclusively that this simple picture is wrong; not always but often. We must therefore consider now what experiments can tell us about the genetic nature of selection limits.

Phenotypic variance. The expected loss of genetic variance should lead to a progressive decline of the observed phenotypic variance. This, however, seldom happens, and sometimes the variance increases instead of declining. A fairly typical example is provided by the experiment with mice which has already been described (Fig. 11.5, and Table 12.2, column (3)). The phenotypic variance of body weight in this experiment is shown in Fig. 12.4. The variance in the large line remained at the same level throughout the experiment, but the variance in the small line showed a large and sudden increase. An example from *Drosophila* is provided by the experiment on abdominal bristle number analysed in Table 12.2 column (1). The phenotypic variance in the base population and in the most extreme of the replicate high and low lines is illustrated by frequency distributions in Fig. 12.5. In this case the variance increased very much during the selection in both directions. Similar increases were found by Jones, Frankham, and Barker (1968), also selecting for abdominal bristles.

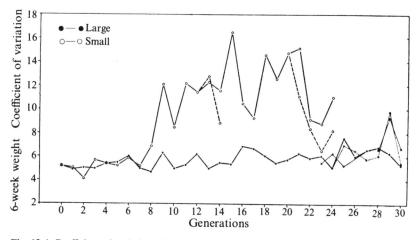

Fig. 12.4. Coefficient of variation of 6-week weight in mice. The thin continuous line starting at generation 23 refers to the unselected control. The broken lines refer to reversed selection and the dotted lines to suspended selection. (*After Falconer, 1955.*)

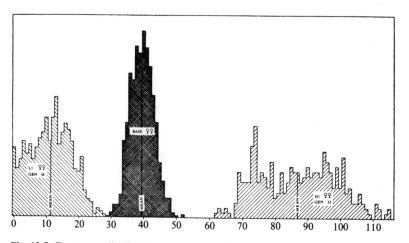

Fig. 12.5. Frequency distributions of abdominal bristle number in *Drosophila melanogaster* (females), in the base population and in the most extreme high and low lines after 35 and 34 generations of selection. (*After Clayton, Morris, and Robertson, 1957.*)

Possible reasons for the variance not declining as expected are the following.

(1) The variance of many characters is not independent of the mean: when the mean changes under selection, the variance automatically changes with it. This is a 'scale effect' which will be more fully discussed in Chapter 17. In Fig. 12.4 the variance is plotted as the coefficient of variation in order to eliminate the scale effect.

(2) The environmental variance may increase. With the approach to fixation of the loci affecting the character, and others linked to them, the frequency of homozygotes will increase. There is evidence, mentioned in Chapter 8 and to be discussed more fully in Chapter 15, that homozygotes are sometimes more variable from environmental causes than are heterozygotes. An increase of

environmental variance from this cause might counterbalance the expected decline of genetic variance.

(3) The genetic variance may not decline as expected, and may even increase. The evidence concerning the genetic variance will now be considered.

Genetic variance.   There are two simple ways of testing for the presence of genetic variance at the limit. First, selection in the opposite direction – reversed selection – can be applied. This tests for the presence of additive variance. Second, the selected line can be inbred. This tests, though not conclusively, for the presence of non-additive genetic variance, for reasons to be explained in Chapter 14. Reversed selection was applied in two of the experiments illustrated in Fig. 12.3. In experiment (c) the high line responsed but the low line did not. In experiment (d) the situation was reversed, the low responding and the high line not responding. In both these experiments therefore, one line seemed to have been fixed but the other retained some additive genetic variance. (The situation in the high line of experiment (d) was rather more complex. After the relaxed selection had been started, the line itself showed a renewed response to continued upward selection, due possibly to recombination but more probably to mutation.) Inbreeding was applied to the high line analysed in Table 12.2, column (4) and led to a decline of litter size, proving the existence of non-additive genetic variance. The same was found in another experiment on litter size in mice (Eklund and Bradford, 1977), and here reversed selection proved the presence also of some additive variance. These few examples illustrate the conclusions from experimental studies of selection limits: some lines lose all additive genetic variance as expected from simple theory; other lines retain additive genetic variance though they seem to have ceased responding. Let us now consider the possible reasons for genetic variance remaining at the selection limit, or what appears to be the limit.

1. The limit may not really have been reached, for two main reasons. First, fertility is often reduced in selected lines, and the selection differential may in consequence be drastically reduced in the later generations. A plot of means against generations may then show a clear approach to a limit, but a plot against the selection differential may reveal little evidence that the line has really ceased responding. Figure 11.5 illustrates this, more particularly in the low line. Second, loci with recessive alleles may remain unfixed. Recessive alleles that are unfavourable will be brought to low frequencies. Selection is then very slow to reduce their frequencies further because most of the variance they contribute is non-additive (see Fig. 8.1). This was thought to be the situation in the two experiments on mouse litter size already referred to (Falconer, 1971; Eklund and Bradford, 1977). In both cases inbreeding and crossing, which was thought to have eliminated the unwanted rare recessives, produced a mean well above the original limit.

2. The limit may be an extrinsic one imposed by the nature of the character or the way in which it is measured. For example, percentages have limits of 0 and 100; *Drosophila* bristle number cannot go below 0. These limits may be reached, or closely approached, with little fixation of the genes concerned. The experiment of Fig. 12.3(b) had extrinsic limits at phototactic scores 1 and 16, though in this case both lines reached selection limits that fell short of the extrinsic limits.

3. The artificial selection may be opposed by natural selection. This should be detectable in a loss of selection differential, as in Example 11.5, provided the selection differential can be appropriately weighted by the number of progeny surviving and measured. It can also be detected by suspending the artificial selection, to see if the line responds in the opposite direction to natural selection acting alone. Selection was suspended, or 'relaxed', in this way in experiments (b), (c), and (d) of Fig. 12.3. In (c) and (d) there was no clear change, showing that natural selection was not seriously opposing the artificial selection in these experiments. But in (b) both lines very clearly responded, showing that natural selection had been opposing the artificial selection in both directions.

4. There may be overdominance at some loci for the character selected. At the selection limit, overdominant genes would be in equilibrium at more or less intermediate frequencies. The variance they give rise to would be non-additive only. There would be no immediate response to reversed selection, but if reversed selection were continued a response would develop as the gene frequencies became changed from their equilibrium values. There is, however, no evidence that this is a common feature of selection limits.

5. Selection may favour heterozygotes through the joint action of artificial and natural selection. This situation, in contrast to the previous one, occurs commonly. The pygmy gene of mice, referred to in Chapter 7, is an example. The gene arose by mutation in a line selected for small size (MacArthur, 1949) and heterozygotes were selected because they are smaller in size. Homozygotes are smaller still but natural selection prevented the fixation of the gene because homozygotes are sterile. Thus, under combined artificial and natural selection, heterozygotes were favoured. When the selection limit is reached under this situation, there is genetic variation due to the gene but no further response. If artificial selection is relaxed, the line responds to the natural selection. If selection is reversed, the artificial and natural selection both act in the same direction and there is an immediate and often rapid response. This may be regarded as an extreme form of asymmetrical response. It leads to the anomaly of a high heritability – about 50 per cent – estimated from the offspring – parent regression, but a realized heritability of zero in one direction and up to 100 per cent in the other direction. The anomaly is only apparent, however, because the estimation of the heritability and the prediction of the response are valid only if natural selection does not interfere with the appearance of the genotypes in their proper Hardy–Weinberg proportions.

The situation described under 5 above has often been found in selected *Drosophila* lines, but with natural selection acting through lethality rather than sterility. In the line selected for high bristle number illustrated in Fig. 12.5, a gene was present which was lethal in homozygotes and which increased bristle number by 22 in heterozygotes, which was 5.8 times the original phenotypic standard deviation. Lethals with large effects were found by Frankham, Jones, and Barker (1968) in 6 out of 9 lines selected for high bristle number. In cases like these it seems probable that the lethal gene does not have so large an effect on the character in the original population, and its effect in heterozygotes is enhanced during the selection, either by interaction with genes at other loci or by a cross-over separating a linked gene whose effect is in the opposite direction. A mechanism of this sort seems to be needed to account for the large increases of variance that are often found in selected lines, as shown for example in Fig. 12.4. It is not known if any of the lethal genes found in the *Drosophila* experiments were present in the original populations. If they were present they would have been at low frequencies because of the deleterious effect on fitness. The appearance of the gene in any particular selected line would then depend very much on the chances of sampling, or on its occurrence later by mutation. Genes of this sort can therefore be expected to cause large differences between replicate lines at the limits, such as those found by Jones, Frankham, and Barker (1968) which were mentioned earlier.

Relevance to practical breeding. It may be thought that experimental studies of long-continued selection are of little relevance to the practice of selection in animal and plant improvement, because the breeder is concerned only with the first five or ten generations. This, however, is not necessarily so. The breeds of animals and varieties of plants which he seeks to improve have already been under selection for more or less the same characters over a long time. They may therefore by now be approaching, if they are not already at, the selection limits. An understanding of the nature of selection limits is therefore very relevant to the exploration of methods of breaking through the limit and achieving further progress. See for example Roberts (1967a,b), Falconer (1971), and Eklund and Bradford (1977).

Note added December 1982. When this chapter was written, it was assumed that mutation was so slow a process that its effect on selection responses must be negligible. This assumption has subsequently been shown to be wrong, on the basis of both experiment and theory.

Lines of *Drosophila* selected over a long period for abdominal bristles were found to have acquired several genes that were not present in the base population, and must therefore have arisen during the course of the selection by some form of mutation (Frankham, 1980). These genes had large enough effects to be detected individually and they caused large differences between replicate lines.

An outline of the theory, worked out by Hill (1982a, b), is briefly as follows. The amount of new variation arising from mutation in each generation is indeed very small (see p. 313). Each new mutant has an initial frequency of $1/2N$ (N being the population size). At such a low gene frequency it contributes very little to the variance and, without selection, most new mutants are soon lost by random drift. Selection, however, can increase the frequency of favourable mutants so that they are not lost and are eventually fixed in the selected line. The mutations arising in the first generation of selection contribute only a very small amount to the response in that generation. But, as their frequencies are increased by continued selection, the variance they produce increases and they contribute more to the response. New mutations are introduced at every generation so that the response per generation attributable to mutation gradually builds up over time and after 20 or more generations reaches a constant value which is not negligible. The final rate of response is proportional to the effective population size, but larger populations take longer to reach the final rate. The response depends also on the mutants' effects on fitness and on the distribution of their effects on the character selected. With reasonable assumptions about these effects the conclusion is that mutation can appreciably increase the response to long-term selection and prolong it indefinitely so that no selection limit is reached. (For references see page 334.)

I3 SELECTION:
III. Information from relatives

In our consideration of selection we have up to now supposed that individuals are measured for the character to be selected and that the best are chosen to be parents in accordance with the individual phenotypic values. An individual's own phenotypic value, however, is not the only source of information about its breeding value; additional information is provided by the phenotypic values of relatives, particularly by those of full or half sibs. With some characters, indeed, the values of relatives provide the only available information. Milk-yield, to take an obvious example, cannot be measured in males, so the breeding value of a male can only be judged from the phenotypic values of its female relatives.

The use of information from relatives is of great importance in the application of selection to animal breeding, for two reasons. First, the characters to be selected are often ones of low heritability, and with these the mean value of a number of relatives often provides a more reliable guide to breeding value than the individual's own phenotypic value. And, second, when the outcome of selection is a matter of economic gain, even quite a small improvement of the response will repay the extra effort of applying the best technique. In this chapter we shall outline the principles underlying the use of information from relatives and the choice of the best method of selection.

If the family structure of the population is taken into account we can compute the mean phenotypic value of each family; this is known as the *family mean*. Suppose, then, that we have a population in which the individuals are grouped in families, which may be full or half sibs, and we have measurements of each individual and of the means of every family. How then is the additional information from the family means to be used? The problem may best be explained by reference to a specific example. Table 13.1 gives some hypothetical but realistic values of litter size in mice. There are 16 individuals whose phenotypic values are entered in the body of the table. The individuals are grouped in four full-sib families, A to D, with 4 individuals in each family. We have to choose the best 4 of these 16 individuals. Basing the choice on the individual phenotypic values we have no difficulty in choosing individuals A1, B1, and A2 with values, 13, 11, 10 respectively. But now there are two with values of 9, B2 in a good family and D1 in a bad family. Which do we choose? The decision rests on whether the differences between families are mainly genetic or mainly environmental. If they are genetic we choose B2, on the grounds that its

Table 13.1 Example of individual values and family means for selection, as explained in the text.

| | Family | | | |
Individual	A	B	C	D
1	13	11	7	9
2	10	9	7	5
3	8	6	6	3
4	5	6	4	3
Family mean	9	8	6	5
Overall mean			7	

better family mean indicates a better breeding value. If, on the other hand, the differences between families are mainly environmental we would choose D1, on the grounds that its low family mean indicates a poor environment and that it has performed well despite this disadvantage. The problem is not only in discriminating between individuals with the same phenotypic values, but is a matter of finding the right weight to be given to the family means. With the correct weighting we might be led to choose A3 with 8 in place of B2 with 9. Application of the principles to be developed shows that this would in fact be the best procedure if these values were litter sizes of mice (See Example 13.1).

To calculate the best weighting of the family means, only three things need be known: the kind of family (whether full or half-sibs), the number of individuals in the families (i.e. the family size), and the phenotypic correlation between members of the families with respect to the character. The information needed to solve what seems a complex problem is thus surprisingly simple; but the explanation of the underlying principles is not so simple. The explanation will be presented in two ways. First we shall extend the concept of heritability as a determinant of the response to selection. This introduces no new principles and leads fairly easily to a solution of the problem posed above; but it is not convenient for the solution of more complex problems found in practice. Then, under the heading of 'Index selection', a more general solution will be briefly explained. This allows information from different sorts of relatives to be combined, for example from parents as well as sibs. It also allows information of a different kind to be used as an aid to selection, in a way to be explained in Chapter 19.

Criteria for selection

The phenotypic value of an individual, P, measured as a deviation from the population mean, is the sum of two parts: the deviation of its family mean from the population mean, P_f, and the deviation of the individual from the family mean, P_w (the within-family deviation); so that

$$P = P_f + P_w \qquad \qquad \ldots [13.1]$$

The procedure of selection, then, varies according to the attention paid, or the weight given, to these two parts. There are three simple procedures that can be followed. First, we may select on the basis of individual values only, as assumed in the last two chapters, giving equal weight to the two components P_f and P_w. This is known as *individual selection*. Second, we may select on the basis of the family mean P_f only, giving zero weight to the within-family deviation P_w. This is known as *family selection*. Applied to Table 13.1, all four individuals in family A would be selected. Third, we may select on the basis of the within-family deviation P_w alone, giving zero weight to the family mean P_f. This is known as *within-family selection*. Applied to Table 13.1, the best individual in each of the four families would be selected.

Instead of one or other of these three simple procedures, we may take account of both components, P_f and P_w, but give them different weights chosen so as to make the best use of the two sources of information. This is known as selection by optimum combination or *combined selection* or, more generally, *index selection*. It represents the general solution for obtaining the maximum rate of response, and the other three simpler methods are special cases in which the weights given to the two sources of information are either 1 or 0. It is therefore in principle always the best method. The appropriate weighting of P_f and P_w will be explained later.

Simple methods
The salient features of the three simpler methods are as follows.

Individual selection. Individuals are selected solely in accordance with their own phenotypic values. This method is usually the simplest to operate and in many circumstances it yields the most rapid response. It should therefore be used unless there are good reasons for preferring another method. *Mass selection* is a term often used for individual selection, especially when the selected individuals are put together *en masse* for mating, as for example *Drosophila* in a bottle. The term 'individual selection' is used more specifically when the matings are controlled or recorded, as with mice or larger animals.

Family selection. Whole families are selected or rejected as units, according to the mean phenotypic value of the family. Individual values are thus not acted on except in so far as they determine the family mean. In other words, the within-family deviations are given zero weight. The families may be of full sibs or half sibs, families of more remote relationship being of little practical significance.

The chief circumstance under which family selection is to be preferred is when the character selected has a low heritability. The efficacy of family selection rests on the fact that the environmental deviations of the individuals tend to cancel each other out in the mean value of the family. So the phenotypic mean of the family comes close to being a measure of its genotypic mean, and the advantage gained is greater when environmental deviations constitute a large part of the phenotypic variance, or in other words when the heritability is low. On the other hand, environmental variation common to members of a family impairs the efficacy of family selection. If this component is large it will tend to swamp the genetic differences between families, and family selection will be correspondingly ineffective. Another important factor in the efficacy of family selection is the number of individuals in the families, or the family size. The larger the family, the closer is the correspondence between mean phenotypic value and mean genotypic value. So the conditions that favour family selection are low heritability, little variation due to common environment, and large families.

There are practical difficulties in the application of family selection, particularly in laboratory populations. They arise from the conflict between the intensity of selection and the avoidance of inbreeding. It is generally desirable to keep the rate of inbreeding as low as possible. If the minimum number of parents is fixed by considerations of inbreeding – say at ten pairs – then under family selection ten families must be selected, since each family represents only one pair of parents in the previous generation. And, if a reasonably high intensity of selection is to be achieved, the number of families bred and measured must be perhaps twice to four times this number. Family selection is thus costly of space, and if breeding space is limited the intensity of selection that can be achieved under family selection may be quite small. The two following methods are variants of family selection.

Sib selection. Some characters, as we have already noted, cannot be measured on the individuals that are to be used as parents, and selection can only be based on the values of relatives. This amounts to family selection but with the difference that now the selected individuals have not contributed to the estimate of their family mean.The difference affects the way in which the response is influenced by family size. Where the distinction is of consequence we shall use the term *sib selection* when the selected individuals are not measured, and family selection when they are measured and included in the family mean. When families are very large the two methods are equivalent, and the term family selection is then to be understood to cover both.

Progeny testing is a method of selection widely applied in animal breeding. We shall not discuss it in detail, except in so far as it can be treated as a form of family selection. The criterion of selection, as the name implies, is the mean value of an individual's progeny. At first sight this might seem to be the ideal method of selection and the easiest to evaluate because, as we saw in Chapter 7, the mean value of an individual's offspring comes as near as we can get to a direct measure of its breeding value, and is in fact the operational definition of breeding value. In practice, however, it suffers from the serious drawback of a much-lengthened generation interval, because the selection of the parents cannot be carried out until the offspring have been measured. The evaluation of selection by progeny testing is apt to be rather confusing because of the inevitable overlapping of generations, and because of a possible ambiguity about which generation is being selected, the parents or the progeny. The progeny, whose mean is used to judge the parents, are ready to be used as parents just when the parents have been tested and await selection. Thus both the selected parents and their progeny are used concurrently as parents. The difficulty of interpretation may be partially overcome by regarding progeny testing as a modified form of family selection. The progenies are families, usually of half sibs, and selection is made between them on the basis of the family means in the manner described above. The only difference is that the selected families are increased in size by allowing their parents to go on breeding. The additional, younger, members of the families do not contribute to the estimates of the family means and are therefore selected by sib selection. Increasing the size of the selected families by unmeasured individuals does not improve the accuracy of the selection, but it reduces the replacement rate and so increases the intensity of selection that can be achieved. This is the principal advantage of progeny testing, but it can only be realized in operations on a large scale, when the danger of inbreeding is not introduced by limitation of space.

Within-family selection. The criterion of selection is the deviation of each individual from the mean value of the family to which it belongs, those that exceed their family mean by the greatest amount being regarded as the most desirable. This is the reverse of family selection, the family means being given zero weight. The chief condition under which this method has an advantage over the others is a large component of environmental variance common to members of a family. Pre-weaning growth of pigs or mice might be cited as examples of such a character. A large part of the variation of individuals' weaning weights is attributable to the mother and is therefore common to members of a family. Selection within families would eliminate this large non-genetic component from the variation operated on by selection. An important practical advantage of selection within families, especially in laboratory experiments, is that it economizes breeding space, for the same reason that family selection is costly of space. If single-pair matings are to be made, then two members of every family must be selected in order to replace the parents. This means that every family contributes equally to the parents of the next generation, a system that we saw in Chapter 4 renders the effective population size twice the actual. Thus when selection within families is practised, the breeding space required to keep the rate of inbreeding below a certain value is only half as great as would be required under individual selection.

Prediction of response

To evaluate the relative merits of the different methods of selection we have to deduce the response expected from each. There is nothing to be added here about individual selection to what was said in Chapter 11. The expected response was given in equation [11.3] as $R = i\sigma_P h^2$, where i is the intensity of selection (i.e. the selection differential in standard deviations), σ_P is the standard deviation and h^2 the heritability of the phenotypic values of individuals. The response expected under family selection or within-family selection is arrived at in an analogous manner. Under family selection, the criterion of selection is the mean phenotypic value of the members of a family, so the expected response to family selection is

$$R_f = i\sigma_f h_f^2 \qquad \qquad \dots [13.2]$$

where i is the intensity of selection, σ_f is the observed standard deviation of family means, and h_f^2 is the heritability of family means. In the same way, the expected response to within-family selection is

$$R_w = i\sigma_w h_w^2 \qquad \qquad \dots [13.3]$$

where σ_w is the standard deviation, and h_w^2 the heritability of within-family deviations.

Heritability. The concept of heritability applied to family means or to within-family deviations introduces no new principle. It is simply the proportion of the phenotypic variance of these quantities that is made up of additive genetic variance. These heritabilities can be expressed in terms of the heritability of individual values (which we shall continue to refer to simply as the heritability, with symbol h^2), the phenotypic correlation between members of families, and the number of individuals in the families, all of which can be estimated by observation. To arrive at the appropriate expressions we have to consider again how the observational components of variance are made up of the causal components, as explained in Chapters 9 and 10 (see in particular Tables 9.5 and 10.4). First let us simplify matters by supposing that all families contain a large number of individuals, so that the means of all families are estimated without error. Consider first the phenotypic variance. The intraclass correlation t between members of families is the between-group component divided by the total variance: $t = \sigma_B^2/\sigma_T^2$. Therefore the between-group component can be expressed as $\sigma_B^2 = t\sigma_T^2$, and the within-group component as $\sigma_W^2 = (1-t)\sigma_T^2$. This expresses the partitioning of the phenotypic variance into its observational components. The total variance, written here as σ_T^2, is the phenotypic variance which we shall write as V_P in the context of causal components. The partitioning of the additive variance between and within families can be expressed in the same way, in terms of the correlation of breeding values, r. Thus the additive variance between families is rV_A and the additive variance within families is $(1-r)V_A$. The dual partitioning is summarized in Table 13.2.

This partitioning of both the additive and the phenotypic variance leads at once to the heritabilities of family means and of within-family deviations, since these heritabilities are simply the ratios of the additive variance to the phenotypic variance. Thus, when the families are large, the heritability of family means is rV_A/tV_P, or $(r/t)h^2$, since V_A/V_P is the heritability of individual values, h^2. The values of r for different relationships were given in Table 9.3; for full sibs it is $\frac{1}{2}$ and for half sibs it is $\frac{1}{4}$. In order to be able to discuss full-sib and half-sib families at the same time in what follows, we shall retain the symbol r in the formulae instead of inserting the appropriate values of $\frac{1}{2}$ or $\frac{1}{4}$.

The foregoing account of the heritabilities of family means and within-family deviations was simplified by the supposition of large families. The simplification is not justified in practice and we must now remove it by considering families of finite size. We shall, however, suppose that all families are of equal size. The number of individuals in a family has to be taken into consideration for the following reason. If selection is based on the family mean, or on the deviations from the family mean, then it is the observed mean that we are concerned with

Table 13.2 Partitioning of the variance between and within families of large size.

Observational component	Additive variance	Phenotypic variance
Between families, σ_B^2	$r\,V_A$	$t\,V_P$
Within families, σ_W^2	$(1-r)\,V_A$	$(1-t)\,V_P$

Table 13.3 Composition of observed variances with families of size n.

Observed variance	Observational components	Causal components	
		Additive	Phenotypic
Of family means, σ_f^2	$\sigma_B^2 + \dfrac{1}{n}\sigma_W^2$	$\dfrac{1 + (n-1)r}{n} V_A$	$\dfrac{1 + (n-1)t}{n} V_P$
Of within-family deviations, σ_w^2	$\sigma_W^2 - \dfrac{1}{n}\sigma_W^2$	$\dfrac{(n-1)(1-r)}{n} V_A$	$\dfrac{(n-1)(1-t)}{n} V_P$

and not the true mean. In other words we are not concerned with the observational components of variance which we have hitherto discussed, but with the variance of the observed means and of the observed within-family deviations. The observed means of groups are subject to sampling variance which comes from the within-group variance. If there are n individuals in a group then the sampling variance of the group-mean is $(1/n)\sigma_W^2$, where σ_W^2 is the component of variance within the group. Thus the variance of observed group-means is augmented by $(1/n)\sigma_W^2$, and the variance of observed deviations within groups is correspondingly diminished by the same amount. The observed variances, with family size n, are therefore made up of the observational components as shown in Table 13.3. The causal components entering into the observed variances can now be found by translating the observational components into causal components from Table 13.2. They are shown in the two right-hand columns of Table 13.3.

To find the heritabilities of family means and of within-family deviations, we have only to divide the additive component by the phenotypic component of the observed variances. Thus the heritability of family means is

$$h_f^2 = \frac{1 + (n-1)r}{1 + (n-1)t} h^2 \qquad \dots [13.4]$$

and the heritability of within-family deviations is

$$h_w^2 = \frac{1-r}{1-t} h^2 \qquad \dots [13.5]$$

Sib selection has to be distinguished from family selection, from which it differs in that the selected individuals are not measured. The appropriate heritability is best deduced by considering it as a regression in the manner of equation [10.2]. In this case it is the regression of the breeding values of the unmeasured individuals on the mean phenotypic value of their measured sibs. The covariance of these is simply the covariance of full or half sibs, i.e., rV_A, and it is not affected by the numbers of either the measured or the unmeasured individuals for reasons explained in Chapter 9. The regression is therefore rV_A/σ_P^2, where σ_P^2 is the observed variance of the family means of the measured individuals as given in Tabe 13.3. Substitution gives the heritability of family means appropriate to sib selection as

$$h_s^2 = \frac{nr}{1 + (n-1)t} h^2 \qquad \dots [13.6]$$

The heritabilities of the different methods of selection, whose derivations have now been explained, are listed in Table 13.4.

Expected responses. To deduce the expected responses is now a simple matter. Family selection will be taken for illustration. The expected response was given in equation [13.2] as $R_f = i\sigma_f h_f^2$. This expression, however, is not much use as it stands, because it does not readily allow a comparison to be made with the other methods. It will be most convenient to cast it into a form that facilitates comparison with individual selection. This can be done by substituting the expression for the heritability of family means, h_f^2, given in equation [13.4], and by putting the standard deviation of observed family means, σ_f, in terms of the standard deviation of individual phenotypic values, $\sigma_P(=\sqrt{V_P})$ from the right-hand column of Table 13.3. The expected response then becomes

$$R_f = i\sqrt{\left[\frac{1+(n-1)t}{n}\right]}\sigma_P\frac{1+(n-1)r}{1+(n-1)t}h^2$$

which reduces to

$$R_f = i\sigma_P h^2\left[\frac{1+(n-1)r}{\sqrt{[n\{1+(n-1)t\}]}}\right]$$

The term $i\sigma_P h^2$ is equivalent to the expected response under individual selection, so the expression within the square brackets is the factor that compares family selection with individual selection. The expression looks very complicated but it contains only three simple quantities: n, which is the family size; r, which is $\frac{1}{2}$ for full-sib and $\frac{1}{4}$ for half-sib families; and t, which is the phenotypic intraclass correlation.

The expected responses under the different methods of selection are listed in Table 13.4, all expressed in this manner which allows the comparisons to be made with individual selection.

Combined selection
Combined selection will be dealt with very briefly here because it will be more fully explained later. The appropriate weighting factors to be used in its application can be deduced as follows. We saw before that the phenotypic value of an individual is made up of two parts, the family mean and the within-family deviation, $P = P_f + P_w$, and that each part gives some information about the individual's breeding value. In Chapter 10 we saw that the heritability is equivalent to the regression of breeding value on phenotypic value (equation [10.2]), so that the best estimate of an individual's breeding value to be derived from its phenotypic value is h^2P (equation [10.4]). This idea can be applied separately to the two parts of the phenotypic value, since these are uncorrelated and supply independent information about the breeding value. Therefore, taking both parts of the phenotypic value into account, the best estimate of an individual's breeding value is given by the multiple regression equation

$$\text{expected breeding value} = h_f^2 P_f + h_w^2 P_w \qquad \ldots [13.7]$$

P_f being measured as a deviation from the population mean, and P_w as a

Table 13.4 Heritability and expected response under different methods of selection.

Method of selection	Heritability	Expected response
Individual	h^2	$R = i\sigma_p h^2$
Family	$h_f^2 = h^2 \dfrac{1 + (n-1)r}{1 + (n-1)t}$	$R_f = i\sigma_p h^2 \dfrac{1 + (n-1)r}{\sqrt{n\{1 + (n-1)t\}}}$
Sib	$h_s^2 = h^2 \dfrac{nr}{1 + (n-1)t}$	$R_s = i\sigma_p h^2 \dfrac{nr}{\sqrt{n\{1 + (n-1)t\}}}$
Within-family	$h_w^2 = h^2 \dfrac{(1-r)}{(1-t)}$	$R_w = i\sigma_p h^2 (1-r)\sqrt{\left[\dfrac{n-1}{n(1-t)}\right]}$
Combined	—	$R_c = i\sigma_p h^2 \sqrt{\left[1 + \dfrac{(r-t)^2}{(1-t)} \times \dfrac{(n-1)}{1 + (n-1)t}\right]}$

i = intensity of selection (selection differential in standard measure): assumed to be equal for
    all methods, but not necessarily so.
σ_p = standard deviation in phenotypic values of individuals.
h^2 = heritability of individual values.
r: with full-sib families, $r = \frac{1}{2}$
    with half-sib families, $r = \frac{1}{4}$
t = correlation of phenotypic values of members of the families.
n = number of individuals in the families.

deviation from the family mean. The weighting factors that make the most efficient use of the two sources of information are therefore the two heritabilities, which are the partial regression coefficients on family mean and within-family deviation respectively. If the values of the two heritabilities from Table 13.4 are inserted in equation [13.7], it will be seen that the term h^2 is common to both weighting factors, and this term may therefore be omitted without affecting the relative weighting. This gives the expected breeding value as

$$E(A) = \left[\frac{1-r}{1-t}\right]P_w + \left[\frac{1 + (n-1)r}{1 + (n-1)t}\right]P_f$$

In practice it is more convenient to work with the individual values in place of the within-family deviations, and to assign them a weight of 1. The family mean is thus used in the manner of a correction, supplementing the information provided by the individual itself. Rearrangement of the appropriate weighting factor for the family mean leads to an index of merit as follows (Lush, 1947):

$$I = P + \left[\frac{r-t}{1-r} \times \frac{n}{1 + (n-1)t}\right]P_f \qquad \ldots [13.8]$$

In this equation, P is the individual value, and P_f is the deviation of the family mean from the population mean, the individual itself being included in the family mean. Note that the weighting of P_f is negative if t is greater than r. This can only occur when there is a large environmental component in the correlation; the family mean is then an indicator of environment rather than of breeding value. The expected response to combined selection, cast in a form suitable for comparison with individual selection, is given in Table 13.4. For its derivation see Lush (1947).

Example 13.1 The operation of combined selection will be illustrated by application of equation [13.8] to the figures in Table 13.1, which are realistic values for litter sizes of mice. The phenotypic values are listed again here in table (i) with the family means as deviations from the overall mean. The full-sib correlation of litter size is about $t = 0.1$. Substituting this, with $r = 0.5$, $n = 4$, in equation [13.8] gives the index of merit as $I = P + 2.46P_f$. The index so calculated for each individual is given in table (ii). The only difference between the individuals selected by combined selection and those selected by individual selection is in the 4th-ranking individual, A3 being chosen instead of B2 or D1.

Table (i)

	A	B	C	D
1	13	11	7	9
2	10	9	7	5
3	8	6	6	3
4	5	6	4	3
P_f	+2	+1	−1	−2
Overall mean		7		

Table (ii)

	A	B	C	D
	17.9	13.5	4.5	4.1
	14.9	11.5	4.5	0.1
	12.9	8.5	3.5	−1.9
	9.9	8.5	1.5	−1.9
		7		

Relative merits of the methods

The merit of any one method of selection relative to any of the others can be worked out from the expected responses given in Table 13.4. The formulae expressing relative merits, however, are very cumbersome, depending in a complicated way on the phenotypic correlation t, the family size n, and whether the families are full or half sibs, i.e., r. The relative merits are summarized graphically in Fig. 13.1. This shows the expected responses of the three simple methods relative to combined selection, which must always be the best method. The relative responses are plotted against the phenotypic correlation t with the

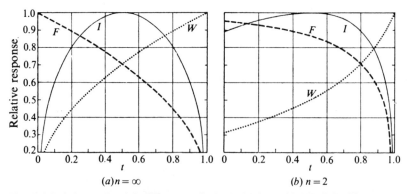

Fig. 13.1. Relative merits of the different methods of selection, with full-sib families. Responses relative to that for combined selection plotted against the phenotypic intraclass correlation, t. I = individual selection; F = family selection; W = within-family selection.

two extremes of family size: very large families in (a) and families of size 2 in (b). The intensity of selection i is assumed to be the same for all methods. The following conclusions can be drawn from the graphs though, for reasons to be explained, they are not immediately applicable to practical breeding operations. First, comparing the three simple methods with combined selection, one or other of the simple methods is seldom more than 10 per cent, and never more than 20 per cent below combined selection in the expected response. Second, comparing the three simple methods, individual selection is best over much of the range t. The reason for this is that individual selection operates on the whole of the additive genetic variance, whereas family selection operates only on the variance between family means, and within-family selection only on the variance within families. Family selection is better than individual selection when the phenotypic correlation t is low. Low sib-correlations imply a low heritability and little resemblance from common environment. These are the conditions that, in general, make family selection better than individual selection. Another factor to be considered is the selection differential, which is likely to be less in family than in individual selection, because the number of families selected cannot be reduced below a certain number for reasons of inbreeding, as was explained earlier. Within-family selection is better than individual selection when the sib-correlation is very high. Very high correlations can only arise from a very large common environment component of variance V_{Ec}; this is the condition that makes within-family selection better than individual selection, but it occurs only rarely. There is, however, another reason for selecting within families, namely the reduced rate of inbreeding which was mentioned earlier. In experimental selection this may be a more important consideration than the rate of response.

The reasons why the comparisons shown in Fig. 13.1 are not immediately relevant to practical breeding operations are the following. Combined selection requires both individual records of performance and pedigree records by which families can be identified. If both are available, the calculations needed to apply combined selection add little to the cost and are justified by quite a small gain in expected response. In animal breeding, individual records are easily obtainable but pedigree records are often costly to obtain. In these circumstances, therefore, the practical choice is between individual selection and combined selection, and the question is whether the gain from combined selection justifies the cost of pedigree records. In plant breeding, on the other hand, families may be easily identified but individual plant records may be difficult to obtain or may not be meaningful. In this case the choice is between family selection and combined selection. For details of family selection applied to plants, see England (1977).

> **Example 13.2** To illustrate how the different methods of selection are compared, their relative merits will be calculated for three characters in mice. The families are assumed to be single litters. Weaning weight has a very large common environment component giving a high full-sib correlation of 0.8. Six-week weight has a smaller, but still large, common environment component giving a correlation of 0.6 (see Example 10.5). Litter size, a character of the adult female, has a low heritability and a small amount of common environment, giving a low full-sib correlation of 0.1. The intensity of selection is assumed to be the same for all methods of selection, though this would not be true in practice. The family size is assumed to be $n = 4$. The families are full sibs,

		Weaning weight	6-week weight	Litter size
t		0.8	0.6	0.1
Weighting of P_f in eq. [13.8]		-0.71	-0.29	$+2.46$
Combined:	R_c/R	1.18	1.01	1.19
Family:	R_f/R	0.68	0.75	1.10
Within family:	R_w/R	0.97	0.68	0.46

so $r = 0.5$. The expected responses relative to individual selection are calculated from Table 13.4, by entering these values of n and r and the appropriate value of t. The relative responses are given in the table. Combined selection would be nearly 20 per cent better than individual selection for weaning weight and litter size, but would have virtually no advantage for 6-week weight. Family selection would be 10 per cent better than individual selection for litter size. (If males were to be selected they would have to be selected by sib selection whatever method was applied to the females). Within-family selection would not be better than individual selection for any of the characters, but it would be for weaning weight if the family size was more than 5.

Example 13.3 A comparison of sib selection with individual selection has been made experimentally with *Drosophila* (Clayton, Morris, and Robertson, 1957). Sib selection was made with both full-sib and half-sib families. The responses were compared with individual selection with intensity $i = 1.40$ as given in Example 11.2. The table gives the data needed to calculate the expected responses, relative to individual selection, from the formulae in Table 13.4. It will be seen that r for the half sibs was a little greater than 0.25. This was because the females mated to the same male were not entirely unrelated to each other. The relative responses expected and observed are given in the right-hand part of the table. The expectation is that sib selection should be less good than individual selection, and so it proved to be. There was, however, some discrepancy between the upward and downward responses, for which the reason is not known.

Data				Relative response, R_s/R		
	Full sibs	Half sibs			Full sibs	Half sibs
i	1.33	1.27	Exp.		0.832	0.614
n	12	20	Obs. up		0.618	0.527
r	0.50	0.275	Obs. down		0.919	0.635
t	0.265	0.121				

Index selection

The optimal procedure for selection uses all the information available about each individual's breeding value, combined into an index of merit. The solution given above for combining the family mean and within-family deviation is not readily applicable to more complex situations when there may be more than two sources of information. There may, for example, be information from the individual, its parents, full sibs, half sibs, and other relatives. Or, if the character is limited to one sex, the information about individuals that cannot themselves be measured will come only from relatives of different sorts. The aim therefore is to combine all the information into an *index* on the basis of which the individuals will be selected. The construction of an index is not easy without the use of matrix methods, particularly if there are more than two sources of information. The technical details are beyond the scope of this book and only a brief account of the principles involved can be given. For more detailed accounts see Henderson (1963), Nordskog (1978), and the review of Lin (1978).

Construction of an index

The index is the best linear prediction of an individual's breeding value and it takes the form of a multiple regression of breeding value on all the sources of information. Consider first the simplest situation where the only information we have is the individual's own phenotypic value P as a deviation from the population mean. Then the predicted, or expected, breeding value is $E(A) = b_{AP}P$, where b_{AP}, is the regression of breeding value on phenotypic value. In this case $b_{AP} = h^2$ (equation [10.2]). Now suppose that there are several pieces of information, P_1, P_2, P_3, etc., where each P is the phenotypic value of an individual or a group of relatives. These pieces of information in the form of phenotypic values will be referred to as 'measurements'. The index of an individual is then

$$I = b_1P_1 + b_2P_2 + b_3P_3 + \ldots \qquad \ldots [13.9]$$

in which the b's are the factors by which each measurement is to be weighted. The problem is to find the best value for each weighting factor. This is done by finding what values will give the maximum correlation r_{IA} between the index and the breeding value. Maximizing r_{IA} is equivalent to minimizing the sum of squared deviations of index values from the linear regression of I on A, i.e., $\Sigma(I - A)^2$. The resulting values of the b's are then the partial regression coefficients of the individual's breeding value on each measurement. The maximizing of r_{IA} is a standard procedure for calculating partial regressions and it need not be explained here. The maximization leads to a set of simultaneous equations, with as many equations as there are measurements, and the solution of these equations gives the values of the b's to be used in equation [13.9]. These index equations for solution are given below (equations [13.10]). The equations for three measurements are given; it will easily be seen how they would be extended or reduced for more or fewer measurements. Each equation relates the phenotypic variances and covariances of the measurements, on the left, to the additive genetic variances and covariances of the individuals measured, on the right. The notation is condensed as follows. P means the phenotypic variance or covariance of the measurements denoted by subscript numbers. For example, P_{11} is the phenotypic variance of measurement 1, and P_{12} is the phenotypic covariance of measurements 1 and 2. The variances and covariances of breeding values are similarly written as A. The equations are

$$\left.\begin{array}{l} b_1P_{11} + b_2P_{12} + b_3P_{13} = A_{11} \\ b_1P_{21} + b_2P_{22} + b_3P_{23} = A_{21} \\ b_1P_{31} + b_2P_{32} + b_3P_{33} = A_{31} \end{array}\right\} \qquad \ldots [13.10]$$

To solve these equations the numerical values of the P's and A's must of course be inserted. The P's and A's can all be expressed in terms of the following parameters: the phenotypic variance V_P, which will be denoted here by σ^2; the heritability of individual values, h^2; the phenotypic correlations between individuals, t; and the coefficients of relationship, r, as in Table 9.3. When 'measurements' are the means of groups of individuals, the number n in the group is also needed. There are standard computer programs for solving the equations. The indices obtained will be illustrated by reference to specific examples, simplified by considering only two measurements.

Individual and one relative. Let measurement 1 be of the individual whose index is to be calculated, and measurement 2 be that of one relative. Then the phenotypic variances of the measurements are the same and $P_{11} = P_{22} = \sigma^2$. The phenotypic covariance is $P_{12} = P_{21} = t\sigma^2$. The additive variance is $A_{11} = h^2\sigma^2$ and the additive covariance is $A_{21} = rh^2\sigma^2$, where r is the coefficient of relationship between the relative and the individual. After dividing all through by σ^2, the equations for solution become

$$b_1 + tb_2 = h^2$$
$$tb_1 + b_2 = rh^2$$

and the solution, after some simplification, is

$$b_1 = \frac{h^2(1 - rt)}{1 - t^2}; \quad b_2 = \frac{h^2(r - t)}{1 - t^2}$$

When individuals are to be selected on the basis of their indices, the absolute values of the b's do not matter, only their relative magnitude. The factor $h^2/(1 - t^2)$, which is common to both b's, can therefore be omitted. This gives a rescaled index I', as follows:

$$I' = I\left(\frac{1 - t^2}{h^2}\right) = (1 - rt)P_1 + (r - t)P_2$$

Further rescaling to give a weight of 1 to the individual's own value would yield

$$I'' = P_1 + \left(\frac{r - t}{1 - rt}\right)P_2$$

It must be noted, however, that if the response to index selection is to be predicted, allowance must be made for any rescaling of the index. The prediction of the response will be explained later.

Mother and one paternal half sister. This exemplifies the selection of males for a female character, such as milk-yield or egg production. The individual whose index is to be calculated is not measured. To make the index equations comparable with those of equations [13.10] we shall regard measurement 1 as being absent, measurement 2 as that of the mother, and measurement 3 as the half sister. The relevant parts of equations [13.10] are therefore

$$b_2P_{22} + b_3P_{23} = A_{21}$$
$$b_2P_{32} + b_3P_{33} = A_{31}$$

The phenotypic variances are again equal, and $P_{22} = P_{33} = \sigma^2$. The mother and half sister will be assumed to be unrelated and uncorrelated environmentally, so $P_{23} = P_{32} = 0$. A_{21} is the additive covariance of the individual with his mother, which is $\frac{1}{2}h^2\sigma^2$. A_{31} is the additive covariance of the individual with his half sib, which is $\frac{1}{4}h^2\sigma^2$. The index equations then reduce directly to the solutions

$$b_2 = \tfrac{1}{2}h^2; b_3 = \tfrac{1}{4}h^2$$

giving a rescaled index of

$$I' = I/h^2 = \tfrac{1}{2}P_2 + \tfrac{1}{4}P_3.$$

Individual and mean of sibs. This is the situation with which the problem was introduced at the beginning of this chapter, and described earlier as 'combined selection'. The individual is again measurement 1, so $P_{11} = \sigma^2$ and $A_{11} = h^2\sigma^2$ as before. The variance of a family mean was given in Table 13.3, from which $P_{22} = [1 + (n - 1)t]\sigma^2/n$. It is usually convenient to include the individual in the family mean. The covariance of the individual with the family mean is then equal to the variance of family means. Thus $P_{12} = P_{21} = P_{22}$. In the same way the additive covariance is equal to the additive variance of family means. Taking the additive variance from Table 13.3 gives $A_{21} = [1 + (n - 1)r]h^2\sigma^2/n$. The index equations are thus

$$b_1 + Kb_2 = h^2$$
$$Kb_1 + Kb_2 = kh^2$$

where $K = \dfrac{1 + (n - 1)t}{n}$ and $k = \dfrac{1 + (n - 1)r}{n}$

Solving for b_1 and b_2 gives the index

$$I = \frac{h^2(1 - k)}{(1 - K)}P_1 + \frac{h^2(k - K)}{K(1 - K)}P_2$$

If different individuals have different numbers of sibs, k and K must be evaluated separately for each individual. If the number of sibs is constant, the index can be rescaled as

$$I' = P_1 + \frac{(k - K)}{(1 - k)K}P_2$$

If k and K are rewritten in terms of r, t, and n, this index becomes the same as equation [13.8].

The usefulness of the information obtainable from sibs depends on the family size n. Because there are usually many more half sibs than full sibs, half-sib families may give more information than full sibs, despite the less close genetic relationship. The relative merits of indices for application to poultry are examined by Osborne (1957).

Other 'measurements' involving means. In applying the quantities denoted by K and k above in other situations, two points need to be noted. First, a group of relatives may be related to each other differently from the way that they are to the individual. Care must then be taken to use the appropriate values of t and r. For example, P_2 might be the mean of a group of full sibs that are half sibs to the individual. Then the t appropriate to P_{22} is the correlation of full sibs, but the t

appropriate to P_{12} and P_{21}, and the r appropriate to A_{21}, are the correlations of half sibs. Second, the 'measurement' of an individual may be the mean of several repeated records. Then the repeatability must be used in place of t.

Efficiency of an index

The correlation r_{IA} between index values and breeding values provides a measure of the efficiency of an index, because, clearly, the higher the correlation the better is the index as a predictor of breeding values. The construction of the index so as to maximize r_{IA} ensures that the index makes the best use of the information, i.e., the 'measurements', in it. Different indices based on different measurements can be compared by means of their correlations r_{IA} to find out, for example, whether a particular measurement is worth including in the index or not. The efficiency r_{IA} is calculated as follows.

First, we need to know the variance of index values. From equation [13.9] this can be seen to be

$$\sigma_I^2 = b_1^2 P_{11} + b_2^2 P_{22} + \ldots + 2b_1 b_2 P_{12} + \ldots \qquad \ldots [13.11]$$

where variances and covariances are written in the notation of equations [13.10]. This expression can be put into a form that is easier to calculate. Rearranging the terms gives

$$\sigma_I^2 = b_1(b_1 P_{11} + b_2 P_{12} + \ldots) + b_2(b_1 P_{21} + b_2 P_{22} + \ldots) + \ldots$$

Substituting for the terms in brackets from equations [13.10] leads to

$$\sigma_I^2 = b_1 A_{11} + b_2 A_{21} + b_3 A_{31} + \ldots \qquad \ldots [13.12]$$

Next we need to know the covariance of index values with breeding values. This is $\text{cov}_{IA} = \sigma_I^2$ for the following reason. The construction of the index so that the weighting factors, the b's, are partial regression coefficients results in the regression of breeding values on index values being unity; i.e., $b_{AI} = 1$. Another way of saying this is that, in its construction, the index is scaled so that 1 unit of the index is equivalent to 1 unit of predicted breeding value. Now, $\text{cov}_{AI}/\sigma_I^2 = b_{AI} = 1$, from which it follows that $\text{cov}_{AI} = \sigma_I^2$. Thus, provided the index has not been rescaled, the correlation is given by

$$r_{IA} = \frac{\text{cov}_{IA}}{\sigma_I \sigma_A} = \frac{\sigma_I}{\sigma_A} \qquad \ldots [13.13]$$

Here σ_I is obtained from equation [13.12] and σ_A is the square-root of the additive genetic variance in the population.

The correlation r_{IA} is a multiple correlation, and its square expresses the fraction of the additive variance that is accounted for by the measurements combined in the index, i.e., $r_{IA}^2 = \sigma_I^2/\sigma_A^2$. Conversely, $(1 - r_{IA}^2)$ is the fraction of the additive variance that is not taken account of in the index. This shows how much room there is for improvement of the index by inclusion of additional measurements.

Response to selection

The response to selection is the mean breeding value of the selected parents, which is predicted from the regression of breeding values on index values as

$R = b_{AI}S$, where S is the selection differential of index values. Putting $b_{AI} = r_{IA}\sigma_A/\sigma_I$ and $S = i\sigma_I$ gives the predicted response as

$$R = i r_{IA}\sigma_A \qquad \qquad \ldots [13.14]$$

This prediction applies to any index, whether rescaled or not, provided r_{IA} refers to the index actually used. If the index has not been rescaled, substitution for r_{IA} can be made from equation [13.13] to give

$$R = i\sigma_I \qquad \qquad \ldots [13.15]$$

The prediction in this form can be arrived at more directly. The unscaled index is a prediction of breeding value, so the predicted breeding value of the selected individuals is their mean index value, which is the selection differential.

Comparison with individual selection. In a practical breeding operation the first question about index selection is whether it is worth the cost of recording pedigrees, without which no information about families is available. The alternative is individual selection and we therefore need to know the efficiency of individual selection for comparison. The efficiency of individual selection is h, the square-root of the heritability, because h is the correlation between breeding values and phenotypic values (see equation [10.3]). Thus the efficiency of index selection relative to individual selection is r_{IA}/h. This can be seen also by comparison of the expected responses in equations [13.14] and [11.4] ($R = ih\sigma_A$).

(Chapter 19 deals with index selection in a different context, and it could be read now by those whose main interest is in selection.)

14 INBREEDING AND CROSSBREEDING: I. Changes of mean value

We turn our attention now to inbreeding, the second of the two ways open to the breeder for changing the genetic constitution of a population. The harmful effects of inbreeding on reproductive rate and general vigour are well known to breeders and biologists, and were mentioned in Chapter 6 as one of the two basic genetic phenomena displayed by metric characters. The opposite, or complementary, phenomenon of hybrid vigour resulting from crosses between inbred lines or between different races or varieties is equally well known, and forms an important means of animal and plant improvement. The production of lines for subsequent crossing in the utilization of hybrid vigour is one of two main purposes for which inbreeding may be carried out. The other is the production of genetically uniform strains, particularly of laboratory animals, for use in bioassay and in research in a variety of fields. Inbreeding in itself, however, is almost universally harmful and the breeder or experimenter normally seeks to avoid it as far as possible, unless for some specific purpose.

In the treatment of inbreeding given in Chapter 3, the consequences were described in terms of the changes of gene frequencies and of genotype frequencies. Here we have to show how these changes of gene and genotype frequencies affect metric characters, and how they can account for the observed effects of inbreeding and crossing. The effects on the mean value will be explained in this chapter and the effects on the variance in the next. In Chapter 16 we shall consider the use of inbreeding and crossing as a means of plant and animal improvement.

The effects of inbreeding to be described do not apply to naturally self-fertilizing plants. Since inbreeding is their normal mating system they cannot be further inbred. They can, however, be crossed and they do then often show hybrid vigour, though less than when inbred lines of outbreeding species are crossed. The improvement of self-fertilizing plants, for which crossing is the first step, will be described briefly in Chapter 16.

Inbreeding depression

The most striking observed consequence of inbreeding is the reduction of the mean phenotypic value shown by characters connected with reproductive capacity or physiological efficiency, the phenomenon known as *inbreeding depression*. Some examples of inbreeding depression are given in Table 14.1, from

Table 14.1 Some examples of inbreeding depression. Approximate *decrease* of mean per 10 per cent increase of inbreeding coefficient: (1) in absolute units; (2) as percentage of non-inbred mean; and (3) as percentage of the original phenotypic standard deviation. The depression given is due only to inbreeding in the individuals on which the characters are measured, except where noted below.

		(1) Units	(2) % of M	(3) % of σ_P
Man				
Height (cm) at age 10; [Schull, 1962]		2.0	1.6	37
IQ score (percentage points); [Morton, 1978]		4.4	4.4	29
Cattle				
Milk-yield (kg): [Robertson, 1954]		135	3.2	17
Sheep [Morley, 1954]				
Fleece weight (kg)		0.29	5.5	51
Body weight at 1 yr (kg)		1.32	3.7	36
Pigs [Bereskin *et al.*, 1968]				
Litter size (no. born alive)	(a)	0.24	3.1	9
Body weight at 154 days (kg)		2.6	4.3	15
Mice				
Litter size: [Bowman and Falconer, 1960]	(b)	0.56	7.2	23
Body weight at 6 wks (g): [White, 1972]		0.19	0.6	7
Maize [Cornelius and Dudley, 1974]	(c)			
Plant height (cm)	(FS)	5.20	2.1	4
	(S)	5.65	2.3	5
Yield of seed (g/plant)	(FS)	7.92	5.6	25
	(S)	9.65	6.8	30

(a) Inbreeding in the mothers; litters non-inbred.

(b) Depression related to inbreeding in the mothers under consecutive full-sib mating; litters one generation more inbred than mothers.

(c) Depression related to inbreeding in the plants measured; inbreeding by consecutive full-sib mating (FS) or selfing (S). *Dr. J. W. Dudley kindly provided the values of σ_P for maize.*

which one can see what sort of characters are subject to inbreeding depression, and – very roughly – the magnitude of the effect. From the results of these and many other studies we can make the generalization that inbreeding tends to reduce fitness. Thus, characters that form an important component of fitness, such as litter size or lactation in mammals, show a reduction on inbreeding; whereas characters that are not closely connected with fitness show little or no change. In *Drosophila*, for example, bristle number and body weight do not change (Kidwell and Kidwell, 1966) but fertility and viability do (Tantawy and Reeve, 1956).

In saying that a certain character shows inbreeding depression, we refer to the average change of mean value in a number of lines. The separate lines are commonly found to differ to a greater or lesser extent in the change they show, as, indeed, we should expect in consequence of random drift of gene frequencies. This matter of differentiation of lines will be discussed later when we deal with changes of variance. It is mentioned here only to emphasize the fact that the changes of mean value now to be discussed refer to changes of the mean value of a number of lines derived from one base population. As in our earlier account of inbreeding we have to picture the 'whole population' consisting of many lines. The population mean then refers to the whole population, and inbreeding depression refers to a reduction of this population mean. Let us now consider the theoretical basis of the change of population mean on inbreeding.

First, we may recall and extend some of the conclusions from Chapter 3, supposing at first that selection does not in any way interfere with the dispersion of gene frequencies. Since the gene frequencies in the population as a whole do not change on inbreeding, any change of the population mean must be attributed to the changes of genotype frequencies. Inbreeding causes an increase in the frequencies of homozygous genotypes and a decrease of heterozygous genotypes. Therefore a change of population mean on inbreeding must be connected with a difference of genotypic value between homozygotes and heterozygotes. Let us now see more precisely how the population mean depends on the degree of inbreeding.

Consider a population, subdivided into a number of lines, with a coefficient of inbreeding F. The expression for the population mean is derived by putting together the reasoning set out in Tables 3.1 and 7.1, in the following way. Table 14.2 shows the three genotypes of a two-allele locus with their genotypic frequencies in the whole population. These frequencies come from Table 3.1, $\bar{p}$ and $\bar{q}$ being the gene frequencies in the whole population. Then the third column gives the genotypic values assigned as in Fig. 7.1. The value and frequency of each genotype are multiplied together in the right-hand column, the summation of which gives the contribution of this locus to the population mean. Thus, referring still to the effects of a single locus, we find that a population with inbreeding coefficient F has a mean genotypic value:

$$M_F = a(\bar{p} - \bar{q}) + 2d\bar{p}\bar{q}(1 - F) \qquad \ldots [14.1]$$
$$= M_0 - 2d\bar{p}\bar{q}F \qquad \ldots [14.2]$$

where M_0 is the population mean before inbreeding, from equation [7.2]. The change of mean resulting from inbreeding is therefore $- 2d\bar{p}\bar{q}F$. This shows that a locus will contribute to a change of mean value on inbreeding only if d is not zero; in other words if the value of the heterozygote differs from the average value of the homozygotes. This conclusion, though demonstrated in detail only for two alleles at a locus, is equally valid for loci with more than two alleles. The following general conclusions can therefore be drawn: that a change of mean value on inbreeding is a consequence of dominance at the loci concerned with the

Table 14.2

Genotype	Frequency	Value	Frequency × Value
A_1A_1	$\bar{p}^2 + \bar{p}\bar{q}F$	$+a$	$\bar{p}^2a + \bar{p}\bar{q}aF$
A_1A_2	$2\bar{p}\bar{q} - 2\bar{p}\bar{q}F$	d	$2\bar{p}\bar{q}d - 2\bar{p}\bar{q}dF$
A_2A_2	$\bar{q}^2 + \bar{p}\bar{q}F$	$-a$	$-\bar{q}^2a - \bar{p}\bar{q}aF$
			Sum $= a(\bar{p} - \bar{q}) + 2d\bar{p}\bar{q} - 2d\bar{p}\bar{q}F$
			$= a(\bar{p} - \bar{q}) + 2d\bar{p}\bar{q}(1 - F)$

character, and that the direction of the change is toward the value of the more recessive alleles. The dominance may be partial or complete, or it may be overdominance; all that is necessary for a locus to contribute to a change of mean is that the heterozygote should not be exactly intermediate between the two homozygotes. Equation [14.2] shows also that the magnitude of the change of mean depends on the gene frequencies. It is greatest when $\bar{p}\bar{q}$ is maximal: that is, when $\bar{p} = \bar{q} = \frac{1}{2}$. Genes at intermediate frequencies therefore contribute more to a change of mean than genes at high or low frequencies, other things being equal.

Now consider the combined effect of all the loci that affect the character. In so far as the genotypic values of the loci combine additively, the population mean is given by summation of the contributions of the separate loci, thus:

$$M_F = \sum a(\bar{p} - \bar{q}) + 2(\sum d\bar{p}\bar{q})(1 - F) \qquad \ldots [14.3]$$
$$= M_0 - 2F\sum d\bar{p}\bar{q} \qquad \ldots [14.4]$$

and the change of mean on inbreeding is $-2F\sum d\bar{p}\bar{q}$.

These expressions show what are the circumstances under which a metric character will show a change of mean value on inbreeding. The chief one is *directional dominance*, which means the dominance of the genes concerned being preponderantly in one direction. If the genes that increase the value of the character are dominant over their alleles that reduce the value, then inbreeding will result in a reduction of the population mean, i.e., a change in the direction of the more recessive alleles. The contribution of each locus, however, depends also on its gene frequencies, those with intermediate frequencies having the greatest effect on the change of mean value.

Another conclusion that can be drawn from equation [14.4] is that when loci combine additively the change of mean on inbreeding should be directly proportional to the coefficient of inbreeding. In other words the change of mean should be a straight line when plotted against F. If there is epistatic interaction between loci, the relation between the mean and the inbreeding coefficient is not linear. The non-linearity is due to the interaction deviation of double, or multiple, heterozygotes. The frequency of double heterozygotes declines in proportion to F^2. Therefore as F increases, the rate of depression of the mean increases if the interaction deviations are on average positive, i.e., favourable, and the rate decreases if they are negative. No other form of interaction affects the linearity, and epistasis without dominance cannot itself cause any inbreeding depression. For the details of how epistasis affects inbreeding depression, see Crow and Kimura (1970, p. 79).

Examples of experimentally observed inbreeding depression are illustrated in Figs 14.1 and 14.2. On the whole, the observed inbreeding depression, as in these examples, does tend to be linear with respect to F, and this might be taken as evidence that epistatic interaction between loci is not of great importance. There are, however, several practical difficulties that stand in the way of drawing firm conclusions from observations of the rate of inbreeding depression. One is that, as inbreeding proceeds and reproductive capacity deteriorates, it soon becomes impossible to avoid the loss of some individuals and of some entire lines. The survivors are then a selected group to which the theoretical expectations no longer apply. Thus precise measurement of the rate of inbreeding depression can

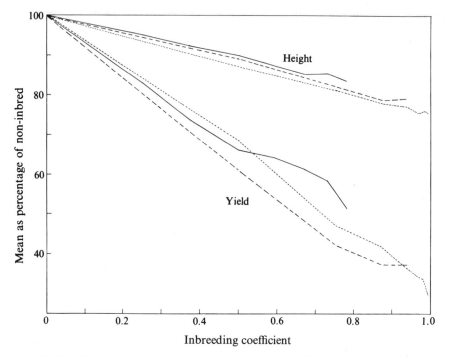

Fig. 14.1. Effects of inbreeding on plant-height and yield of seed in maize (*Zea mays*). The dotted and dashed lines refer to consecutive selfing; the continuous lines refer to consecutive full-sib mating. No selection was practised. *Data from Hallauer and Sears (1973) (dotted lines), and Cornelius and Dudley (1974) (continuous and dashed lines).*

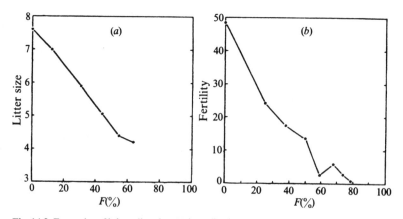

Fig. 14.2. Examples of inbreeding depression affecting fertility. (*a*) Litter-size in mice (*Data from Bowman and Falconer, 1960*). Mean number born alive in 1st litters, plotted against the coefficient of inbreeding of the litters. The first generation was by double-first-cousin mating; thereafter by full-sib mating. No selection was practised. (*b*) Fertility in *Drosophila subobscura*. Mean number of adult progeny per pair per day, plotted against the inbreeding coefficient of the parents. Consecutive full-sib matings. (*Based on Hollingsworth and Maynard Smith, 1955.*)

generally be made only over the early stages, before the inbreeding coefficient reaches high levels. Another difficulty, met with particularly in the study of mammals, arises from maternal effects. Maternal qualities are among the most sensitive characters to inbreeding depression. The effect of inbreeding on another character that is influenced by maternal effects is therefore twofold: part being attributable to the inbreeding of the individuals measured and part to the inbreeding in the mothers. When continuous inbreeding is practised, the mothers and the offspring have different coefficients of inbreeding in the early generations, so the relationship between the character measured and the coefficient of inbreeding cannot be expressed in any simple way. In consequence of these difficulties, reliable conclusions cannot easily be drawn from the exact form of the depression observed under continuous inbreeding.

The effect of selection

The neglect of selection during inbreeding is an unrealistic omission because natural selection cannot be wholly avoided even in laboratory experiments, and because deliberate inbreeding is usually accompanied by some artificial selection for characters subject to inbreeding depression. There are two effects of selection which must be distinguished: delay in the approach to fixation, and reduction in the inbreeding depression. Delay in the approach to fixation was discussed at the end of Chapter 5. It results in the actual proportion of homozygotes being less than would be predicted by the inbreeding coefficient calculated from the pedigree or the population size. Reduction of the inbreeding depression can occur without a delay in the approach to homozygosis if selection leads to the better allele being fixed at more of the loci than would occur by chance. To make effective use of selection in this way it is essential for a large number of lines to be inbred in parallel. Selection is then applied between the lines, so that the worst lines are eliminated and the best retained. The reason for this, which is explained in the next chapter, is that the genetic variation is progressively reduced within lines and increased between lines.

How effective can selection be in counteracting inbreeding depression? This is an important practical question because, if highly inbred lines are to be used for any practical purpose, the individuals must be reasonably viable and fertile. Selection has undoubtedly been successful in reducing, or even completely overcoming, the inbreeding depression of litter size in mice. If the graph in Fig. 14.2(a) is extrapolated to complete inbreeding at $F = 100$ per cent, the predicted litter size of highly inbred mice would be about 2. This is what would be expected of inbreeding without selection. In fact, most of the existing inbred strains have mean litter sizes well above 2, showing that selection in their development was successful in reducing the inbreeding depression. (Twenty strains listed by Green (1968) have a mean litter size of 5.7 with a range from 4.1 to 7.2.) Details of the making of one highly inbred strain with a mean better than the original population are given in the following example.

Example 14.1 The experiment on litter size illustrated in Fig. 14.2(a) was started with 20 lines. All females born in first litters were subsequently mated to provide the mean litter size. Only first litters were reared, so lines became eliminated when they failed to produce one offspring of each sex in their first litters. No other selection between lines was applied. Selection within lines was applied in 10 of the lines (these are not included in Fig. 14.2). This selection had virtually no effect on the inbreeding depression. At generation 3 ($F = 55\%$), 1 of the 20 lines was lost, and at generation 6 ($F = 76\%$), only 3 lines remained. Over the next 5 generations, these 3 had a mean litter size of 6.6, an improvement of over 2 mice per litter attributable to selecting the best 3 out of 20 lines. One of the lines survived indefinitely and became a 'standard' inbred strain known as JU. Over the three years after it had had more than 20 generations of inbreeding, its mean litter size was 9.0 in first litters (McCarthy, 1965), which was better than the original population before inbreeding. The experiment is described by Bowman and Falconer (1960) and Falconer (1960a). Similar results have been obtained in two other experiments with mice (Falconer, 1971; Eklund and Bradford, 1977).

The three experiments cited in Example 14.1 prove that by selection it is possible to get highly inbred lines of mice that are at least as good as the original population in respect of the character selected. This provides strong, though not conclusive, evidence that overdominant loci are not an important cause of inbreeding depression of litter size in mice. If an overdominant locus is at its equilibrium gene frequency in the population before inbreeding, it is not possible to have a homozygous line that is as good as the non-inbred population. But if the frequency of the better homozygote is below its equilibrium value in the non-inbred population, fixing the better homozygote in an inbred line may increase the mean (Minvielle, 1979). The ways in which selection favouring heterozygotes affects the inbreeding depression are not straightforward. They depend on whether the gene frequencies start at their equilibrium values or not and, if they do, on whether the equilibrium frequency is intermediate or extreme, i.e., on whether the two homozygotes are nearly equal or very different in fitness (Hill and Robertson, 1968). When the initial frequency is the equilibrium value, selection reduces the inbreeding depression by delaying the approach to homozygosis if the two homozygotes are nearly equal in fitness; but it reduces the rate and the total amount of inbreeding depression if the two homozygotes are very different in fitness, because it then causes the better homozygote to be fixed preferentially.

Heterosis

Complementary to the phenomenon of inbreeding depression is its opposite, 'hybrid vigour' or *heterosis*. When inbred lines are crossed, the progeny show an increase of those characters that previously suffered a reduction from inbreeding. Or, in general terms, the fitness lost on inbreeding tends to be restored on crossing. The amount of heterosis is the difference between the crossbred and inbred means. That the phenomenon of heterosis is simply inbreeding depression in reverse can be seen by consideration of how the population mean depends on the coefficient of inbreeding, as shown in equation [14.4]. Consider, as before, a population subdivided into a number of lines, inbred without selection so that the mean gene frequencies are not changed. If the lines are crossed at random, the average inbreeding coefficient in the crossbred progeny reverts to that of the base population and, if the gene frequencies have not changed, the frequencies of the genotypes are the same as in the base population. Thus if a number of crosses are made at random between the lines, the mean value of any character in the crossbred progeny is expected to be the same as the population mean of the base population. In other words, the heterosis on crossing is expected to be equal to the depression on inbreeding. Furthermore, if the population is continued after the crossing by random mating among the crossbred and subsequent generations, the coefficient of inbreeding will remain unchanged, and the population mean is consequently expected to remain at the level of the base population. We may thus make the following generalization on theoretical grounds: that, in the absence of selection, inbreeding followed by crossing of the lines in a large population is not expected to make any permanent change in the population mean.

Example 14.2 An experiment with mice (Roberts, 1960) was designed to test the theoretical expectation that, in the absence of selection, the heterosis on crossing should be equal to the depression on inbreeding. The character studied was litter size. Thirty lines taken from a random-bred population were inbred by 3 consecutive generations of full-sib mating, bringing the coefficient of inbreeding up to 50 per cent in the litters and 37.5 per cent in the mothers. No selection was practised during the inbreeding, and only 2 of the 30 lines were lost as a consequence of their inbreeding depression.

	F in		Mean
	litters	mothers	litter size
Before inbreeding	0	0	8.1
Inbred	0.50	0.375	5.7
Crossbred litters	0	0.50	6.2
Crossbred litters and mothers	0	0	8.5

After the third generation of inbreeding, crosses were made at random between the lines, and in the next generation crosses between the F_1's were made so as to give crossbred mothers with non-inbred young. The mean litter sizes observed at the different stages are given in the table. The inbreeding depression was 2.4 and the heterosis 2.8; the two are equal within the limits of experimental error.

Single crosses

The foregoing theoretical conclusions refer to the average of a large number of crosses between lines derived from a single base population. In practice, however, one is often interested in a somewhat different problem, namely the heterosis shown by a particular cross between two lines, or between two populations which may have no known common origin. To refer the changes of mean value to changes of inbreeding coefficient would be inappropriate under these circumstances, and the theoretical basis of the heterosis is better expressed in terms of the gene frequencies in the two lines. We may recall from Chapter 3 that inbreeding leads to a dispersion of gene frequencies among the lines, the lines becoming differentiated in gene frequency as inbreeding proceeds; and the coefficient of inbreeding is a means of expressing the degree of differentiation (equation [3.14]). In turning from the inbreeding coefficient to the gene frequencies as a basis for discussion we are therefore turning from the general, or average, consequence of crossing, to the particular circumstances in two lines.

Let us, then, consider two populations, referred to as the 'parent popu-lations', both random-bred though not necessarily large. The parent populations are crossed to produce an F_1 or 'first crossbred generation', and the F_1 individuals are mated together at random to produce an F_2 or 'second crossbred generation'. The amount of heterosis shown by the F_1 or the F_2 will be measured as the deviation from the mid-parent value, i.e., as the difference from the mean of the two parent populations. First consider the effects of a single locus with two alleles whose frequencies are p and q in one population, and p' and q' in the other. Let the difference of gene frequency between the two populations be y, so that $y = p - p' = q' - q$. The algebra is then simplified by writing the gene frequencies p' and q' in the second population as $(p - y)$ and $(q + y)$. Let the genotypic values be a, d, $-a$, as before. They are assumed to be the same in the two populations, epistatic interaction being disregarded. We have to find the mean of each parent population and the mid-parent value; then the mean of the F_1 and the mean of the F_2. The parental means, M_{P_1} and M_{P_2}, are found from equation [7.2]. They are

$$M_{P_1} = a(p - q) + 2dpq$$
$$M_{P_2} = a(p - y - q - y) + 2d(p - y)(q + y)$$
$$= a(p - q - 2y) + 2d[pq + y(p - q) - y^2]$$

The mid-parent value is

$$M_{\bar{P}} = \tfrac{1}{2}(M_{P_1} + M_{P_2})$$
$$= a(p - q - y) + d[2pq + y(p - q) - y^2] \qquad \dots [14.5]$$

When the two populations are crossed to produce the F_1, individuals taken at random from one population are mated to individuals taken at random from the other population. This is equivalent to taking genes at random from the two populations, as shown in Table 14.3. The F_1 is therefore constituted as follows:

	Genotypes		
	A_1A_1	A_1A_2	A_2A_2
Frequencies	$p(p - y)$	$2pq + y(p - q)$	$q(q + y)$
Genotypic values	a	d	$-a$

The mean genotypic value of the F_1 is therefore:

$$M_{F_1} = a(p^2 - py - q^2 - qy) + d[2pq + y(p - q)]$$
$$= a(p - q - y) + d[2pq + y(p - q)] \qquad \dots [14.6]$$

The amount of heterosis, expressed as the difference between the F_1 and the mid-parent values, is obtained by subtracting equation [14.5] from equation [14.6]:

$$H_{F_1} = M_{F_1} - M_{\bar{P}}$$
$$= dy^2 \qquad \dots [14.7]$$

Thus heterosis, just like inbreeding depression, depends for its occurrence on dominance. Loci without dominance (i.e., loci for which $d = 0$) cause neither inbreeding depression nor heterosis. The amount of heterosis following a cross between two particular lines or populations depends on the square of the

Table 14.3 Frequencies of zygotes in the F_1.

		Gametes from P_1	
		A_1	A_2
		p	q
Gametes $\}$	A_1 $p - y$	$p(p - y)$	$q(p - y)$
from P_2	A_2 $q + y$	$p(q + y)$	$q(q + y)$

difference of gene frequency (y) between the populations. If the populations crossed do not differ in gene frequency there will be no heterosis, and the heterosis will be greatest when one allele is fixed in one population and the other allele in the other population.

Now consider the joint effects of all loci at which the two parent populations differ. In so far as the genotypic values attributable to the separate loci combine additively, we may represent the heterosis produced by the joint effects of all the loci as the sum of their separate contributions. Thus the heterosis in the F_1 is

$$H_{F_1} = \sum dy^2 \qquad \qquad \dots [14.8]$$

Three conclusions can be drawn from equation [14.8]:

(1) If some loci are dominant in one direction and some in the other, their effects will tend to cancel out, and no heterosis may be observed, in spite of the dominance at the individual loci. The occurrence of heterosis on crossing is therefore, like inbreeding depression, dependent on directional dominance, and the absence of heterosis is not sufficient ground for concluding that the individual loci show no dominance.

(2) The amount of heterosis is something specific to each particular cross. The genes by which two lines differ will not be the same for all pairs of lines, so different pairs of lines will have different values of $\sum dy^2$ and will show different amounts of heterosis.

(3) If the lines crossed are highly inbred, and so completely homozygous, the differences of gene frequency between them can only be 0 or 1. The heterosis as shown by equation [14.8] is then the sum of the dominance deviations d of those loci that have different alleles in the two lines.

Before we go on to consider the F_2 it is perhaps worth noting that the formulation of the heterosis in terms of the square of the difference of gene frequency, in equations [14.7] and [14.8], is quite in line with the previous formulation of the inbreeding depression in terms of the coefficient of inbreeding. If we think of a population subdivided into many lines, and we suppose that pairs of lines are taken at random, then the mean squared difference of gene frequency between the pairs of lines will be equal to twice the variance of gene frequency among the lines, i.e., $\overline{y^2} = 2\sigma_q^2$. The relationship between the variance of gene frequency and the coefficient of inbreeding was given in equation [3.14] as $\sigma_q^2 = \bar{p}\bar{q}F$. Therefore $\overline{y^2} = 2\bar{p}\bar{q}F$, showing that the mean amount of heterosis in crosses between random pairs of lines is equal to the inbreeding depression as given in equation [14.2], though of opposite sign.

Now consider the F_2 of a particular cross of two parent populations, the F_2 being made by random mating among the individuals of the F_1. In consequence of the random mating, the genotype frequencies in the F_2 will be the Hardy–Weinberg frequencies corresponding to the gene frequency in the F_1. The mean genotypic value of the F_2 is then easily derived by application of equation [7.2]. The gene frequency in the F_1, being the mean of the gene frequencies in the two parent populations, is $(p - \frac{1}{2}y)$ for one allele, and $(q + \frac{1}{2}y)$ for the other. Putting these gene frequencies in place of p and q respectively in equation [7.2] gives the mean genotypic value of the F_2 as

$$M_{F_2} = a(p - \tfrac{1}{2}y - q - \tfrac{1}{2}y) + 2d(p - \tfrac{1}{2}y)(q + \tfrac{1}{2}y)$$
$$= a(p - q - y) + d[2pq + y(p - q) - \tfrac{1}{2}y^2] \qquad \ldots [14.9]$$

The amount of heterosis shown by the F_2 is the difference between the F_2 and mid-parent values. So, from equations [14.5] and [14.9],

$$H_{F_2} = M_{F_2} - M_{\bar{P}}$$
$$= \tfrac{1}{2}dy^2$$
$$= \tfrac{1}{2}H_{F_1} \qquad \ldots [14.10]$$

We find therefore that the heterosis shown by the F_2 is only half as great as that shown by the F_1. In other words, the F_2 is expected to drop back half-way from the F_1 value toward the mid-parent value. At first sight this conclusion may seem to contradict the one arrived at earlier, when we were considering crosses between many lines, the F_1 and F_2 means then being equal. The difference between the two situations is that an F_2 made by random mating among a large number of different crosses has the same inbreeding coefficient as the F_1. But an F_2 made from an F_1 derived from a single cross has inevitably an increased inbreeding coefficient. If the inbreeding coefficient is worked out in the manner described in Chapter 5, it will be found to be half the inbreeding coefficient of the parent lines. The change of mean from F_1 to F_2 may therefore be regarded as inbreeding depression. It cannot be overcome by having a large number of parents of the F_2 because the restriction of population size that causes the inbreeding has already been made in the single cross of only two lines, or parent populations. There need, however, be no further rise of the inbreeding coefficient in the F_3 and subsequent generations. Provided, therefore, that there is no other reason for the gene frequency to change, the population mean will be the same in the generations following as in the F_2.

That the heterosis expected in the F_2 is half that found in the F_1 is equally true when the joint effects of all loci are considered, provided that epistatic interaction is absent. The conclusion for a single locus was based on the principle that Hardy–Weinberg equilibrium is attained by a single generation of random mating. This, however, is not true with respect to genotypes at more than one locus considered jointly, for reasons that were explained in Chapter 1. Therefore if there is epistatic interaction, the population mean will not reach its equilibrium value in the F_2, but will approach it in subsequent generations more or less rapidly according to the closeness of the linkage between the interacting loci. The existence of epistatic interaction is intimately connected with the scale of measurement, but this matter will not be discussed until Chapter 17. Here we need only note that, for reasons connected with the scale of measurement, the halving of the heterosis in the F_2 expected on theoretical grounds is not often found at all exactly in practice, though the F_2 usually falls somewhere between the F_1 and mid-parent values. Some examples from plants of the heterosis observed in the F_1 and F_2 generations are illustrated in Fig. 14.3. It will be noticed that with some of the characters shown the F_1 and F_2 are lower in value than the mid-parent, and the heterosis is consequently negative in sign. This is in no way inconsistent with our definition of heterosis as the difference between the F_1 or F_2 and the mid-parent value. The sign of the difference depends simply on the nature of the measurement. For example, the character 'days to first fruit', represented in graphs (e) to (h), shows heterosis of negative sign: but if the character were called 'speed of development' and expressed as a reciprocal of time

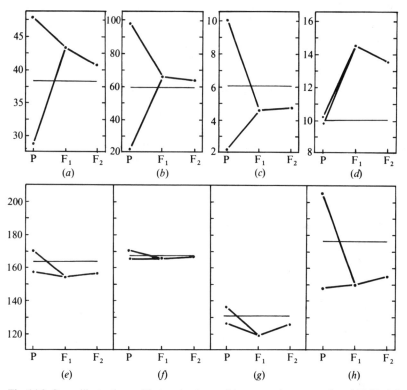

Fig. 14.3. Some illustrations of heterosis observed in crosses between pairs of highly inbred strains of plants. The points show the mean values of the two parent strains, the F_1 and the F_2 generations. The mid-parent values are shown by horizontal lines. Graph (a) refers to tobacco, *Nicotiana rustica*. (*Data from Smith, 1952*). All the other graphs refer to tomatoes, *Lycopersicon* (*Data from Powers, 1952*). The characters represented are:

  (a) Height of plant (in)
  (b) Mean weight of one fruit (g)
  (c) Number of locules per fruit
  (d) Mean weight per locule (g)
  (e)-(h) Mean time in days between the planting of the seed and the ripening of the first fruit, in
      4 different crosses.

the heterosis would be positive in sign. Not all the crosses in Fig. 14.3 provide heterosis that would be useful to a breeder. If the mean of the character is the only criterion of value, a cross is of no use unless the F_1 is better than both of the two parental lines. The term heterosis is sometimes used to mean useful heterosis, that is to say the amount by which the F_1 exceeds the better parent line.

Maternal effects. The relative amount of heterosis observed in the F_1 and F_2 generations may be complicated by maternal effects. A character subject to a maternal effect, such as litter size, has two components belonging to different generations. Each component is expected to follow the general pattern of heterosis in the F_1 and F_2 described above, but the two components are one generation out of step with each other. Thus the heterosis observed in the F_1 is attributable to the non-maternal part, the maternal effect being still at the inbred level. In the F_2, however, the non-maternal part will lose half the heterosis as explained above, but the maternal effect will now show the full effect of its heterosis since the mothers are now in the F_1 stage. This rather complicated situation may perhaps be more readily grasped from the diagrammatic representation in Fig. 14.4. This shows how the heterosis appears in two steps, the first due to the non-maternal component and the second due to the maternal component. These two steps were illustrated in Example 14.2.

Wide crosses. We have seen that the amount of heterosis shown by a particular cross depends, among other things, on the differences of gene frequency between the two populations crossed. This would seem to indicate that the amount of heterosis would increase with the degree of genetic differentiation between the two populations and would be limited only by the barrier of

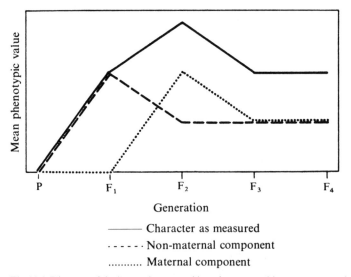

Fig.14.4. Diagram of the heterosis expected in a character subject to a maternal effect, when two lines are crossed and the F_2 and subsequent generations are made by random mating. The maternal and non-maternal components of the character separately are here supposed to show equal amounts of heterosis, and to combine by simple addition to give the character as it is measured.

interspecific sterility. This, however, is not true. Populations that are widely differentiated through adaptations to local conditions may fail to show heterosis and may suffer a reduction of fitness in the F_2 generation, as has been shown by studies of *Drosophila* populations (Wallace and Vetukhiv, 1955). The failure of wide crosses to show the heterosis that might have been expected can be attributed to epistatic interaction, which we have so far assumed to be absent. Adaptation to widely different local conditions involves many different charac- ters because the fitness of an organism depends on the harmonious interrelations of all its functions. Genes at many loci are therefore selected for their joint effects on fitness; the combinations of genes selected in this way are said to be 'coadapted'. In other words, some of the adaptation comes from epistatic interactions. When two populations adapted to different conditions are crossed, the hybrids are adapted to neither. Furthermore, the favourable epistatic combinations of genes are lost by segregation in the F_2. The following example shows how heterosis for yield in maize is related to the width of the cross.

Example 14.3 (*Data from Moll et al.*, 1965). Varieties, or populations, of maize are adapted to different geographical regions. Two varieties from each of four regions

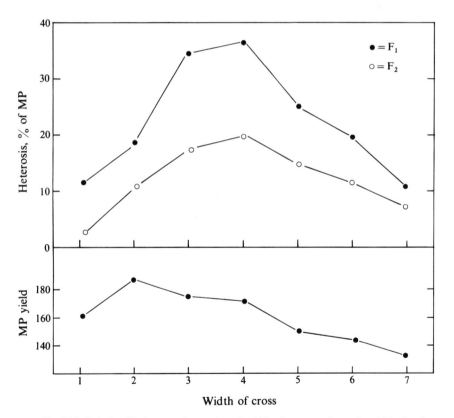

Fig. 14.5. Relationship between heterosis and width of cross as shown by yield of maize, as explained in Example 14.3. The heterosis is expressed as the percentage difference from the mid-parent. The mid-parent yields in g per plant are shown in the lower graph. (*Data from Moll et al.*, 1965.)

were chosen and all of the 28 possible crosses were made. F_2 generations were made by random mating among the individuals of each F_1. The varieties and the crosses were grown in each of the four regions of origin. The regions were: (A) South-east USA, (B) Midwest USA, (C) Puerto Rico, (D) Mexico. The degree of diversity in the crosses was assessed in seven grades according to the known ancestral relationships of the varieties and the climatic conditions of the regions. Grade 1 was crosses between the varieties from the same region. The other grades were: $2 = A \times B$; $3 = A \times C$; $4 = B \times C$; $5 = A \times D$; $6 = B \times D$; $7 = C \times D$. Figure 14.5 shows the heterosis in the F_1's and F_2's as percentages of the mean of the parental varieties, plotted against the degree of diversity. The mean parental yields are also shown. The greater heterosis was shown by the crosses of intermediate 'width'. The widest crosses gave much less heterosis though it was still positive even in the F_2 generations.

15 INBREEDING AND CROSSBREEDING: II. Changes of variance

The effect of inbreeding on the genetic variance of a metric character is apparent, in its general nature, from the description of the changes of gene frequency given in Chapter 3. Again, we have to imagine the whole population, consisting of many lines. Under the dispersive effect of inbreeding, or random drift, the gene frequencies in the separate lines tend toward the extreme values of 0 or 1, and the lines become differentiated in gene frequency. Since the mean genotypic value of a metric character depends on the gene frequencies at the loci affecting it, the lines become differentiated, or drift apart, in mean genotypic value. And, since the genetic components of variance diminish as the gene frequencies tend toward extreme values (see Fig. 8.1), the genetic variance within the lines decreases. The general consequence of inbreeding, therefore, is a redistribution of the genetic variance; the component appearing between the means of lines increases, while the component appearing within the lines decreases. In other words, inbreeding leads to genetic differentiation between lines and genetic uniformity within lines.

The subdivision of an inbred population into lines introduces an additional observational component of variance, the between-line component, and it is not surprising that this adds a considerable complication to the theoretical description of the components of genetic variance. Only a brief description of the main outlines will be given here. For detailed treatment see Wright (1969), and Weir and Cockerham (1977). Similarly, when lines are crossed, the variance can be partitioned into components between and within crosses. The variance of crosses will be described at the end of this chapter. The redistribution of genetic variance is not the only effect of inbreeding; experiments have shown that the environmental variance is sometimes also affected. The greater sensitivity of inbred individuals to environmental sources of variation was mentioned earlier, in Chapter 8. This phenomenon interferes with the experimental study of the changes of variance, and until it is better understood we cannot put much reliance on the theoretical expectations concerning the genetic variance being manifest in the observed phenotypic variance. Another matter concerning inbreeding to be considered is the genetic stability of highly inbred lines, which is important in connection with the use of 'standard' inbred strains for experimental purposes.

Inbreeding

Redistribution of genetic variance

The redistribution of variance arising from additive genes (i.e., genes with no dominance) is easily deduced. This is because with additive genes the proportion in which the original variance is distributed within and between lines does not depend on the original gene frequencies. When there is dominance, however, we cannot deduce the changes of variance without a knowledge of the initial gene frequencies. This not only adds to the mathematical complexity, but it renders a general solution impossible. We shall first consider the case of additive genes, and then very briefly indicate the conclusions arrived at for dominant genes. The effect of selection will not be specifically discussed. We need only note that natural selection will tend to render the actual state of dispersion of gene frequencies less than that indicated by the inbreeding coefficient computed from the population size or pedigree relationships.

No dominance.   What follows refers to the variance arising from additive genes: it does not apply to the additive variance arising from genes with dominance. The conclusions therefore apply, strictly speaking, only to characters with no non-additive variance. They serve, however, to indicate the general effect of inbreeding on variance, and may be taken as a fair approximation to what is expected of characters that show little non-additive genetic variance. The description to be given refers to slow inbreeding, and is not strictly true of rapid inbreeding by sib-mating or self-fertilization. The redistribution of the variance under rapid inbreeding is, however, not very different except in the first few generations.

Consider first a single locus. When there is no dominance the genotypic variance in the base population, given in equation 8.5 is

$$V_G = V_A = 2p_0q_0a^2$$

The variance within any one line is $V_G = 2pqa^2$, where p and q are the gene frequencies in that line. The mean variance within lines is

$$V_{Gw} = 2\overline{(pq)}a^2$$

where $\overline{(pq)}$ is the mean value of pq over all lines. Now, $2\overline{(pq)}$ is the overall frequency of heterozygotes in the whole population, which, by Table 3.1, is equal to $2p_0q_0(1 - F)$, where F is the coefficient of inbreeding. Therefore

$$V_{Gw} = 2p_0q_0a^2(1 - F)$$
$$= V_G(1 - F)$$

and this remains true when summation of the variances is made over all loci. Thus the within-line variance is $(1 - F)$ times the original variance, and as F approaches unity the within-line variance approaches zero.

Now consider the between-line variance. This is the variance of the true means of lines, and would be estimated from an analysis of variance as the between-line component. For a single locus, still with no dominance, the mean genotypic value of a line with gene frequency p and q is obtained from equation

[7.2] as

$$M = a(p - q)$$
$$= a(1 - 2q)$$

Thus we want to find the variance of $(a - 2aq)$. Epistasis is assumed to be absent, so a is constant, i.e., the same in all lines. Therefore

$$\sigma_M^2 = \sigma_{(2aq)}^2$$
$$= 4a^2\sigma_q^2$$
$$= 4a^2 p_0 q_0 F \qquad \text{(from equation [3.14])}$$
$$= 2FV_G$$

and this also remains true when summation is made over all loci. Thus the between-line genetic variance is $2F$ times the genetic variance in the base population.

Table 15.1 Partitioning of the variance in a population with inbreeding coefficient F, when the variance V_G in the base population is due to genes with no dominance.

Between lines	$2FV_G$
Within lines	$(1 - F)V_G$
Total	$(1 + F)V_G$

The partitioning of the genetic variance into components as explained above is summarized in Table 15.1. The total genetic variance in the whole population is the sum of the within-line and between-line components, and is equal to $(1 + F)$ times the original genetic variance. (This is true also of close inbreeding.) Thus

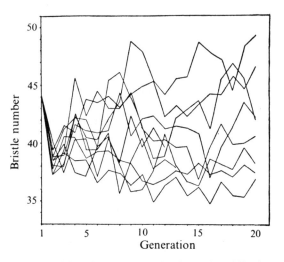

Fig. 15.1. Differentiation between lines by random drift, shown by abdominal bristle number in *Drosophila melanogaster*. The graphs show the mean bristle number in each of 10 lines during full-sib inbreeding without artificial selection. (*After Rasmuson, 1952.*)

when inbreeding is complete the genetic variance in the population as a whole is doubled, and all of it appears as the between-line component. The increasing variance between lines is illustrated from experimental data in Fig. 15.1.

The genetic variance within lines, before inbreeding is complete, is partitioned within and between the families of which the lines are composed. Under slow inbreeding with random mating within the lines, it is partitioned equally within and between full-sib families. The covariance of relatives within the lines is just as described in Chapter 9, each line being a separate random-breeding population with a total genetic variance of $(1 - F)V_G$, on the average. From this we can deduce what the heritability is expected to be within any one line. It will be $(1 - F)V_G/[(1 - F)V_G + V_E]$, and this reduces to

$$h_t^2 = \frac{h_0^2(1 - F_t)}{1 - h_0^2 F_t} \qquad \qquad \dots [15.1]$$

where h_t^2 and F_t are the heritability within lines and the inbreeding coefficient at time t, and h_0^2 is the original heritability in the base population. This shows how the heritability is expected to decline with the inbreeding in a small population. The formula, however, is applicable only to characters with no non-additive variance, and in the absence of selection. The operation of natural selection renders the reduction of the heritability less than expected, especially under slow inbreeding. This point has been demonstrated experimentally with *Drosophila* (Tantawy and Reeve, 1956).

Dominance. The changes in the components of variance arising from additive genes will have been seen to be independent of the gene frequencies in the base population. When we consider genes with any degree of dominance, however, we find that the changes of variance on inbreeding depend on the initial gene frequencies, and this makes it impossible to give a general solution in terms of the genetic variance present in the base population. We shall therefore do no more than give the conclusions arrived at by Robertson (1952) for the case of fully dominant genes, when the recessive allele is at low frequency. This is the situation most likely to apply to variation in fitness arising from deleterious recessive genes, though the effects of selection are here disregarded. Figure 15.2 shows the redistribution of variance arising from recessive genes at a frequency of $q = 0.1$ in the base population. Figure 15.2(*a*) refers to full-sib mating with only one family in each line, and Fig 15.2(*b*) refers to slow inbreeding. A surprising feature of the conclusions is that the within-line variance at first increases, reaching a maximum when the coefficient of inbreeding is a little under 0.5, and it remains at a fairly high level until the coefficient of inbreeding approaches 1. The reason, in general terms, for the apparent anomaly that the variation within lines increases during the first stages of inbreeding can be seen from a consideration of the relationship between the gene frequency and the variance arising from a dominant gene shown in Fig. 8.1(*b*). The gene frequency is taken to start at a value of 0.1, and on inbreeding it will increase in some lines and decrease in others, the increase being on the average equal in amount to the decrease. But examination of the graph shows that an increase of gene frequency by a certain amount will increase the variance more than a decrease of the same amount will reduce it. Therefore, on

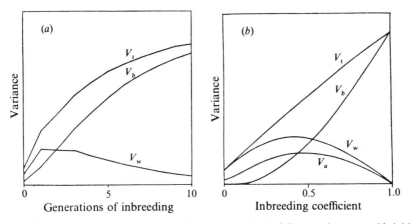

Fig. 15.2. Redistribution of variance arising from a single fully recessive gene with initial frequency $q_0 = 0.1$. (a) with full-sib mating, (b) with slow inbreeding. (*After Robertson, 1952.*)
V_t = total genetic variance.
V_b = between-line component.
V_w = within-line component.
V_a = additive genetic variance within lines.

average, the variance within the lines will increase in the early stages of inbreeding. This increase of variance would be detectable in practice only if a substantial part of the genetic variance were due to recessive genes at low frequencies.

Environmental variance

Several times in previous chapters we have referred to the fact that the environmental component of variance may differ according to the genotype; in particular, that inbred individuals often show more environmental variation than non-inbred individuals. This fact has been revealed by many experiments in which the variances of inbreds and of hybrids have been compared. Any difference of phenotypic variance between highly inbred lines and the F_1 between them (i.e., the 'hybrid') must be attributed to a difference of the environmental component, because the genetic variance is negligible in amount in the hybrids as well as in the inbred lines. The greater susceptibility of inbreds than of hybrids to environmental sources of variation is not a universal phenomenon, but it has been observed in a wide variety of characters and organisms. Some examples are cited in Table 15.2; others are given by Lerner (1954) and Wright (1977). The phenomenon has been extensively studied in behavioural characters of rats and mice; these have been reviewed by Hyde (1973), who found that 14 out of 19 behavioural characters showed the phenomenon.

Before we consider the possible reasons for the differences of environmental variance, there are two points in connection with the phenomenon that should be mentioned. The first is a technical matter. If the mean of the character differs between inbreds and hybrids, as it frequently does, then it may be difficult to decide on a proper basis for the comparison of the variances. It is necessary to be sure that any difference of variance found is not simply a reflection of the difference of mean. In other words, a measure of variation must be found that is uncorrelated with the mean. The coefficient of variation is often, though not

Table 15.2 Examples of characters with phenotypic variance greater in inbreds than in hybrids. The values are phenotypic variances averaged over the inbred lines and over the F_1's where more than one cross was made. $(C.V.)^2$ = squared coefficient of variation.

	Inbreds	Hybrids
Drosophila melanogaster		
(Robertson and Reeve, 1952)		
Wing-length: 6 inbreds, 6 F_1's.	2.35	1.24
Mice (McLaren and Michie, 1956)		
Duration of "Nembutal" anaesthesia		
(log minutes).		
2 inbreds, 1 F_1	0.0665	0.0165
Mice (Yoon, 1955)		
Age at opening of vagina (days).		
3 inbreds, 2 F_1's	51.7	17.4
Mice (Chai, 1957)		
Weight (g) at ages given $(C.V.)^2$ ⎰ Birth	119	59
2 inbreds, 1 F_1 ⎱ 21 days	98	47
60 days	24	19
Rats (Livesay, 1930)		
Weight at 90 days $(C.V.)^2$, 3 inbreds, 2 F_1's	522	170
Maize (Shank and Adams, 1960)		
Plant height $(C.V.)^2$	44	30
Ear weight $(C.V.)^2$	412	198
10 inbreds, 5 F_1's		

always, an appropriate measure. The problem is basically a matter of the choice of scale, and will be considered again in Chapter 17. The second point concerns the nature of the environmental variation that is being measured. There is a distinction to be made between variation induced by the environment and adaptive responses by the organism to the particular environmental circumstances. The distinction is necessary when one considers the possible relation of variation to fitness. In the first case, the presumption is that there is an optimal phenotype that is the same for all the environmental circumstances under consideration. The body temperature of mammals is an obvious example. Insensitivity to environmental variables is then an aspect of fitness: the fittest individuals are those that can regulate their development, or their physiological processes, so as to achieve the optimal phenotype despite sub-optimal values of the environmental variables. The restriction of variation in this way is called *homeostasis*. The environmental factors causing the variation are not necessarily external, but include also internal causes of developmental variation. In the second case, when the organism responds adaptively, the character has different optima in different environments. The output of sweat in relation to variation of ambient temperature is an example. With characters of this sort the relation with fitness is reversed: individuals with the ability to vary are fitter than those without the ability.

What then is the cause of some characters being more variable in inbreds than in hybrids? The phenomenon can be looked on as a manifestation of inbreeding depression or heterosis (Mather, 1953). On this view, a character showing the phenomenon is one for which homeostasis is beneficial, and inbreeding has reduced this component of fitness, causing environmental variance to be increased in inbreds and correspondingly reduced when hybrids are compared with inbreds. This interpretation implies that there are genes that affect variability, i.e., homeostasis, independently of any effect they may have on the mean of the character, and that these genes must have some degree of directional dominance for increased homeostasis. Another interpretation of the phenomenon is in biochemical terms. Different allelic forms of enzymes often have their maximal enzymic activity at different values of environmental variables such as temperature or pH. Heterozygotes, with both forms of the enzyme, may therefore maintain an adequate level of enzyme activity over a wider range of environmental variation than homozygotes, with only one form of the enzyme, can do. In so far as the enzyme activity is reflected in a measured character, then, heterozygotes are less sensitive to environmental variables than are homozygotes. On this view, the stability is not necessarily related to fitness, but is simply a biochemical consequence of heterozygosity at some loci that affect the mean of the character. These loci are consequently overdominant with respect to their effect on variability, though not necessarily in their effect on the mean of the character or on fitness.

Uniformity of inbred strains

Inbred strains of laboratory animals, particularly mice, are widely used as experimental material in testing and assay, and in many other areas of biological research. The inbred strains are single lines and are used because uniformity is desired. For some purposes, when for example the absence of antigenic differences is necessary, it is genetic uniformity that is required. Abundant experience has shown that the highly inbred strains of laboratory mice fully satisfy this requirement of genetic uniformity. For other purposes, however, it is not genetic uniformity alone that matters, but phenotypic uniformity. The less variable the animals, the smaller the number that need be used to attain a given degree of precision in measuring their response to a treatment. The value of inbred animals in this respect therefore depends on how much of the variance of the character is genetic in origin, because only the genetic variance is removed by inbreeding. It depends also on how much the environmental variance is affected by inbreeding; the increased environmental variance, as exemplified in Table 15.2, may sometimes offset the reduced genetic variance so that an inbred strain is more variable phenotypically than a non-inbred stock. The way to obtain genetic uniformity without increased environmental variation is, of course, to use the F_1 of a cross between two inbred strains. This has the added advantage that the F_1 individuals are usually healthier, more viable and more fertile than the inbreds, though it does not reduce the cost of production since the inbreds have to be maintained as parents. Another point in connection with the use of inbred or F_1 animals may be mentioned. An inbred strain or the F_1 of two inbred strains has a unique genotype; and that of an inbred, moreover, is one that cannot occur in a natural population. Testing the response to any treatment on one inbred strain or one hybrid is therefore testing it on one genotype. There may be appreciable differences of response between genotypes, and consequently results obtained with one strain or cross cannot be relied on to be applicable to other strains.

Subline divergence. The use of standard strains facilitates the comparison of results from different laboratories, and many inbred strains of laboratory animals are widely dispersed in different laboratories. The dispersion of a strain inevitably leads to it becoming split into sublines and it is important to know to what extent the sublines become differentiated genetically. The main question concerns the origin of new variation by mutation and its fixation by the continued inbreeding. Differentiation can, of course, occur by errors of parentage, or by residual segregation if the sublines have been separated before fixation was complete. Several studies have shown, however, that differentiation does occur when both these causes have been excluded (see Grewal, 1962, and Example 15.1 below). The rate of origin of allelic differences that are selectively neutral was shown in Chapter 4 to be equal simply to the mutation rate to neutral alleles, irrespective of the population size. To repeat the argument in the present context: in a sib-mated line there are four representatives of each autosomal locus, so the rate of occurrence of mutations at a particular locus is $4u$, where u is the mutation rate per gamete per generation. But a mutation that has occurred has a $\frac{1}{4}$ chance of becoming fixed. So the rate of allelic substitution per generation is equal to the mutation rate. Mutants that are not strictly neutral can become fixed, especially in very small populations. In Chapter 4 it was stated that a mutant is 'effectively neutral' if the coefficient of selection against it is less than $1/2N_e$, which is about 0.2 in a sib-mated line (see Chapter 5). Thus if, for example, the rate of mutation to alleles with coefficients of selection against them of less than about 20 per cent were 10^{-5}, and if the total number of loci were 10,000, then in any one subline there would be, on average, 1 allelic substitution every 10 generations ($10^{-5} \times 10^4 \times t = 1$; $t = 10$). If two sublines were kept with contemporaneous generations they would come to differ, on average, at 1 locus for every 5 generations that elapse. This is because after 5 elapsed generations they are separated by 10 generations in which mutation can occur. It is not possible to translate this theoretical rate of allelic differentiation into an effect on a metric character because we do not know the number of relevant loci; the number, that is, at which an allelic difference would affect the character. It may seem that the effect of mutation would be trivial, but several experimental studies have revealed detectable, and even quite large, differences between sublines that had not been very long separated (see McLaren and Michie, 1954; Grewal, 1962; Festing, 1973). One such study is described in the following example. The conclusion to be drawn from the experimental work is that differences between sublines that have been separated for more than a few generations are by no means negligible.

Example 15.1 This is a brief description of a study of two inbred mouse strains by Bailey (1959). Six measurements of various skeletal dimensions were studied in two sublines of the C57BL/6 strain and two sublines of the BALB/cAn strain. The strains had been separated into sublines after 30 and 78 generations respectively of full-sib mating, so differentiation by residual segregation was not a serious possibility. The C57BL sublines had been separated from each other by 46 generations and the BALB sublines by 9 generations. (The generations of separation are counted down both sides of the pedigree, to give the number of generations over which mutations accumulate to give the differentiation.) In both strains four of the six characters differed significantly between the sublines. The amount of the differentiation can be expressed in the following way. If a strain were subdivided into a number of sublines, any two of which were separated by t generations, what would be the variance between the sublines and what fraction would this be of the variance between unrelated inbred strains? This fraction, averaged over the six characters, was 0.2 per cent per generation of separation in the C57BL strain and 0.5 per cent per generation in the BALB strain. To see what these figures mean, consider the following question. How long would it take for the variance between sublines to reach 10 per cent of the variance between unrelated inbred strains? For the C57BL strain this would mean $0.1 = 0.002t$, whence $t = 50$ generations of separation. If the generations were contemporaneous in all sublines this would require 25 elapsed generations in each line. With roughly 3 generations per year, this amount of differentiation would occur in about 8 years. The corresponding time required for the BALB strain would be only about 3 years.

Crossing

In the first part of this chapter we saw how the genetic variance of a metric character is distributed between and within inbred lines. We now have to consider the complementary problem of how the variance is distributed between and within crosses. The variance between crosses is important in predicting what improvement can be expected from inbreeding and crossing; this will be explained in the next chapter. The variance after crossing presents a simpler problem than the variance after inbreeding for the following reason. The gametes produced by inbred lines are not different from the gametes produced by a non-inbred population, provided selection has not changed gene frequencies. This means that any F_1 individual has a genotype that could, in principle, be found in a random-breeding population; and, conversely, any genotype in the original random-breeding population could be found among the crosses. Consequently the total genetic variance after crossing is the same as that in the base population. The F_1 individuals of the same cross can be regarded as a 'family' with a degree of relationship dependent on the inbreeding coefficient of the parental lines. The covariance of these 'families' is the variance between crosses. If the parental lines are fully inbred, all members of the same F_1 have identical genotypes and the variance between crosses is equal to the total genetic variance in the base population. If the parental lines are not fully inbred, the covariance of the 'families' can be worked out from the coefficients of relationship derived from coancestries. This will now be explained by consideration of random crosses.

Variance between crosses

Assume that a large number of lines have been inbred without selection from the same base population and all to the same inbreeding coefficient. Crosses are then made at random between the lines. Assume further that each cross is made from many individuals of the parent lines, these parental individuals not being related to each other within their lines. This means that the genetic variance within the lines is fully represented in the crosses. Each cross, then, is to be regarded as a family with a certain degree of relationship between its members. The component of variance between crosses is the covariance of these related

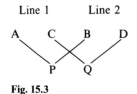

Fig. 15.3

individuals. The composition of the covariance of relatives was given in equation [9.13] in terms of the genetic components and the coefficients of relationship, r and u. The coefficient r concerns the additive variance and u the dominance variance. The coefficients r and u can be derived from the coancestries as shown by equations [9.11] and [9.12]. The coancestries concerned in a cross are as follows. The relevant individuals are shown in Fig. 15.3, where they are given letters to correspond with those in Fig. 5.4. A and C are two individuals of one parental line; B and D two individuals of the other parental line. P and Q are two F_1 individuals produced by crossing A with B, and C with D, respectively. There are many other crosses producing other F_1's with pedigree relationships like P and Q. We need the covariance appropriate to the relationship of P with Q expressed in the coancestry f_{PQ}. First, by equation [9.11], $r = 2f_{PQ}$, and by equation [5.2], $f_{PQ} = \frac{1}{4}(f_{AC} + f_{AD} + f_{BC} + f_{BD})$. The two parental lines are unrelated, so $f_{AD} = f_{BC} = 0$. The two lines have the same inbreeding coefficient, so $f_{AC} = f_{BD} = F$, where F is the inbreeding coefficient of the next generation of the lines, i.e., the generation contemporaneous in terms of generations with the F_1. Thus $f_{PQ} = \frac{1}{2}F$, and this gives

$$r = F \qquad \qquad \dots [15.2]$$

Next, by equation [9.12], $u = f_{AC}f_{BD} + f_{AD}f_{BC}$, which reduces to

$$u = F^2 \qquad \qquad \dots [15.3]$$

The covariance can now be expressed in terms of the inbreeding coefficient by substituting equation [15.2] and [15.3] into equation [9.13]. This gives the variance between crosses as

$$\sigma_X^2 = FV_A + F^2V_D + F^2V_{AA} + F^3V_{AD} + F^4V_{DD} + \dots \qquad \dots [15.4]$$

In this expression V_A and V_D are the additive and dominance variances in the base population; V_{AA}, V_{AD} and V_{DD} are the interaction components as explained in Chapter 8; and F is the inbreeding coefficient, not of the parents used in the crosses, but of the next generation of the lines. The remainder of the genetic variance appears within lines, i.e., $(1 - F)V_A + (1 - F^2)V_D$, etc. The variance between crosses is the variance of the true means, which would be estimated as the component of variance from an analysis of variance. Equation [15.4] gives the genetic content of this component. It corresponds to what would be estimated experimentally only if there are no environmental differences between the crosses. If the variance of the observed means of crosses were to be calculated, this would be increased by a fraction of the within-cross variance depending on the number of individuals measured in each cross for the reasons explained in connection with family selection in Chapter 13 (see Table 13.3).

The main point of interest in equation [15.4] is that the contributions of the different components of genetic variance are differently related to the inbreeding coefficient. The contribution of the additive variance increases linearly with F, but those of the dominance and interaction components increase in proportion to the square or higher powers of F. The consequence is that if the character is one with much non-additive variance, the crosses become differentiated much more rapidly in the late stages of inbreeding, as F approaches 1, than they do in the early stages. When $F = 0.5$, each cross is genetically equivalent to a full-sib family in the base population; when $F = 1$, the whole of the genetic variance appears between crosses. The practical importance of the way in which the between-cross variance is related to F will be considered in the next chapter.

Combining ability

The variance between crosses was derived above on the assumption that a large number of lines were crossed at random. The 'large number' implies that each line was used in only one cross. If, in contrast, each line is crossed with several others, the variance between crosses can be partitioned in a way that has great importance for understanding the use of cross-breeding for improvement. We shall assume, for the sake of explanation, that large numbers of lines, crosses, and individuals are used, so that all means are estimated without error. Crossing each line to several others provides an additional measure in the mean performance of each line in all its crosses. This mean performance of a line, when expressed as a deviation from the mean of all crosses, is called the *general combining ability* of the line. It is the average value of all F_1's having this line as one parent, the value being expressed as a deviation from the overall mean of crosses. Any particular cross, then, has an 'expected' value which is the sum of the general combining abilities of its two parental lines. The cross may, however, deviate from this expected value to a greater or lesser extent. This deviation is called the *specific combining ability* of the two lines in combination. In statistical terms, the general combining abilities are main effects and the specific combining ability is an interaction. The true mean X of a cross between lines P and Q can thus be expressed as

$$X - \bar{X} = GCA_P + GCA_Q + SCA_{PQ} \qquad \ldots [15.5]$$

where $\bar{X}$ is the mean of all crosses, and GCA and SCA are the general and specific combining abilities respectively. In practice another term, E, must be added to the right-hand side to represent sampling error in estimating X.

The terms on the right-hand side of equation [15.5] are uncorrelated with each other, so the total between-cross variance (excluding error variance) is made up as follows:

$$\sigma_X^2 = \sigma_{GCA}^2(M) + \sigma_{GCA}^2(F) + \sigma_{SCA}^2 \qquad \ldots [15.6]$$

Where (M) and (F) refer to the general combining abilities of lines used as male and as female parents respectively. If the lines are not distinguished by sex or in any other way, then equation [15.6] becomes

$$\sigma_X^2 = 2\sigma_{GCA}^2 + \sigma_{SCA}^2 \qquad \ldots [15.7]$$

The two components into which the total between-cross variance can be partitioned are the variance of general combining ability and the variance of specific combining ability. These are observational components of variance in the sense explained in Chapter 9, and are estimated from an analysis of variance. Their importance lies in the fact that the causal components of genetic variance contribute to them differently, as we shall now see.

A set of crosses with one line as common male parent can be regarded as a family analogous to a paternal half-sib family. The covariance of these families is the variance of general combining abilities of male lines, $\sigma^2_{GCA}(M)$. This covariance is found from the coefficients of relationship in the same way as before. Figure 15.3 will serve to illustrate the relevant pedigree. Individuals A and C are two members of the common male line, but B and D are now members of two different lines to which A and C are crossed, producing the two F_1's, P and Q. The coancestries are now all zero except $f_{AC} = F$. This gives $r = \frac{1}{2}F$ and $u = 0$. Substitution into equation [9.13] as before gives

$$\sigma^2_{GCA}(M) = \tfrac{1}{2}FV_A + \tfrac{1}{2}F^2 V_{AA} + \cdots$$

The same argument applies to lines used as female parents and therefore, provided there are no maternal or sex-linked effects, $\sigma^2_{GCA}(F) = \sigma^2_{GCA}(M)$. The variance of specific combining ability is what is left of the total variance between crosses as given in equation [15.4]. We thus arrive at the following composition of the components of variance of crosses:

General combining ability

$$\left.\begin{array}{ll}
\text{of male parents:} & \sigma^2_{GCA}(M) = \tfrac{1}{2}FV_A + \tfrac{1}{2}F^2 V_{AA} + \cdots \\
\text{of female parents:} & \sigma^2_{GCA}(F) = \tfrac{1}{2}FV_A + \tfrac{1}{2}F^2 V_{AA} + \cdots \\
\text{Specific combining ability:} & \sigma^2_{SCA} = F^2 V_D + F^3 V_{AD} + F^4 V_{DD} + \cdots
\end{array}\right\} \cdots [15.8]$$

From this it can be seen that differences of general combining ability are due to the additive variance and $A \times A$ interactions in the base population; and differences of specific combining ability are attributable to the non-additive genetic variance. Consequently the variance of general combining ability increases linearly with F (apart from the interaction component), while the variance of specific combining ability increases with higher powers of F. It is therefore the specific, and not the general, combining ability that is expected to increase in variance more rapidly as the inbreeding reaches high levels.

The components of genetic variance in equation [15.8] are those of a random-breeding population with all gene frequencies equal to those in the lines crossed and with coupling and repulsion linkages in equilibrium. This random-breeding population can be regarded as being the base population, real or hypothetical, from which the lines were derived without selection. Or, alternatively, it can be regarded as a synthetic population made by random mating among the crosses and then bred by random mating for long enough to reach linkage equilibrium. Which of these viewpoints is to be adopted affects the details of the analysis of variance. In the first case the lines are regarded as a sample of the population and are therefore random factors: in the second case the lines are the whole population and are fixed factors (see Griffing, 1956b, for details).

Estimation of combining abilities. A method of estimating general combining abilities that is convenient for use with plants is known as the *polycross* method. A number of plants from all the lines to be tested are grown together and allowed to pollinate naturally, self-pollination being prevented by the natural mechanism for cross-pollination, or by the arrangement of the plants in the plot. The seed from the plants of one line are therefore a mixture of random crosses with other lines, and their performance when grown tests the general combining ability of that line. The general combining abilities measured are those of lines used as female parents. If the variances of general combining ability are assumed to be the same for male and female parents, the variances of general and specific combining ability can be estimated and interpreted as in equation [15.8].

The general combining ability of a line can be estimated by crossing it with individuals from the base population instead of with other inbred lines. This method is known as *top-crossing*. It is equivalent to crossing with a random set of lines inbred from the base population without selection because, as noted earlier, the gametes from inbred lines are not different in genetic content from those of the base population.

A commonly used experimental design for crossing inbred lines is the *diallel cross*, in which each line is crossed with every other line. The estimation of the general combining abilities of the lines is explained and illustrated in Example 15.2 below. The analysis of a diallel cross for the purpose of estimating the variances of general and specific combining ability is rather complicated because it depends on whether reciprocal crosses are included, and on the assumption made about the population to which the genetic components of variance refer. The theory underlying diallel crosses is explained by Griffing (1956a) and the analytical procedure by Griffing (1956b).

Example 15.2 The calculation of general and specific combining abilities will be illustrated by data from Sprague and Tatum (1942) on crosses between inbred lines of maize. Ten lines were used and each was crossed with each of the other nine, but reciprocal crosses were not made. There were therefore $\frac{1}{2}n(n-1) = 45$ crosses, n being the number of lines. The yields, in bushels per acre, are given in table (i); these are the mean yields of each cross. (100 bushels per acre of maize = 6.725 tonnes per hectare.) The column headed T gives the total yield of each line in all nine crosses, obtained by summing down the appropriate column and along the row as indicated by the arrows. Note that each cross contributes to two totals, so $\Sigma T = 2\Sigma X$, where X is the yield of a cross.

If there were a very large number of lines, the general combining ability of a line would be calculated simply as the deviation of its mean, $T/(n-1)$, from the overall mean, $\bar{X}$. With a small number of lines, however, this is not valid because each of the other lines contributes a fraction, $1/(n-1)$, of its general combining ability to the mean of the line in question. Thus, for example, the mean $\bar{A}$ of line A in all its crosses is

$$\bar{A} - \bar{X} = G_A + \frac{1}{n-1}(G_B + G_C + \ldots + G_N)$$

where the G's are the general combining abilities of the lines A to N as indicated by the subscripts. Now, $\Sigma G = 0$, so $(G_B + G_C + \ldots + G_N) = -G_A$, and so

$$\bar{A} - \bar{X} = \left(\frac{n-2}{n-1}\right)G_A$$

Table (i)

	B	C	D	E	F	G	H	I	J	T	GCA
A	86	84	98	98	92	92	97	81	88	816	3.75
B	↳	91	105	102	86	92	79	80	90	811	3.125
C	.	↳	87	80	65	84	93	77	83	744	− 5.25
D	.	.	↳	97	100	101	97	91	80	856	8.75
E	.	.	.	↳	97	83	93	78	83	811	3.125
F	.	.	.	.	↳	80	93	76	70	759	− 3.375
G	.	.	.	.	.	↳	90	74	72	768	− 2.25
H	.	.	.	.	.	.	↳	91	96	829	5.375
I	.	.	.	.	.	.	.	↳	78	726	− 7.50
J	.	.	.	.	.	.	.	.	↳	740	− 5.75

| Sums | | | | | | | | | $2\sum X = \sum T = 7860$ | | 0.000 |
| Overall mean | | | | | | | | | $\bar{X} = 87.333$ | | |

from which

$$G_A = \left(\frac{n-1}{n-2}\right)(\bar{A} - \bar{X})$$

It is more convenient to work with the totals than with the means. Substituting $\bar{A} = T_A/(n-1)$ and $\bar{X} = \sum T/n(n-1)$ leads to

$$G_A = \frac{T_A}{n-2} - \frac{\sum T}{n(n-2)}$$

The general combining abilities of the lines in the table are entered in the column headed GCA. Formulae appropriate to other designs of diallel cross are given by Simmonds (1979, p. 112).

The 'expected' value of each cross can now be calculated, 'expected' meaning the value that would be predicted from the two general combining abilities, in the absence of any knowledge about the specific combining ability. To take the best cross, BD, as an example, $E(X_{BD}) = 3.125 + 8.75 + 87.333 = 99.21$. The difference between the observed and expected values estimates the specific combining ability of the two lines in combination: $SCA_{BD} = 105 - 99.21 = + 5.79$. The value of the specific combining ability so obtained is subject to the sampling error in estimating X_{BD}. Figure 15.4 shows a plot of the observed yields against the expected yields. If there were no deviations from expectation, the points would lie on the diagonal line with a slope of 1. The vertical distance of any point from the diagonal is the specific combining ability together with the sampling error of the yield of the cross. (If the lines were not highly inbred, there would also be error in estimating the specific combining abilities, this error being due to the sampling of genotypes from the lines.)

For the purpose of illustration, the variances of general and specific combining ability will be calculated on the supposition that the lines were a random sample from a population of lines, though in fact they were not randomly selected. The analysis of variance for estimating the components is given in table (ii). The sums of squares for GCA and SCA were calculated from the data in table (i); the mean square for error is the value stated in the paper. The variance of the true means of crosses, i.e., after deducting the variance due to error, is made up of 54.5 per cent due to general

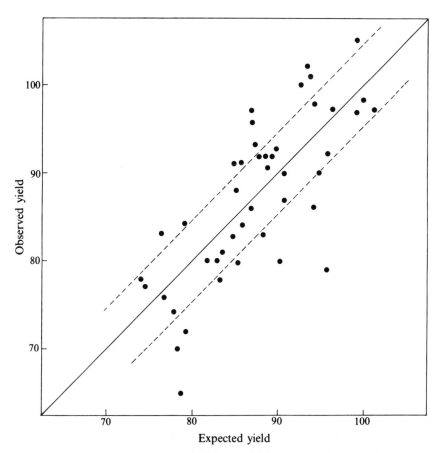

Fig. 15.4. Observed and expected yield (bushels per acre) in crosses between ten lines of maize. The expected yield of each cross is the sum of the general combining abilities of the two parental lines. Deviations from the regression line are due to specific combining ability and error in estimating the mean yield of the cross. The two dashed lines show the positions of deviations from the regression of ± 2 standard errors of cross-means. (*Data from Sprague and Tatum, 1942.*)

combining ability and 45.5 per cent due to specific combining ability. The proportion attributed to specific combining ability is large because these lines were a selected group, and the variation in general combining ability among them was consequently less that would be expected in a random sample.

Table (ii)

Source	d.f.	SS	MS	Expectation of MS
GCA	9	2,179	242.11	$\sigma_E^2 + \sigma_{SCA}^2 + 8\sigma_{GCA}^2$
SCA	35	1,617	46.20	$\sigma_E^2 + \sigma_{SCA}^2$
Error			5.36	σ_E^2

$$2\sigma^2_{GCA} = 48.98 = 54.5\%$$

$$\sigma^2_{SCA} = 40.84 = 45.5\%$$

16 INBREEDING AND CROSSBREEDING: III. Applications

The crossing of inbred lines to produce hybrids plays a major role in the improvement of some plants, most notably maize. Crossing is also widely used in animal breeding, though highly inbred lines of farm animals are not available because of the severe loss of fertility from inbreeding depression. Animal crosses are therefore made between mildly inbred lines or between different breeds. The principles underlying the use of inbreeding and crossing for improvement will be explained in this chapter. We shall be concerned mainly with outbreeding plants, but animals and naturally self-fertilizing plants will be considered briefly in separate sections. Technical details will not be given; for these the reader should consult a textbook of plant breeding, e.g., Simmonds (1979). Two simplifications will be made. First, it will be assumed that the only criterion of merit in plant breeding is yield, though in practice other characters have to be taken into consideration as well as yield. Second, the complications arising from genotype × environment interactions will not be discussed. Obviously an improved hybrid must perform well in a range of different environments associated with different years and different localities. It will be assumed that this requirement is included in the assessment of merit. Crossing highly inbred lines is used also as a method of genetic analysis both with plants and laboratory animals. These analyses of crosses and the later generations derived from them are fully described by Mather and Jinks (1971, 1977) and will not be dealt with here.

Applied to outbreeding organisms, the purpose of crossing is, of course, to produce superior crossbred, or F_1, individuals. Consider first a set of lines all derived from the same random-breeding base population. The crosses must then be superior not only to the inbred lines but to the outbred population from which the lines were derived. Something more than heterosis is therefore sought, since heterosis is the superiority over the inbred lines. It was shown in Chapter 14 that when lines are inbred without selection the mean of all their crosses is expected to be equal to the mean of the outbred population from which they were derived. Therefore inbreeding and crossing alone cannot produce any improvement; there must be selection at some stage if any improvement is to be made.

The lines that are crossed are, however, usually derived from different base populations. Some of the superiority of the crosses then comes from heterosis. If the two base populations differ in gene frequencies, a cross between them will show heterosis, as explained in Chapter 14. In the same way, the mean of crosses between sets of inbred lines derived without selection from the two base populations will be superior to the mean of the two base populations. Some improvement would therefore be achieved even without any selection. With animals, the lines crossed already exist and there is no existing base population from which they were derived. All the gain from crossing is therefore heterosis, and the only selection is in the choice of which lines or breeds to cross. There is nothing to add here to the account of heterosis given in Chapter 14. We are concerned therefore with the selection by which most of the improvement is achieved in plants. It will be assumed for simplicity that all the lines to be crossed are derived from the same base population.

Some improvement can be expected from the effects of natural selection. It eliminates lethal and severely deleterious genes during the inbreeding and, in so far as these genes affect the desired character, an improvement of the crossbred mean over that of the base population is to be expected. But this improvement will not be very great, because the deleterious genes eliminated will have been at low frequencies in the base population – and the more harmful, the lower the frequency – so that their effect on the population mean will be small. It has been calculated, on the basis of assumptions about the number of loci concerned and their mutation rates, that an improvement of 5 per cent in fitness is the most that could be expected from the elimination of deleterious recessive genes (Crow, 1948, 1952). Most of the improvement, therefore, must come from artificial selection applied to the economically desirable characters. The methods of applying this artificial selection are the main topic for consideration. There is, however, another question that must be considered at the same time. Developing inbred lines and evaluating their crosses is a long and costly process. What are the advantages of this method over straightforward selection applied to the original outbreeding population? This question can be partly answered now.

The crossing of inbred lines produces no genotypes that could not occur in the base population. But whereas the best genotypes occur only in certain individuals in the base population, they are replicated in every individual of certain crosses. It is in this replication of a desirable genotype that the chief merit of the method lies. When a good cross has been found, its genotype can be produced in any required number of individuals and, by repeating the cross, in successive generations. Furthermore, the genetic identity of the F_1 individuals gives them a phenotypic uniformity which is an economic benefit, particularly for mechanical harvesting. For example, uniformity of ripening time means that all individuals are ready for harvesting at the same time. Though the genotype of a cross might be found in an individual of the base population, the replication of the genotype in the cross allows the genotypic value to be measured with little error; whereas the genotypic value of an individual in the base population is only crudely measured by its phenotypic value. Further, it is the genotypic value that is measured in the cross and can be reproduced indefinitely, as long as the inbred lines are maintained; whereas only the breeding value can be reproduced by selection of individuals in a non-inbred population. Therefore the condition under which inbreeding and crossing are likely to be a better means of improvement than selection without inbreeding is when much of the genetic variance of the character is non-additive.

Selection for combining ability

Ultimately the breeder is looking for the pair of inbred lines among all those available that will give the best cross. In other words, the selection is ultimately to be applied to the crosses. The amount of improvement that can be made by selection among a number of crosses depends on the amount of variation between the crosses, and on the intensity of selection as described in Chapter 11. To get a high intensity of selection requires a large number of crosses, and to get the maximum amount of variation between the crosses requires the lines to be inbred to a high level, as was shown in the previous chapter. Time and space can, however, be saved by applying some preliminary selection to the lines. This can be done in two ways. First, the lines to be used for the cross finally selected must themselves be reasonably productive as inbreds. Lines are therefore selected first for their own performance. A line's inbred performance is correlated with its performance in crosses to some extent depending on how much of the variance is due to additive genes. The correlation, however, is rather small – about 0.1 for yield in maize (Gama and Hallauer, 1977) – so the improvement of the crosses expected from selection of the lines for their performance as inbreds is not very great. Second, the value of a cross is made up of two parts, as explained in the previous chapter: the general and specific combining abilities of the two parent lines. The general combining abilities of the lines can be tested in a variety of ways, outlined in the previous chapter, without the necessity of making all possible crosses. Furthermore, a useful guide to the general combining ability can be obtained from lines that are not yet fully inbred.

The improvement made by the preliminary selection of the lines for their general combining ability comes from the additive variance in the base population. Any further improvement, making use of the non-additive genetic variance, must come from selection for specific combining ability. Here there is no way of selecting the lines by preliminary tests; the crosses must be made, from which to select the best. Since the variance of specific combining abilities is proportional to the square or higher powers of the inbreeding coefficient (equation [15.8]), the lines must have reached a high level of inbreeding before much can be gained from selection for specific combining ability.

Relative importance of general and specific combining abilities.   How much of the improvement is expected to come from general combining ability and how much from specific combining ability? If the intensity of selection applied to each is the same, the relative amount of improvement due to each will be proportional to their variances. If the lines are fully inbred, the variance of general combining ability is equal to the additive variance in the base population and the variance of specific combining ability is equal to the non-additive variance (equation [15.8]). So if the variance components in the base population are known, the relative amount of improvement from the two combining abilities can be predicted. In an open-pollinated variety of maize the additive variance of yield as a proportion of the total was 0.149, and the non-additive variance was 0.032 (Gardner, 1977). The proportion of additive variance in the total genetic variance was thus $0.149/0.181 = 0.823$. Therefore if inbreeding and crossing were applied to this population, and selection was applied to the crosses, about 80 per cent of the improvement would be expected to come from general combining ability and only about 20 per cent from specific combining ability. If these variance components are characteristic of maize populations, it seems that by far the largest part of the improvement in hybrid maize comes from general combining ability and ultimately from additive variance in the base population. This raises the question of how far yield could be improved by selection in a random-breeding population without inbreeding. Several experiments have shown that selection in open-pollinated varieties is effective. In one, the population responded for 14 generations with a total improvement of 40 per cent in yield (Gardner, 1977). An improved open-pollinated variety does not have the uniformity which is an important feature of inbred-crosses, but prior selection in the base populations is an effective way of increasing the general combining abilities of inbred lines subsequently made from them.

Synthetic populations.   When inbred lines have been made and selected for their combining abilities, no further improvement can be made to the crosses of those lines. To achieve further improvement a new set of inbred lines must be made from an improved base population. The new base population may have been improved by selection without inbreeding, or it can be constructed from the selected inbred lines. Crossing a number of selected inbred lines and allowing the F_1 and later generations to cross-pollinate at random creates a new *synthetic* population. The improved general combining ability of the lines, being based on additive variance, is retained in the synthetic population. Segregation in the F_2 and later generations then allows a new set of inbred lines to be made with gene combinations different from those of the lines used to construct the population. In this way, further improvement of combining abilities can be achieved. The hybrid maize currently in use is the product of two or more such cycles of inbreeding and crossing. In addition to the improvement of the hybrids, the yield of the lines as inbreds is also improved, which is important economically because the hybrid seed sold for commercial growing must be produced by an inbred parent.

Three-way and four-way crosses; backcrosses
    The practical difficulties associated with the low productivity of inbred lines can be overcome by the use of 3-way and 4-way crosses, though with some loss of performance and of uniformity in the crosses. These crosses were widely used for the production of hybrid maize until the improved inbreds mentioned above were available. In a *3-way cross* the F_1 of two lines is used as female or seed parent, in which high productivity is required, and the F_1 is then crossed with a third line. In a *4-way cross*, or *double-cross*, two F_1's of different pairs of lines are crossed. If lines derived from different base populations are available, then in order to make use of the inter-population heterosis the final cross is made between lines of different origin. For example, if lines A and B are from one origin and lines P and Q from another origin, a 4-way cross is made as A × B and P × Q, followed by AB × PQ. The performance of 3-way and 4-way crosses can be reliably predicted from the performances of single crosses of the constituent lines, provided there is no epistatic interaction. Consider for example the 3-way cross (A × B) × P. Let these letters represent alleles at a single locus carried by the corresponding lines. The F_1 of A × B then has the genotype AB, which when crossed with line P gives

two genotypes, AP and BP in equal proportions. These are the genotypes of crosses of lines A × P and B × P respectively. Therefore the mean performance of these two crosses predicts the performance of the 3-way cross. In the same way, the performance of the 4-way cross (A × B) × (P × Q) is predicted by the mean of A × P, A × Q, B × P, and B × Q. If more than one locus is considered, however, segregation in the F_1 parents produces genotypes in the final cross that could not appear in any single cross of the lines used. Therefore if there is epistatic interaction, the single crosses will not predict the final cross accurately.

The population produced by any particular 3- or 4-way cross is a mixture of genotypes, all of which could in principle have been produced by single crosses, but 3- and 4-way crosses are expected to differ from single crosses in the following ways: (1) If the lines crossed have been selected, and if any of the consequent superiority of their single crosses is due to epistatic interactions, some of this superiority will be lost in the 3- and 4-way crosses. (2) There is genetic variation within crosses and a consequent loss of phenotypic uniformity. (3) The variance between crosses is reduced and the best 3- or 4-way cross is consequently not as good as the best single cross. For experimental comparisons of 3- and 4-way crosses with single crosses, illustrating these consequences in maize, see Weatherspoon (1970), and Otsuka, Eberhart, and Russell (1972).

Another way of avoiding the low productivity of inbred lines is by a *backcross*. Here only two lines are involved, the F_1 being mated to one of the lines used in the first cross, i.e., (A × B) × B. The genotypes in the progeny of the backcross are AB and BB in equal proportions. Therefore, in the absence of epistatic interactions, the mean of the backcross is equal to the mean of the F_1 and the line used in the second cross. Consequently there is less heterosis in backcrosses than in 3-way or 4-way crosses.

Crosses in animals. Crossing is widely used in animal production, most of the animals produced for meat being the progeny of either a 3-way cross or a backcross. As was noted earlier, the lines crossed are not deliberately inbred. They have, however, been previously selected for desirable characters and have become mildly inbred with a consequent reduction of some desirable characters, particularly fertility. The purpose of the crossing is partly to make use of heterosis to improve fertility and partly to combine the different characteristics for which the lines were previously selected. For meat production a desirable quality in the final product is rapid growth and what is desired of the final cross is to produce large numbers of rapidly growing individuals. This requires good fertility in the mother coupled with good growth rate in the progeny. Accordingly, the first cross (A × B) is made to produce F_1's with good fertility, which comes from heterosis. These F_1 (AB) individuals are used as mothers and crossed to a third line (C) with good growth rate to produce the (A × B) × C progeny. Or if no suitable third line is available, the F_1 is backcrossed to one of the lines used in the first cross. The improved growth of the final progeny comes partly from heterosis and partly from the additive effects of the sire line. Their growth rate may not always be as good as the best of the lines, but the increased numbers produced by the fertile AB mothers makes the crossing economically advantageous. The expected gains from different types of cross are reviewed by Dickerson (1969).

Example 16.1 A 3-way cross of sheep breeds will serve to illustrate the gain from combining the heterosis of fertility with the superior growth of the sire-line. The data come from Sidwell, Everson, and Terrill (1962, 1964). The table gives the pure-bred and 3-way cross performances of the three characters: (a) fertility as the number of lambs weaned per ewe mated; (b) growth rate as the weight per lamb at weaning; and (c) the economically important character total weight of lamb weaned per ewe mated, which is the product (a) × (b). The weaning weight of the cross was not as good as the sire-line itself, but the larger number of lambs weaned by the F_1 females made the 3-way cross superior to the best pure breed in respect of the total weight of the weaned lambs.

	Production per ewe mated		
	(a) No. of lambs weaned	(b) Weaning wt. (kg) per lamb	(c) Total wt. (kg) weaned
Pure breeds			
A = Shropshire	0.80	23.0	18.4
B = Southdown	0.79	19.1	15.1
C = Hampshire	1.00	29.2	29.2
Mid-parent ($\frac{1}{4}$A + $\frac{1}{4}$B + $\frac{1}{2}$C)	0.90	25.1	22.6
3-way cross (A × B) × C	1.25	27.5	34.4
Heterosis, % above mid-parent	39	10	52
Superiority over best breed (%)	+ 25	− 6	+ 18

Reciprocal recurrent selection (RRS)

The specific combining ability of a cross cannot be measured without making and testing that particular cross. Therefore to achieve a reasonably high intensity of selection for specific combining ability, a large number of crosses must be made and tested. Is no short-cut possible? Could the superior combining ability not be, as it were, built into the lines by selection? From the causes of heterosis explained in Chapter 14 it is clear that what is wanted is a pair of lines that differ widely in the gene frequencies at all loci that affect the character and that show dominance. It should therefore be possible to build up these differences of gene frequency in two lines by selection. Instead of the differences of gene frequency being produced by the random process of inbreeding, they would be produced by the directed process of selection, which would be both more effective and more economical. Furthermore, both general and specific combining ability would be selected for simultaneously. Selection for combining ability in this way is known as *reciprocal recurrent selection*. Its theoretical basis has been examined by Comstock, Robinson, and Harvey (1949) and Dickerson (1952). In outline, the procedure is as follows.

The start is made from two populations, preferably two already known to give some heterosis when crossed. These two populations, whose combining ability is to be improved, will be referred to as lines A and B. Crosses are made reciprocally, a number of A ♂♂ being mated to B ♀♀, and a number of B ♂♂ to A ♀♀. The crossbred progeny are then measured for the character to be improved and the parents are judged from the performance of their progeny. The best parents are selected and the rest discarded, together with all the crossbred progeny, which are used only to test the combining ability of the parents. The selected individuals must then be remated, to members of their own line, to produce the next generation of parents to be tested. These are crossed again as before and the cycle repeated. Deliberate inbreeding is avoided because random changes of gene frequencies are not desired.

An essential prerequisite is that there should be some difference of gene frequency between the two lines at the beginning, or else selection for combining ability will be unable to produce a differentiation of the lines. Any locus at which the gene frequencies are the same in the two lines will be in equilibrium, though an unstable equilibrium. Any shift in one direction or the other will give the selection something to act on and the difference will be increased. The initial difference between the lines may be obtained by starting from two different breeds or varieties, choosing two that already cross well; or by deliberate inbreeding, up to perhaps 25 per cent, and relying on random differentiation of gene frequencies.

Evidence about the practical value of reciprocal recurrent selection is conflicting. It is used by some commercial breeders of poultry for egg production (see Krosigk *et al.*, 1973), and has given promising results with maize (Eberhart, 1977). On the other hand, direct comparisons with other methods of selection in poultry and in laboratory animals have not been encouraging (see Calhoon and Bohren, 1974; McNew and Bell, 1976).

Overdominance

The question of whether inbreeding and crossing is a better method of improvement than selection without inbreeding hinges on overdominance as a property of the genes concerned. Overdominance for fitness was discussed in Chapter 2 as a mechanism for the maintenance of polymorphism, and the different ways in which true overdominance and pseudo-overdominance due to linkage can arise were explained. Here we are concerned with overdominance for the character to be improved and in practice it matters little how the overdominance arises. Both methods of improvement involve selection, as we have already seen, so the essential distinction is in the crossing. Crossing two lines in which different alleles are fixed gives an F_1 in which all individuals are heterozygotes; and this is the only way of producing a group of individuals that are all heterozygotes. In a non-inbred population no more than 50 per cent of the individuals can be heterozygotes for a particular pair of alleles. Consequently, if heterozygotes of a particular pair of alleles are superior in merit to homozygotes, inbreeding and crossing will be a better means of improvement than selection without inbreeding. Furthermore, it is only when there is overdominance with respect to the desired character, or combination of characters, that inbreeding and crossing can achieve what selection without inbreeding cannot. Under any other conditions of dominance the best genotype is one of the homozygotes and all individuals can, in theory, be made homozygous by selection, without the disadvantages attendant on inbreeding and much more simply than by methods dependent on crossing. It was stated earlier in this chapter that the potentialities of inbreeding and crossing are greatest when there is much non-additive genetic variance and little additive. Now we see that this is only part of the truth: in theory, and leaving all practical considerations aside, inbreeding and crossing can surpass selection without inbreeding only when there is at least some degree of overdominance of the genes concerned.

A variety of experimental work on both plants and animals, some of which has been mentioned in earlier chapters, suggests that overdominance for the characters studied is not an important property of the genes. This is true even of yield in maize for which inbreeding and crossing has been so successful (Eberhart, 1977). There seems, therefore, to be little theoretical justification for believing that inbreeding and crossing is a better way of increasing the mean of the desired character. Its advantages are mainly in the uniformity rather than in the improved mean.

Naturally self-fertilizing plants

Self-fertilizing plants usually show some heterosis when crosses are made. The reason for this is presumably that deleterious genes arise by mutation and some are fixed by the inbreeding despite natural selection against them. To make use of the heterosis for commercial growing is, however, not easy because making the crosses is usually difficult technically, and to produce hybrid seed requires crossing on a large scale in every generation. There are several ways of overcoming the technical difficulties of crossing but these have been successful with only a few crops (see Simmonds, 1979, p. 231). The purpose of the crossing is therefore usually to generate segregation. After a cross has been made, the F_1 and subsequent generations are allowed to self-fertilize, producing a new set of inbred lines which become differentiated by recombination and random drift. The aim is to find one or more of these recombinant lines that is better than either of the parental lines. The amount of improvement to be expected from any particular cross can be predicted from the additive genetic variance in the F_2 and the intensity of selection that will be applied to the recombinant lines (see Jinks and Pooni, 1976; Pooni and Jinks, 1979).

17 SCALE

The choice of a suitable scale for the measurement of a metric character has been mentioned several times in the foregoing chapters. The explanation of what is involved in the choice of a scale and a discussion of the criteria of suitability have, however, been deferred till this point because these are matters that cannot be properly appreciated until the nature of the deductions to be made from the data are understood. In other words, the choice of a scale has to be made in relation to the object for which the data are to be used. The data from any experimental or practical study are obtained in the form most convenient for the measurement of the character. That is to say, the phenotypic values are recorded in grams, centimetres, days, numbers, or whatever unit of measurement is most convenient. The point at issue is whether these raw data should be transformed to another scale before they are subjected to analysis or interpretation. A transformation of scale means the conversion of the original units to logarithms, reciprocals, or some other function, according to what is most appropriate for the purpose for which the data are to be used.

It is tempting to suppose that each character has its 'natural' scale, the scale on which the biological process expressed in the character works. Thus, growth is a geometrical rather than an arithmetical process, and a geometric scale would appear to be the most 'natural'. For example, an increase of 1 g in a mouse weighing 20 g has not the same biological significance as an increase of 1 g in a mouse weighing 2 g: but an increase of 10 per cent has approximately the same significance in both. For this reason a transformation to logarithms would seem appropriate for measurements of weight. This, however, is largely a subjective judgement, and some objective criterion for the choice of a scale is needed. Different criteria, however, are often inconsistent in the scale they indicate and, moreover, the same criterion applied to the same character may indicate different scales in different populations. Therefore the idea that every character must have its 'natural' and correct scale is largely illusory.

There are, broadly speaking, three main reasons for making a scale transformation: (1) to make the distribution normal; (2) to make the variance independent of the mean; and (3) to remove or reduce non-additive interactions. The criterion for the choice of a scale is in each case the empirical one of achieving the particular objective. When a scale transformation is called for but is not made, certain phenomena arise, called *scale effects*, which disappear when the appropriate transformation is made. The objectives noted above might equally well be stated as being the removal of these scale effects. We shall discuss in particular the logarithmic transformation which converts an arithmetic to a geometric scale. This is probably the commonest and most useful transformation. Other transformations are described by Wright (1968, Ch. 10). The general principles, outlined by reference to the log transformation, apply equally to other transformations.

Distribution and variance

Consider first the distribution of phenotypic values. Figure 17.1 shows three distributions plotted as if from the original data on an arithmetic scale. They would all three be symmetrical and normal if the data were first transformed to logarithms. There are two points of importance to notice. First, the degree of departure from normality depends on the amount of variation in relation to the mean. This may be seen from a comparison of the two upper graphs, (a) and (b), which are not very noticeably asymmetrical, with the lower graph, (c), which is. The relationship between the amount of variation and the mean, which determines the degree of departure from normality, is best expressed as the coefficient of variation, i.e., the ratio of standard deviation to mean, often multiplied by 100 to bring it to a percentage. The coefficient of variation of the two upper graphs is 20 per cent, while that of the lower graph is 50 per cent. Thus, a transformation to logarithms does not make an appreciable difference to the shape of the distribution unless the coefficient of variation is fairly high – that is, above about 20 per cent or so. Consequently, statistical procedures which do not

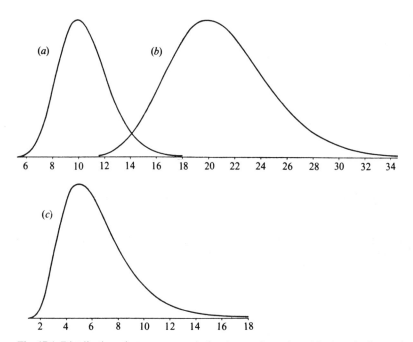

Fig. 17.1. Distributions that are symmetrical and normal on a logarithmic scale shown plotted on an arithmetic scale. Explanation in text.

rely on a strictly normal distribution, such as the analysis of variance, can be carried out on the untransformed data when the coefficient of variation is not above about 20 per cent. Transformations to other scales are also less necessary when the coefficient of variation is low than when it is high.

The second point to notice in Fig. 17.1 is that the variance, when computed in arithmetic units, increases when the mean increases. This may be seen in graphs (a) and (b). These both have the same variance in logarithmic units, but different means. The mean – or strictly speaking the mode – of (b) is double that of (a) and the standard deviation in arithmetic units is correspondingly doubled. Though the distributions are not very noticeably skewed and a transformation does not seem to be very strongly indicated, yet in consequence of the difference of mean the variances differ very greatly. Here, then, is one of the commonest scale effects, namely a change of variance following a change of the population mean. The two graphs (a) and (b) in Fig. 17.1 might well represent two populations which have diverged by some generations of two-way selection, if the character were something like body size measured in units of weight. Such characters are commonly found to increase in variance when the mean increases, and to decrease in variance when the mean decreases. Figure 17.2 shows an example from an experiment with mice, the character being weight at 60 days. Note that none of the three distributions considered separately seems to be sufficiently asymmetrical or non-normal to need a scale transformation on this criterion; but

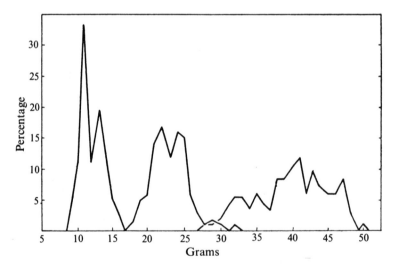

Fig. 17.2. Distributions of body weight of male mice at 60 days. Centre: base population before selection. Left and right: small and large strains after 21 generations of two-way selection. (*Based on MacArthur, 1949.*)

	Small	Unselected	Large
Mean	11.97	23.16	39.85
Standard deviation	1.71	2.56	5.10
Coeff. of variation, %	14.3	11.1	12.8

to make the variance independent of the mean, a transformation is very obviously required.

Phenomena such as the change of variance discussed above are called scale effects if they disappear when the measurements are appropriately transformed: in other words, if their cause can be attributed to the scale of measurement. But they are none the less real, though labelled as a scale effect or removed by transformation. The large mice, for example, are really more variable than the small when their weights are measured in grams. What is gained by recognizing this as a scale effect is that there is no need to look deeper into the genetic properties of the character for an explanation.

A convenient test for the appropriateness of a logarithmic transformation is provided by the proportionality of standard deviation and mean, which we noted in connection with graphs (a) and (b) in Fig. 17.1. If two distributions have the same variance on a logarithmic scale then the coefficients of variation in arithmetic units will be the same. Thus, constancy of the coefficient of variation indicates constancy of variance on a logarithmic scale. And, if variances are to be compared, we may simply compare the coefficients of variation instead of expressing the variances in logarithmic units. The standard deviations and coefficients of variation of the distributions shown in Fig. 17.2 are given in the legend to the figure. The coefficients of variation, though not identical, are much more alike than the standard deviations, and this shows that the changes of variance that have resulted from the selection can be attributed, in large part at least, to the scale of measurement.

When a logarithmic transformation is required, it is not always necessary to convert each individual measurement. Conversion of the mean and of the variance can conveniently be made by the following formulae (Wright, 1968, p. 229):

$$(\overline{\log x}) = \log \bar{x} - \tfrac{1}{2}\log (1 + C^2) \qquad \ldots [17.1]$$

$$\sigma^2_{(\log x)} = 0.4343 \log (1 + C^2) \qquad \ldots [17.2]$$

The first converts the mean of arithmetic values to the mean of logarithmic values, and the second converts the variance as computed from the arithmetic values to the variance as it would be computed from logarithmic values. In these formulae C is the coefficient of variation in the form $\sigma/\bar{x}$ computed from arithmetic values, and the logarithms are to the base 10. The formulae are accurate only if the distribution really is normal on the logarithmic scale.

When conclusions about variances depend critically on eliminating any scale effect, it may be necessary to find the empirical relationship between variance, or standard deviation, and mean. This can be done if several populations with different means are available, and if there are no reasons other than the scale effect for thinking that their variances would differ. Then the regression equation relating the standard deviation to the mean gives the expected standard deviation in another population with a particular mean. If the regression is linear the regression equation is $\sigma = a + b\bar{x}$, where σ is the expected standard deviation in a population with a mean of $\bar{x}$; a is the intercept and b is the regression coefficient. A scale on which the variance would be independent of the mean would be

$X = \log (x + a/b)$, where x is the original measurement and X its transformed value (Wright, 1968, p. 232).

Let us return to the consequences of selection and pursue them a little further. If the variance changes with the change of mean as a result of selection, so also will the selection differential and the response. The response per generation of a character such as we have been considering would therefore be expected to increase with the progress of selection in the upward direction, and to decrease correspondingly in the downward direction. The response to two-way selection would then be asymmetrical. An example of an asymmetrical response which can most probably be attributed to a scale effect in this way is shown in Fig. 17.3. Plotted in arithmetic units, as in (a), the response is much greater in the upward than in the downward direction. A transformation to logarithms, shown in (b), renders the response much more nearly symmetrical. This does not do away with the fact that the character as measured increased much more than it decreased under selection. But it accounts for the asymmetry without the need for more elaborate hypotheses. A convenient way of eliminating scale effects from the graphical presentation of a response to selection is to plot the response in the form of the realized heritability, as explained in Chapter 11 and illustrated in Fig. 11.5. The realized heritability, which is the ratio of response to selection differential, is very little influenced by scale effects (Falconer, 1954).

Interactions

We turn now to what is perhaps a more fundamental effect of a scale transformation – its effect on the apparent nature of the genetic variance. To understand this we must go back to a single locus and consider the effect, or mode of action, of the genes. Imagine a locus with two alleles whose mode of action is geometric, the genotypic value of A_2A_2 being 50 per cent greater than A_1A_2, and that of A_1A_2 being also 50 per cent greater than A_1A_1. Thus on the logarithmic

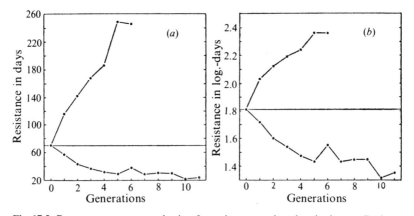

Fig. 17.3. Response to two-way selection for resistance to dental caries in rats. Resistance is measured in days and plotted on an arithmetic scale in (a), and on a logarithmic scale in (b). The arithmetic means were converted to logarithmic means by formula [17.1]. The coefficient of variation was high – about 50 per cent – and was approximately constant. The reason why the upward selection has not covered so many generations as the downward is simply that the increased resistance lengthened the generation interval. (*Data from Hunt, Hoppert, and Erwin, 1944.*)

scale there is no dominance, the heterozygote being exactly mid-way between the two homozygotes. Now suppose the genotypic values are measured in arithmetic units, such as grams, and that A_1A_1 has a value of 10 units. Then A_1A_2 will be 15 units and A_2A_2 22.5 units. On the arithmetic scale, therefore, A_1 is partially dominant to A_2, the heterozygote no longer falling mid-way between the homozygotes. Thus the degree of dominance is influenced by the scale of measurement, and so also is the proportionate amount of dominance variance. This effect of a scale transformation, however, is normally rather small. A gene that causes a 50 per cent difference between the genotypic values, such as we have considered, would be a major gene, easily recognizable individually. But even so, the degree of dominance on the arithmetic scale is not very great. Minor genes with effects of perhaps 1 per cent or 10 per cent would be scarcely influenced in their dominance.

In the same way that the dominance is affected by the scale, so also is the epistatic interaction between different loci. Loci with geometric effects would combine without interaction if the genotypic values were measured in logarithmic units. But when measured in arithmetic units there would be interaction deviations due to epistasis. Thus the amount of interaction variance is also influenced by the scale of measurement. The following example illustrates the dependence of interaction on scale.

Example 17.1 The pygmy gene in mice is a major gene affecting body size, homozygotes being much reduced in size. The effect of this gene was studied in different genetic backgrounds (King, 1955). The gene was transferred from the strain selected for small size where it arose, to a strain selected for large size, by repeated backcrosses. The mean difference between pygmy homozygotes and normals (i.e., heterozygotes and normal homozygotes together) was measured in the two strains and during the transference, the comparisons being made between pygmies and normals in the same litters. The results are shown in Fig. 17.4. The difference between pygmies and normals increases with the weight of the normals. In the background of the small strain the pygmies were about 7 g smaller than normals, but in the background of the large strain they were about 14 g smaller. Thus the pygmy gene shows epistatic interaction with the other genes that affect body size. But if the effect of the gene is expressed as a proportion, it is constant and independent of the other genes present. Pygmies are about half the weight of their normal litter-mates, no matter what the actual weights are. Thus if the comparisons are made in logarithmic units there is no epistatic interaction.

In general, therefore, a scale transformation may remove or reduce the variance attributable to epistatic interaction, and this variance might then be labelled as a scale effect. A transformation which removes or reduces interaction variance may be useful if conclusions are to be drawn from an analysis that depends for its validity on the absence of interaction. Interactions between genotype and environment may also arise from a scale effect, and a transformation may be useful for removing or reducing them. Interactions, whether epistatic or genotype × environment, however, cannot always be removed or even reduced by a transformation of scale. For example, no meaningful transformation can remove an interaction that causes a reversal of order, such as was illustrated in Fig. 8.2.

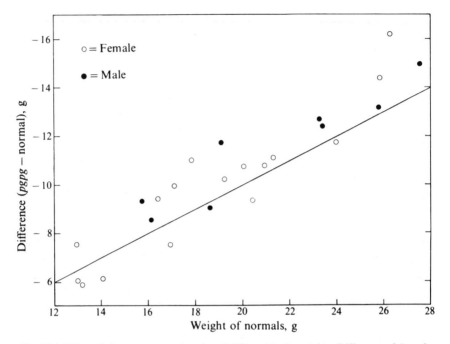

Fig. 17.4. Effect of the pygmy gene in mice of different body weights. Difference of 6-week weight between pygmies and their normal litter-mates plotted against the weight of the normals in the same litter. The straight line is not a fitted regression, but shows the relationship: weight of pygmy = 0.5 × weight of normal. See Example 17.1. (*Data from King, 1955.*)

There are two ways by which a suitable scale for removing interactions can be found. The first is by comparing the effects of a specific factor and finding a scale that makes the effect additive. This was illustrated in Example 17.1 above. The effects of a single gene were compared in different genetic backgrounds, and it was found that on a logarithmic scale the gene added the same amount on all genetic backgrounds. The specific factor whose effects are to be made additive can be environmental rather than genetic. The second test of a suitable scale for removing epistatic interaction can be applied when two populations with different means are available and can be crossed. It was shown in Chapters 14 and 16 that in the absence of epistatic interaction the means of the F_2 and backcross generations were expected to be as follows:

$$\left.\begin{array}{l} F_2 = \frac{1}{2}(F_1 + \bar{P}) \\ B_1 = \frac{1}{2}(F_1 + P_1) \\ B_2 = \frac{1}{2}(F_1 + P_2) \end{array}\right\} \quad \ldots [17.3]$$

where $\bar{P}$ is the mean of the two parental populations and all the other symbols are the means of the corresponding generations. A scale is chosen which brings the observed means closest to their expectations. For details, see Mather and Jinks (1971, p. 71).

Conclusions
In this chapter we have outlined some of the scale effects most commonly met

with, and have indicated the circumstances under which a transformation of scale may be helpful to the interpretation of results and the drawing of conclusions. Transformations of scale, however, should not be made without good reason. The first purpose of experimental observations is the description of the genetic properties of the population, and a scale transformation obscures rather than illuminates the description. If epistasis, for example, is found, this is an essential part of the description, and it is better labelled as epistasis than as a scale effect. The transformation of scale is essentially a statistical device to be employed for the purpose of simplifying the analysis of the data, or to make possible the drawing of valid conclusions from the analysis. It is sometimes helpful also in the interpretation of results. If epistasis, for example, were found to disappear on transformation to a logarithmic scale we could conclude that the effects of different loci combined by multiplication rather than by addition. Or, if there were good reasons for attributing a difference of variance to a scale effect we should not need to invoke more complicated genetic explanations. The choice of scale, however, raises troublesome problems in connection with the interpretation of results. Logical justification of a scale transformation can only come from some criterion other than the property about which the conclusions are to be drawn. If there is no independent criterion the argument becomes circular, and the distinction between a scale effect and some other interpretation becomes meaningless. There is also a more fundamental difficulty: the scale appropriate for one population may not be appropriate for another, and the scale appropriate to the genetic and environmental components of the variation may be different. This difficulty is strikingly illustrated by an analysis of the character 'weight per locule' in a number of crosses between varieties of tomato (Powers, 1950). By the same criterion – normality of the distribution – this character was found to require an arithmetic scale in some crosses and a geometric scale in others; and, moreover, in the F_2 generations of some crosses the genetic variation required one scale while the environmental variation required another.

18 THRESHOLD CHARACTERS

There are many characters of biological interest or economic importance which vary in a discontinuous manner but are not inherited in a simple Mendelian manner. Familiar examples are susceptibility to disease, where there are two phenotypic classes – affected or not-affected – and litter size of the larger mammals that usually bear one young at a time but sometimes two or three. There are also discontinuous anatomical differences, such as the number of vertebrae of mice, whose genetics has been extensively studied. Characters of this sort appear at first sight to be outside the realm of quantitative genetics; yet when they are subjected to genetic analysis they are found to be inherited in the same way as continuously varying characters.

Liability and threshold

The clue to understanding the inheritance of such characters lies in the idea that the character has an underlying continuity with a *threshold* which imposes a discontinuity on the visible expression, as depicted in Fig. 18.1. When the underlying variable is below this threshold level the individual has one form of phenotypic expression, e.g., is 'normal'; when it is above the threshold the individual has the other phenotypic expression, e.g., is 'affected'. The underlying continuous variable has been called the *liability* in the context of human diseases as threshold characters, and this term will be used here. The continuous variation of liability is both genetic and environmental in origin, and may be thought of as the concentration of some substance, or the rate of some developmental process – of something, that is to say, that could in principle be measured and studied as a metric character in the ordinary way. It may be a compound of several different physiological or developmental processes but it is not necessary to know how these are combined to give the liability, or even to know what they really are.

That the idea of an underlying variable is a realistic one can be appreciated by thinking of litter size. The litter size of mice or pigs, though in reality obviously discontinuous, can be treated as a continuous variable because there are a large enough number of classes. The litter size of cows has only two classes, single and twin births, more than two calves being exceedingly rare. But there is no reason to think that the physiological causes of twinning in cattle are different from those of litter size in mice or pigs. The underlying variable in both cases is made up mainly of the levels of circulating gonadotrophic hormones, which determine the number of eggs shed, the intra-uterine factors that affect embryonic survival and, in the case of cattle, the factors determining monozygotic twinning.

Two classes, one threshold

Let us first consider characters which have only two phenotypic classes with a single threshold separating them. The two classes will be referred to as normal and affected. On the phenotypic level, or visible scale, individuals can have only

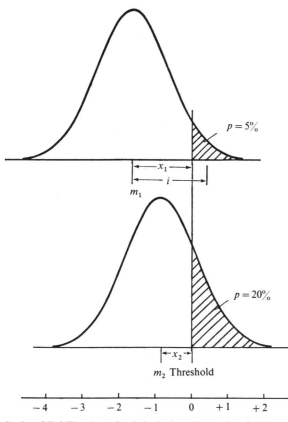

Fig. 18.1. Two populations or groups with different incidences, p, of a threshold character and consequently different mean liabilities. The variance of liability is the same in the two groups and the means differ by 0.8 standard deviations of liability. x is the normal deviate of the threshold from the mean; i is the mean deviation of affected individuals from their group mean.

two possible values which might be designated 0 for normal and 1 for affected. Groups of individuals, however, such as families or the population as a whole, can have any value, in the form of the proportion or percentage of individuals that are affected. This is referred to as the *incidence* or, in the context of human diseases, the *prevalence*. The incidence is quite adequate as a simple description of the population or group, but the percentage scale in which the incidence is expressed is inappropriate for many purposes, because on a percentage scale variances differ according to the mean. For genetic analyses, therefore, incidences must be converted to mean liabilities. In order to make this transformation it is necessary to define the liability as being normally distributed. This definition carries with it the requirement that if we could measure the liability directly, it would be possible to render its distribution normal by some scale transformation. We shall return to this requirement later; meantime we assume that it can be met. With liability being normally distributed, then, the unit of liability is its standard deviation σ. The mean liability is then related to the incidence by the (single-tailed) normal deviate x, which is the deviation of the threshold from the mean in standard-deviation units of liability. Values of x for different incidences are tabulated in Appendix Table A.

Comparison of means. Consider two populations or groups with different incidences, as shown in Fig. 18.1. By how much do they differ in mean liability? For comparing different groups, the threshold must be defined as being fixed; that is to say, it is at the same level of liability in all groups. The mean of any group is then expressed as a deviation in σ units from the threshold. In other words, the threshold is taken as the origin or zero-point on the scale of liability. The upper group in Fig. 18.1 has an incidence $p_1 = 0.05$, which gives $x_1 = 1.6\sigma_1$. The mean liability is thus $m_1 = -1.6\sigma_1$. (Care must be taken with the signs of mean liabilities. If the threshold is above the mean, the mean is below the threshold and therefore negative.) Similarly, the lower group in Fig. 18.1, with an incidence of 0.20, has a mean of $m_2 = -0.8\sigma_2$. Note that the means are expressed in units of their own population's standard deviation. We can go no further with the comparison of the means unless we assume that the two standard deviations are the same. If this can be accepted as a reasonable assumption then $\sigma_1 = \sigma_2 = \sigma$, and $m_2 - m_1 = 0.8\sigma$; the means of the two groups differ by 0.8 standard deviations of liability.

Heritability of liability. Suppose that the upper distribution in Fig. 18.1 represents a parental generation from which affected individuals are selected as parents. When these parents are bred they produce offspring with the lower distribution. Knowing the incidence in the parental generation and in the progeny, we have all that is needed to calculate the regression of offspring on mid-parent values of liability, and from this the heritability of liability. Consider first the response to the selection. This is the difference in mean between the parental and progeny generations which was given above as 0.8σ assuming the variances of the two generations to be equal. In fact the variances will not be quite the same, but the small error introduced will be neglected for the moment. Now consider the selection differential. This is the mean liability of the affected individuals in the parent generation as a deviation from their population mean. The proportion of individuals used as parents may be less than the incidence: in other words, not all affected individuals may be used as parents. But so long as all parents are affected, the mean of those used is expected to be the same as the mean of all affected individuals. The mean of the affected individuals in standard deviation units is equivalent to the intensity of selection, i, corresponding to the incidence as the proportion selected. Values of i are tabulated in Appendix Table A. With the incidence of $p = 0.05$ the intensity of selection is $i = 2.1$, and the selection differential is therefore $S = 2.1\sigma$ (see equation [11.5]). The regression of offspring on mid-parent values is the ratio of response to selection differential (equation [11.1]) and is $R/S = 0.8\sigma/2.1\sigma = 0.38$. Finally, provided there is no environmental resemblance between offspring and their parents, the heritability of liability is equal to the regression of offspring on mid-parent values.

The calculation of the regression and heritability explained above by reference to parents and offspring can be applied to any sort of relationship. Suppose that the upper distribution in Fig. 18.1 represents a population, and the lower distribution represents any specified sort of relatives, e.g., full sibs, of affected individuals. Then the regression calculated is that of an individual on his relative. Since the variances of the two groups – the population and the relatives – are approximately equal, the regression is approximately equal to the correlation. Thus, with the incidences in Fig. 18.1 the correlation of full sibs in respect of liability would be 0.38. In the absence of resemblance due to common environment and of dominance, the heritability would be estimated as twice the correlation of full sibs, or 0.76.

The calculations explained above can be summarized in the following formulae. The correlation of liability between relatives of any specified sort is given by

$$t = \frac{m_R - m_P}{i} = \frac{x_P - x_R}{i} \qquad \dots [18.1]$$

where the subscripts P and R refer to the population and the relatives respectively, m is the mean as a deviation from the threshold, x is the normal deviate of the threshold from the mean, and i is the mean deviation of affected individuals from the population mean. The signs are appropriate to a scale that assigns a higher liability to affected than to normal individuals. The heritability is then obtained from the correlation as

$$h^2 = t/r \qquad \dots [18.2]$$

as in equation [10.5], r being the coefficient of relationship as given in Table 9.3. For first-degree relatives r is $\frac{1}{2}$; for second-degree, $\frac{1}{4}$; when the relatives are offspring of parents both of which are affected, r is 1.

The error introduced by assuming the variance to be the same in the relatives of affected individuals as it is in the population as a whole leads to the correlation estimated by equation [18.1] being too low by a factor of 5 or 10 per cent. A modified formula that takes account of the unequal variances is the following (Reich, James, and Morris, 1972):

$$t = \frac{x - x_R \sqrt{[1 - (x^2 - x_R^2)(1 - (x/i))]}}{i + x_R^2(i - x)} \qquad \dots [18.3]$$

where x and i without subscript refer to the population, and x_R refers to the relatives; the sign of the square root is taken to make t between 0 and 1.

Example 18.1 Cryptorchidism is a congenital defect of males that occurs in some herds of pigs. The data in the table refer to one herd reported by Mikami and Fredeen (1979). The incidence in the herd, i.e., the population, was 3.9 per cent and the incidence among the full sibs of affected males was 11.6 per cent. The corresponding values of x and i taken from Appendix Table A are given in the table. The full-sib correlation is calculated by equation [18.1] as follows:

$$t = \frac{1.762 - 1.195}{2.165} = 0.26$$

Assuming no common environment and no dominance, the heritability of liability, by equation [18.2], is

$$h^2 = 2t = 0.52$$

	Numbers		Incidence		
	Affected	Total	p%	x	i
Population (P)	44	1,129	3.9	1.762	2.165
Full sibs (R)	25	215	11.6	1.195	

Calculated by the more accurate formula in equation [18.3], the correlation is $t = 0.28$ giving $h^2 = 0.56$.

The estimate of the heritability was used to calculate the effectiveness of different methods of selection aimed at reducing the incidence. Half-sib family selection was found to be the best, and individual selection the worst. To reduce the incidence from 5 per cent to 1 per cent would require about 3 generations by culling entire half-sib families, but over 50 generations by culling only affected males.

The calculation of the heritability of liability has been widely applied in the study of the inheritance of human diseases. The data are collected by questioning patients with a particular disease about the disease status of their relatives. The correlation in respect of liability can then be calculated as above. The interpretation of the correlation in terms of the heritability, however, is subject to the uncertainties about resemblances due to common environment that were emphasized in Chapter 10. Estimates obtained for the heritability of liability, assuming no environmental resemblance, range from 85 per cent for schizophrenia to 35 per cent for congenital heart diseases (Emery, 1976, p. 54). Knowledge of the heritability is useful in genetic counselling for calculating recurrence risks in families because it allows all the information about the family to be combined correctly. For further details about the application to human diseases, see Falconer (1965a, 1967), and Curnow and Smith (1975).

Adequacy of the liability model

The definition of liability as a normally distributed continuous variable must be examined more closely. It implies, as noted earlier, the assumption that the underlying variables that combine to give the liability could be made normal by a scale transformation. This in turn implies that the distribution of liability is unimodal. If in reality it is bimodal or multimodal, no reasonable scale transformation could make its distribution normal. The calculations of correlations and heritabilities would then be invalid. A bimodal or trimodal distribution could arise in two main ways; first, if there was a single gene whose effect on liability was fairly large in relation to the residual variation, and second, if there were an environmental factor with an effect large in relation to the other variation. Such an environmental factor affecting liability to a disease might be exposure to a pathogen. Thus the genetic analyses in terms of liability are valid only if liability is multifactorial, which means that there are many causes of variation, all with relatively small effects, and the genetic control is by genes at more than one or a few loci.

There is no means of knowing in advance whether these requirements for valid analyses are met, because liability cannot be measured to see if its distribution is unimodal. One can, however, see whether the results obtained are reasonable and consistent. For example, a heritability in excess of 100 per cent would obviously be unacceptable, and would suggest a single major gene. Also, in the absence of resemblance due to common environment, the heritability estimated from different sorts of relatives should be the same. On the whole, the results obtained have been reasonably consistent and have given no strong reason to doubt the adequacy of the liability model. A method of analysis which tests the consistency of different sorts of relatives, and at the same time detects a single major gene, and environmental resemblance if present, is known as complex segregation analysis (see Morton and MacLean, 1974; Lalouel *et al.*, 1977), but it is too complicated for description here. A different test of adequacy can be made with characters that have two thresholds, which will be described below.

Scale relationships

It is possible to assign arbitrary values, 0 and 1, to the two phenotypic classes of a threshold character and to calculate the correlation between relatives in respect of these values. To do this is like using the phenotypic expression as a very coarsely graduated instrument for measuring the liability; an instrument, in fact, with only one graduation mark. This introduces a large amount of measurement error which appears as environmental variance if components of variance are estimated. The amount of variance due to measurement error on the (0, 1) scale depends on the incidence; it is least with an incidence of 0.5 and becomes larger with lower or higher incidences. In consequence, a correlation calculated on the (0, 1) scale varies with the incidence, becoming reduced as the incidence decreases or increases from 0.5. Transformation to the liability scale as described above removes this variance due to measurement error and renders correlations, and heritabilities derived from them, independent of the incidence. There is a simple relationship by which correlations or heritabilities can be converted from one scale to the other (Dempster and Lerner, 1950). If t_c is a correlation on the continuous scale of liability and t_{01} the same correlation calculated on the (0, 1) scale, the two are related by

$$t_c = t_{01} \left(\frac{1 - p}{i^2 p} \right) \qquad \ldots [18.4]$$

where p is the incidence of the phenotype scored as 1. The two heritabilities are related in the same way.

Three classes, two thresholds

Genetic analysis of threshold characters can be taken further if there are three phenotypic classes, provided the classes can be logically ordered with respect to liability. That is to say, provided there are biological reasons for believing that one class is intermediate in liability between the other two. An example would be single, twin and triplet births. Then the two thresholds separating the three classes are at different levels of liability. The two thresholds mark fixed points on the liability scale, which are the same in all groups. The difference in liability between the two thresholds therefore provides a fixed unit of liability which is independent of the standard deviation. This makes it possible to compare the standard deviations or variances of different groups, and to compare the means of groups that are expected to have different variances. The idea is most easily explained by a numerical example.

Consider the two populations illustrated in Fig. 18.2. They have different means and different variances. The thresholds T_1 and T_2 are fixed points on the liability scale and the difference between them is 1 *threshold unit* (t.u.) of liability. The scale of liability is shown at the bottom with the zero at the position of T_1. The first step is to express each standard deviation in terms of threshold units. This is done from the incidences as follows. Let p' be the proportion of individuals above T_1, i.e., the incidence of the intermediate and extreme classes together, and let p'' be the proportion of individuals above T_2, i.e., the incidence of the extreme class. Let x' and x'' be the deviations of T_1 and T_2 from the population mean in standard deviations. Then the difference between the thresholds, for any population, is $(x'' - x')\sigma$, where σ is the standard deviation of the population. The difference between the thresholds is by definition 1 t.u., so $(x'' - x')\sigma = 1$ t.u., and

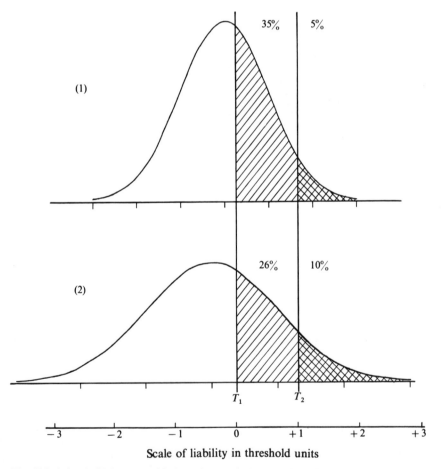

Fig. 18.2. A threshold character with three phenotypic classes and two thresholds. Distributions of liability in two populations with different means and different variances.

the standard deviation in threshold units is given by

$$\sigma = \frac{1}{x'' - x'} \text{ t.u.} \qquad \qquad \dots [18.5]$$

The calculation of the standard deviations of the populations in Fig. 18.2 is as follows, where the values of x are found from Appendix Table A:

Population	$p'(\%)$	$p''(\%)$	x'	x''	1 t.u.	σ
(1)	40	5	0.25	1.64	$1.39\sigma_1$	0.72 t.u.
(2)	36	10	0.36	1.28	$0.92\sigma_2$	1.09 t.u.

Thus the standard deviation of population (2) is found to be 1.5 times that of population (1). The means as deviations from T_1 in threshold units are found as follows;

$$m_1 = -x'_1\sigma_1 = -0.25 \times 0.72 \text{ t.u.} = -0.18 \text{ t.u.}$$
$$m_2 = -x'_2\sigma_2 = -0.36 \times 1.09 \text{ t.u.} = -0.39 \text{ t.u.}$$

Thus the mean liability of population (2) is 0.21 threshold units below that of population (1).

When the variance can be estimated in the manner described above, it becomes possible to study crosses between lines with different incidences and to compare the variances of F_1 and F_2 generations. This provides another test of the adequacy of the liability model, and in particular of the interpretation of one class as being intermediate in liability between the other two. The following example illustrates such a study of a cross of inbred lines of mice. The means and variances of the parental lines, F_1, F_2, and the two backcrosses agree very well with what would be expected of a metric character; in this case there is no reason to doubt the validity of the threshold model.

Example 18.2 The number of presacral vertebrae in mice varies between 25 and 27, presacral being defined as being anterior to the first vertebra that is fused to the sacrum. The character therefore reflects the longitudinal position at which the sacrum is fused to the vertebral column. Usually only two numbers are present in any one inbred strain, and we consider here two inbred strains with 25 and 26 but in very different proportions. A third phenotypic class is provided by the few individuals which are asymmetrical, having 25 on one side and 26 on the other, through having the last vertebra fused to the sacrum on one side only. There is clearly some doubt about the asymmetrical mice being intermediate in liability; they might, instead, be less well regulated in development. But treating them as intermediate seems to be justified by the genetic analysis of crosses. The data here refer to one of several crosses described by Green (1962). The strains were C3H having 13 per cent of individuals with 26 vertebrae, and C57BL having 90 per cent. The incidences of the different generations are given in the table with the means and standard deviations calculated as above. The means are deviations in threshold units from the threshold separating '25' from 'asymmetrical'. The mid-parent value, and the expected means of the F_2 and backcross generations calculated from equations [17.3], are also shown in the table. On the whole, the results agree well with expectations: the F_1 is intermediate between P_1 and P_2; the F_2 has its mean near the F_1 but has a greatly increased variance; the backcross means are between the F_1 and parental means and have variances between those of the F_1 and F_2. The only disturbing anomaly is the very greatly different variances of the two parental lines.

	No. of mice	$p'\%$	$p''\%$	m	$\bar{P}$ E(m)	σ	$\sigma^2 = V_P$
P_1	282	12.8	8.5	-4.8⎫	-0.7	4.2	18.0
P_2	619	96.4	89.8	$+3.4$⎭		1.9	3.6
F_1	532	50.0	31.8	0.0	—	2.1	4.5
F_2	206	56.8	46.1	$+0.6$	-0.3	3.7	13.8
B_1	205	33.7	21.5	-1.1	-2.4	2.7	7.4
B_2	194	75.3	60.8	$+1.7$	$+1.7$	2.4	5.9

The degree of genetic determination, V_G/V_P, in the F_2 can be estimated from the difference of variance between the F_2 and the F_1 as follows:

$$
\begin{aligned}
F_2 &: V_G + V_E = 13.8 \\
F_1 &: \qquad\quad V_E = 4.5 \\
F_2 - F_1 &: V_G \qquad\quad = 9.3 \\
V_G/(V_G + V_E) &= 0.68
\end{aligned}
$$

Thus 68 per cent of the variation of liability in the F_2 was genetic. This, again, is a very reasonable result.

From what has been said in this chapter it will be clear that threshold characters do not provide ideal material for the study of quantitative genetics, because the genetic analyses to which they can be subjected are limited in scope and subject to assumptions that one would be unwilling to make except under the force of necessity. If a continuously varying character that is closely correlated with liability can be found, it would clearly be better to analyse this as a metric character instead of the threshold character. For example, 'time of survival' might be used instead of 'resistant versus susceptible'; or an abnormality might be graded in degrees of severity.

Selection for threshold characters

The application of selection to a threshold character does not involve the theoretical difficulties of genetic analyses. It has some practical importance in connection with reducing the incidence of abnormalities and with changing the response of experimental animals to treatments such as, for example, increasing or decreasing drug resistance. We shall consider individual selection applied to a character with two visible classes.

The response to selection depends in the usual way on the selection differential. But the selection differential does not depend primarily on the proportion selected, as with a continuously varying character, but on the incidence, for the following reason. We may breed exclusively from those individuals in the desired phenotypic class, but we cannot discriminate between those with high and those with low liabilities. The selected individuals are therefore a random sample from the desired class and their mean is the mean of the desired class, irrespective of whether we select all of the desired class or only a portion of it. Thus selecting a smaller proportion than the incidence gives no advantage. If, on the other hand, the proportion that has to be selected is greater than the incidence, we shall be forced to use some individuals of the undesired class. Their mean liability will be below the population mean, so the use of undesired individuals as parents will apply some negative selection. (The mean of the undesired class is easily calculated as $-ip/(1-p)$, where i is the mean of the desired class whose incidence is p.)

With some characters the incidence can be altered, and this provides a means of increasing the selection differential and so of improving the response to selection. If the character is, for example, a reaction to some treatment, the treatment can be increased or reduced in intensity so that the incidence is altered. This changed incidence is best regarded as a shift in the threshold relative to the mean liability of the population. In other words, under the changed treatment there is a different threshold of liability for expression of the character. When the level of the threshold can be controlled in this way, the maximum speed of progress under selection will be attained by adjusting the threshold so that the incidence is kept as nearly as possible equal to the minimum proportion that must be selected for breeding. The progress made can be assessed by subjecting the population, or part of it, to the original treatment under which the threshold is at its original level. The situation is different, however, if family selection or progeny testing is applied. The criterion of selection is then a continuous variable, namely the mean liability of the family, and the best discrimination between families will be obtained when the sampling variance of the liability is minimal. The sampling variance of the mean liability is given by $(1/n)(1-p)/pi^2$, where n is the total number in the sample (see Falconer, 1965a), and this is minimal when the incidence p is 50 per cent.

Genetic assimilation.   A very interesting result of the application of this principle of changing the threshold by environmental means is the phenomenon known as 'genetic assimilation' (Waddington, 1953). If a threshold character appears as a result of an environmental stimulus, and selection is applied for this character, it may eventually be made to appear spontaneously, without the necessity of the environmental stimulus. In this way, what was originally an 'acquired character' becomes by perfectly orthodox principles of selection an 'inherited character' (Waddington, 1942). In such a situation there are two thresholds, one spontaneous and the other induced, as shown in Fig. 18.3. The spontaneous threshold is at first outside the range of variation of the population, so that there is no variation of phenotype and no selection can be applied (Fig. 18.3(a)). The induced threshold, however, is within the range of liability covered by the population, and it allows individuals toward one end of the distribution to be picked out by selection. In this way the mean genotypic value of the population is changed. If this change goes far enough, some individuals will eventually cross the spontaneous threshold and appear as spontaneous variants (Fig. 18.3(b)). When the spontaneous incidence becomes high enough, selection may be continued without the aid of the environmental stimulus, and the spontaneous incidence may be further increased (Fig. 18.3(c)).

Example 18.3 An experimental demonstration of genetic assimilation in *Drosophila melanogaster* is described by Waddington (1953). The character was the absence of the posterior cross-vein of the wing. In the base population no flies with this abnormality

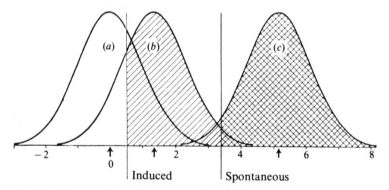

Fig. 18.3. Diagram illustrating the genetic assimilation described in Example 18.3. The distributions of liability are marked in standard deviations from the original population mean. The vertical lines show the positions of the induced and spontaneous thresholds, and the arrows mark the population means at the following three stages of selection.

(a) before selection: incidence – induced =   30%, spontaneous =   0%
(b) after some selection: incidence – induced =   80%, spontaneous =   2%
(c) after further selection: incidence – induced = 100%, spontaneous = 95%

were present, but treatment of the puparium by heat shock caused about 30 per cent of cross-veinless individuals to appear. Selection in both directions was applied to the treated flies, and after 14 generations the incidence of the induced character had risen to 80 per cent in the upward selected line, and fallen to 8 per cent in the downward selected line. At this time cross-veinless flies began to appear in small numbers among untreated flies of the upward-selected line, and by generation 16 the spontaneous incidence was between 1 and 2 per cent. Selection was then continued without treatment, the population being subdivided into a number of lines. The best four of the lines, selected without further treatment, reached spontaneous incidences ranging from 67 per cent to 95 per cent. The distributions in Fig. 18.3 illustrate the progress of the upward selection. Graph (b) shows a spontaneous incidence of 2 per cent and an induced incidence of 80 per cent and thus corresponds approximately with generation 16. On the assumption of constant variance, the change of mean at this stage amounted to 1.36 standard deviations. Graph (c) shows a spontaneous incidence of 95 per cent and represents the line that finally showed the greatest progress. Its mean liability is 5.15 standard deviations above that of the initial population.

19 CORRELATED CHARACTERS

This chapter deals with the relationships between two metric characters, in particular with characters whose values are correlated – either positively or negatively – in the individuals of a population. Correlated characters are of interest for three chief reasons. Firstly, in connection with the genetic causes of correlation through the pleiotropic action of genes: pleiotropy is a common property of major genes, but we have as yet had little occasion to consider its effects in quantitative genetics. Secondly, in connection with the changes brought about by selection: it is important to know how the improvement of one character will cause simultaneous changes in other characters. And thirdly, in connection with natural selection: the relationship between a metric character and fitness is the primary agent that determines the genetic properties of that character in a natural population. This last point, however, will be discussed in the next chapter.

Genetic and environmental correlations

In genetic studies it is necessary to distinguish two causes of correlation between characters, genetic and environmental. The genetic cause of correlation is chiefly pleiotropy, though linkage is a cause of transient correlation, particularly in populations derived from crosses between divergent strains. Pleiotropy is simply the property of a gene whereby it affects two or more characters, so that if the gene is segregating it causes simultaneous variation in the characters it affects. For example, genes that increase growth rate increase both stature and weight, so that they tend to cause correlation between these two characters. Genes that increase fatness, however, influence weight without affecting stature, and are therefore not a cause of correlation. The degree of correlation arising from pleiotropy expresses the extent to which two characters are influenced by the same genes. But the correlation resulting from pleiotropy is the overall, or net, effect of all the segregating genes that affect both characters. Some genes may increase both characters, while others increase one and reduce the other; the former tend to cause a positive correlation, the latter a negative one. So pleiotropy does not necessarily cause a detectable correlation. The environment is a cause of correlation in so far as two characters are influenced by the same differences of environmental conditions. Again, the correlation resulting from environmental causes is the overall effect of all the environmental factors that vary; some may tend to cause a positive correlation, others a negative one.

The association between two characters that can be directly observed is the correlation of phenotypic values, or the *phenotypic correlation*. This is determined from measurements of the two characters in a number of individuals of the population. Suppose, however, that we knew not only the phenotypic values of the individuals measured, but also their genotypic values and their environmental deviations for both characters. We could then compute the correlation between the genotypic values of the two characters and the correlation between the environmental deviations, and so assess independently the genetic and environmental causes of correlation. And if, in addition, we knew the breeding values of the individuals, we could determine also the correlation of breeding values. In principle there are also correlations between dominance deviations, and between the various interaction deviations. To deal with all these correlations would be unmanageably complex, but fortunately is not necessary since the practical problems can be quite adequately dealt with in terms of two correlations. These are the *genetic correlation*, which is the correlation of breeding values, and the *environmental correlation*, which is not strictly speaking the correlation of environmental deviations, but the correlation of environmental deviations together with non-additive genetic deviations. In other words, just as the partitioning of the variance of one character into two components, additive genetic versus all the rest, was adequate for many purposes, so now the covariance of two characters need only be partitioned into these same two components. The 'genetic' and 'environmental' correlations thus correspond to the partitioning of the covariance into the additive genetic component versus all the rest. The methods of estimating these two correlations will be explained later. The first problem to be considered is how the genetic and environmental correlations combine together to give the directly observable phenotypic correlation.

The following symbols will be used throughout this chapter:

X and Y: the two characters under consideration.
r_P the phenotypic correlation between the two characters X and Y.
r_A the genetic correlation between X and Y (i.e., the correlation of breeding values).
r_E the environmental correlation between X and Y (including non-additive genetic effects).
cov   the covariance of the two characters X and Y, with subscripts P, A, or E, having the same meaning as for the correlations.
σ^2 and σ variance and standard deviation, with subscripts P, A, or E, as above, and X or Y according to the character referred to; e.g. σ_{PX}^2 = phenotypic variance of character X.
h^2 the heritability, with subscript X or Y, according to the character.
e^2 $= 1 - h^2$.

(The symbol r_G is often used for the genetic correlation but, since the correlation referred to is almost always the correlation of breeding values, the symbol r_A will be used here for the sake of consistency with previous chapters.)

A correlation, whatever its nature, is the ratio of the appropriate covariance to the product of the two standard deviations. For example, the phenotypic correlation is

$$r_P = \frac{\text{cov}_P}{\sigma_{PX}\sigma_{PY}}$$

and the phenotypic covariance can be written as

$$\text{cov}_P = r_P\,\sigma_{PX}\,\sigma_{PY}$$

The phenotypic covariance is the sum of the genetic and environmental covariances, i.e.,

$$\text{cov}_P = \text{cov}_A + \text{cov}_E$$

Writing these covariances in terms of the correlations and standard deviations as above gives

$$r_P\sigma_{PX}\sigma_{PY} = r_A\sigma_{AX}\sigma_{AY} + r_E\sigma_{EX}\sigma_{EY}$$

Now note that $\sigma_A = h\sigma_P$ and $\sigma_E = e\sigma_P$. Substituting these gives

$$r_P\sigma_{PX}\sigma_{PY} = r_A h_X \sigma_{PX} h_Y \sigma_{PY} + r_E e_X \sigma_{PX} e_Y \sigma_{PY}$$

Dividing through by $\sigma_{PX}\sigma_{PY}$ leads to

$$r_P = h_X h_Y r_A + e_X e_Y r_E \qquad\qquad \dots [19.1]$$

This shows how the genetic and environmental causes of correlation combine together to give the phenotypic correlation. If both characters have low heritabilities, then the phenotypic correlation is determined chiefly by the environmental correlation: if they have high heritabilities, then the genetic correlation is the more important. The dual nature of the phenotypic correlation makes it clear that the magnitude and even the sign of the genetic correlation cannot be determined from the phenotypic correlation alone.

A few examples of genetic and environmental correlations are given in Table 19.1. In some cases the genetic and environmental correlations are different in magnitude, or even in sign. In other cases the two correlations are of the same sign and not very different magnitude, and this is the more usual situation. A large difference, and particularly a difference of sign, shows that genetic and environmental sources of variation affect the characters through different physiological mechanisms.

The genetic correlation expresses the extent to which two measurements reflect what is genetically the same character. For example, the lengths of the two wings of *Drosophila* must obviously be measures of the same character, wing length. But wing length and thorax length, though both measures of body size, are not quite the same character; the genetic correlation between them is about 0.75 (Reeve and Robertson, 1953). In this connection the genetic correlation has a bearing on the interpretation of the repeatability of multiple measurements. In Chapter 8 it was said, without explanation, that the repeatability has a precise genetic interpretation only if the different measurements are of the same genetic character. The meaning of this requirement can now be seen to be that the genetic correlation between the measurements must be 1. If the two characters X and Y in

Table 19.1 Some examples of phenotypic, genetic, and environmental correlations. The estimates quoted refer to particular populations in particular circumstances; they should not be taken as generally applicable.

	r_P	r_A	r_E
Man (Grundbacher, 1974)			
Serum immunoglobulin levels, IgG: IgM	0.20	0.07	0.31
Cattle (Barker and Robertson, 1966)			
Milk-yield: butterfat % (1st lactation)	− 0.26	− 0.38	− 0.18
Milk yield in 1st: 2nd lactations	0.40	0.75	0.26
Pigs (Smith, King, and Gilbert, 1962)			
Weight gain: backfat thickness	0.00	0.13	− 0.18
Weight gain: efficiency	0.66	0.69	0.64
Poultry (Emsley, Dickerson, and Kashyap, 1977)			
Body weight: egg weight	0.33	0.42	0.23
Body weight: egg production	0.01	− 0.17	0.08
Egg weight: egg production	− 0.05	− 0.31	0.02
Mice (Rutledge, Eisen, and Legates, 1973)			
Body weight: tail length	0.45	0.29	0.56
Drosophila melanogaster (Sheridan et al., 1968)			
Bristle number, abdominal: sternopleural	0.14	0.41	0.06

equation [19.1] are the same, the expression for the phenotypic correlation reduces to $r_P = h^2 + e^2$, which is equivalent to the repeatability in equation [8.12], since here e^2 is the proportion of variance due to the general environment (V_{Eg}) and the non-additive genetic variance. The repeatability nevertheless remains a useful concept even though the genetic correlation may often be somewhat less than 1.

There are some pairs of characters for which no phenotypic correlation exists. These are characters that cannot both be measured on the same individual. The age at sexual maturity in males and in females is an example of two such characters. Though no phenotypic correlation can be measured, the two characters may nevertheless be correlated genetically and environmentally, and both these correlations can be estimated.

Estimation of the genetic correlation

The estimation of genetic correlations rests on the resemblance between relatives in a manner analogous to the estimation of heritabilities described in Chapter 10. Therefore only the principle and not the details of the procedure need be described here. Instead of computing the components of variance of one character from an analysis of variance, we compute the components of covariance of the two characters from an analysis of covariance which takes exactly the same form as the analysis of variance. Instead of starting from the squares of the individual values and partitioning the sums of squares according to the source of variation, we start from the product of the values of the two characters in each individual and partition the sums of products according to the source of variation. This leads to estimates of the observational components of covariance, whose interpretation in terms of causal components of covariance is exactly the same as that of the components of variance given in Table 10.4. Thus, in an analysis of half-sib families the component of covariance between sires estimates $\frac{1}{4}\text{cov}_A$, i.e., one-quarter of the covariance of breeding values of the two characters. For the estimation of the correlation, the components of variance of each character are also needed. Thus the between-sire components of variance estimate $\frac{1}{4}\sigma^2_{AX}$ and $\frac{1}{4}\sigma^2_{AY}$. Therefore the genetic correlation is obtained as

$$r_A = \frac{\text{cov}_{XY}}{\sqrt{(\text{var}_X \text{ var}_Y)}} \qquad \dots [19.2]$$

where var and cov refer to the components of variance and covariance.

The offspring–parent relationship can also be used for estimating the genetic correlation. To estimate the heritability of one character from the resemblance between offspring and parents, we compute the covariance of offspring and parent for the one character by taking the product of the parent or mid-parent value and the mean value of the offspring. To estimate the genetic correlation between two characters we compute what might be called the 'cross-covariance', obtained from the product of the value of X in parents and the value of Y in offspring. This 'cross-covariance' is half the genetic covariance of the two characters, i.e., $\frac{1}{2}\text{cov}_A$. The covariances of offspring and parents for each of the characters separately are also needed, and then the genetic correlation is given by

$$r_A = \frac{\text{cov}_{XY}}{\sqrt{(\text{cov}_{XX} \text{ cov}_{YY})}} \qquad \dots [19.3]$$

where cov_{XY} is the 'cross-covariance', and cov_{XX} and cov_{YY} are the offspring–parent covariances of each character separately. The genetic correlation can also be estimated from responses to selection in a manner analogous to the estimation of realised heritability. This will be explained in the next section.

Data that provide estimates of genetic correlations provide also estimates of the heritabilities of the correlated characters, and of the phenotypic correlations. The environmental correlation can then be found from equation [19.1]. If highly inbred lines are available, the environmental correlations can be estimated directly from the phenotypic correlation within the lines, or preferably within the F_1's of crosses between the lines.

Estimates of genetic correlations are usually subject to rather large sampling errors and are therefore seldom very precise. Furthermore, genetic correlations are strongly influenced by gene frequencies (Bohren, Hill, and Robertson, 1966), so they may differ markedly in different populations. For these reasons the examples quoted in Table 19.1 must be regarded as approximate values and not necessarily valid for other populations. The sampling variance of genetic correlations is treated by Reeve (1955b) and by Robertson (1959b). The standard error of an estimate is given approximately by the following formula:

$$\sigma_{(r_A)} = \frac{1 - r_A^2}{\sqrt{2}} \sqrt{\left[\frac{\sigma_{(h_X^2)} \sigma_{(h_Y^2)}}{h_X^2 h_Y^2} \right]} \qquad \dots [19.4]$$

where σ denotes standard error. Since the standard errors of the two heritabilities appear in the numerator, an experiment designed to minimize the sampling variance of an estimate of heritability, in the manner described in Chapter 10., will also have the optimal design for the estimation of a genetic correlation.

Correlated response to selection

The next problem for consideration concerns the response to selection: if we select for character X, what will be the change of the correlated character Y? The expected response of a character Y, when selection is applied to another character X, may be deduced in the following way. The response of character X – i.e. the character directly selected – is equivalent to the mean breeding value of the selected individuals. This was explained in Chapter 11. The consequent change of character Y is therefore given by the regression of the breeding value of Y on the breeding value of X. This regression is

$$b_{(A)YX} = \frac{\text{cov}_A}{\sigma_{AX}^2} = r_A \frac{\sigma_{AY}}{\sigma_{AX}}$$

The response of character X, directly selected, by equation [11.4], is

$$R_X = i h_X \sigma_{AX}$$

Therefore the correlated response of character Y is

$$CR_Y = b_{(A)YX} R_X \qquad \qquad \text{...} [19.5a]$$

$$= i h_X \sigma_{AX} r_A \frac{\sigma_{AY}}{\sigma_{AX}}$$

$$= i h_X r_A \sigma_{AY} \qquad \qquad \text{...} [19.5b]$$

Or, by putting $\sigma_{AY} = h_Y \sigma_{PY}$, the correlated response becomes

$$CR_Y = i h_X h_Y r_A \sigma_{PY} \qquad \qquad \text{...} [19.6]$$

Thus the response of a correlated character can be predicted if the genetic correlation and the heritabilities of the two characters are known. And, conversely, if the correlated response is measured by experiment, and the two heritabilities are known, the genetic correlation can be estimated. If the heritability of character Y is to be estimated as the realized heritability from the response to selection, then it is necessary to do a double selection experiment. Character X is selected in one line and character Y in another. Then both the direct and the correlated responses of each character can be measured. This type of experiment provides two estimates of the genetic correlation (by equation [19.6]), one from the correlated response of each character; and the two estimates should agree if the theory of correlated responses expressed in equation [19.6] adequately describes the observed responses. A joint estimate of the genetic correlation can be obtained from such double selection experiments, without the need for estimates of the heritabilities, from the following formula which may be easily derived from equations [11.4] and [19.5b]:

$$r_A^2 = \frac{CR_X}{R_X} \frac{CR_Y}{R_Y} \qquad \qquad \text{...} [19.7]$$

Example 19.1 In a study of wing length and thorax length in *Drosophila melanogaster*, Reeve and Robertson (1953) estimated the genetic correlation between these two measures of body size from the responses to selection. There were two pairs of selection lines; one pair was selected for increased and for decreased thorax length, and the other pair for increased and for decreased wing length. In each line the correlated response of the character not directly selected was measured, as well as the response of the character directly selected. Two estimates of the genetic correlation were obtained by equation [19.7], one from the responses to upward selection and the other from the responses to downward selection. In addition, estimates of the genetic correlation in the unselected population were obtained from the offspring–parent covariance and also from the full-sib covariance. The four estimates were as follows:

Method	Genetic correlation
Offspring–parent	0.74
Full sib	0.75
Selection, upward	0.71
Selection, downward	0.73

The agreement between the estimates from selection and the estimates from the unselected population shows that the correlated responses were very close to what would have been predicted from the genetic analysis of the unselected population.

Close agreement between observed and predicted correlated responses, such as was shown in the above example, cannot always be expected and, indeed, is not often found, particularly if the genetic correlation is low. Furthermore, double selection experiments are often inconsistent in the estimates of the genetic correlation that they give. There are two reasons for the low predictability and the inconsistency of correlated responses. The first is the low precision of estimates of the genetic correlation in the base population, resulting from the large sampling errors already mentioned. The second reason is the sensitivity of genetic correlations to gene frequency changes (Bohren, Hill, and Robertson, 1966); the genetic correlation, and therefore the correlated response, can change rapidly during the course of the selection as a result of the selection itself and of random drift. For these reasons there must be some lack of confidence in applying the theory of correlated responses in practice. We shall, however, pursue the practical implications of the theory a little further in the next section, but with the caution that the theory cannot always be relied on to work well in practice.

Correlated selection differential.   When selection is applied to one character X, any phenotypically correlated character Y will have a *correlated*, or *apparent*, *selection differential* on it. In other words, the individuals selected for X will have a mean value of Y that is different from the population mean. At first sight it might seem that some use could be made of this correlated selection differential for predicting the correlated response, or for estimating the heritability of the correlated character, in a manner analogous to equation [11.7] ($h^2 = R/S$). Unfortunately, however, the correlated selection differential is of no use for either of these purposes. The reason is briefly as follows. Let S'_Y be the correlated selection differential on Y. S'_Y is related to the selection differential on X by the phenotypic regression of Y on X: $S'_Y = b_{(P)YX} S_X$. Writing the correlated response in the form of equation [19.5a] gives

$$\frac{CR_Y}{S'_Y} = \frac{b_{(A)} R_X}{b_{(P)} S_X} \qquad \ldots [19.8a]$$

Substituting $b = r\sigma_Y/\sigma_X$ and $R_X/S_X = h_X^2$ leads to

$$\frac{CR_Y}{S_Y'} = \frac{r_A}{r_P}h_X h_Y \qquad \ldots [19.8b]$$

Thus, without knowing the genetic correlation r_A, it is not possible to use the correlated selection differential S_Y', either to estimate the heritability of character Y or to predict the correlated response. Equation [19.8] can also be written in the form

$$\frac{CR_Y}{S_Y'} = \frac{\text{cov}_{(A)}}{\text{cov}_{(P)}} \qquad \ldots [19.8c]$$

which is analogous to the direct response, $R/S = V_A/V_P$.

Indirect selection

Consideration of correlated responses suggests that it might sometimes be possible to achieve more rapid progress under selection for a correlated response than from selection for the desired character itself. In other words, if we want to improve character X, we might select for another character Y, and achieve progress through the correlated response of character X. We shall refer to this as *indirect selection*; that is to say, selection applied to some character other than the one it is desired to improve. And we shall refer to the character to which selection is applied as the *secondary character*. The conditions under which indirect selection would be advantageous are readily deduced. Let R_X be the direct response of the desired character, if selection were applied directly to it; and let CR_X be the correlated response of character X resulting from selection applied to the secondary character Y. The merit of indirect selection relative to that of direct selection may then be expressed as the ratio of the expected responses, CR_X/R_X. Taking the expected correlated response from equation [19.5b] and the expected direct response from equation [11.4], we find

$$\frac{CR_X}{R_X} = \frac{i_Y h_Y r_A \sigma_{AX}}{i_X h_X \sigma_{AX}}$$

$$= r_A \frac{i_Y}{i_X} \frac{h_Y}{h_X} \qquad \ldots [19.9]$$

If the same intensity of selection can be achieved when selecting for character Y as when selecting for character X, then the correlated response will be greater than the direct response if $r_A h_Y$ is greater than h_X. Therefore indirect selection cannot be expected to be superior to direct selection unless the secondary character has a substantially higher heritability than the desired character, and the genetic correlation between the two is high; or unless a substantially higher intensity of selection can be applied to the secondary than to the desired character. The circumstances most likely to render indirect selection superior to direct selection are chiefly concerned with technical difficulties in applying selection directly to the desired character. Three such technical difficulties may be mentioned briefly.

1. If the desired character is difficult to measure with precision, the errors of measurement may so reduce the heritability that indirect selection becomes advantageous.

2. If the desired character is measurable in one sex only, but the secondary character is measurable in both, then a higher intensity of selection will be possible by indirect selection. Other things being equal, the intensity of selection would be twice as great by indirect as by direct selection; but a better plan would be to select one sex directly for the desired character and the other indirectly for the secondary character.

3. The desired character may be costly to measure, as for example the efficiency of food-conversion. Then it may be economically better to select for an easily measured correlated character, such as growth rate.

For a detailed evaluation of indirect selection, see Searle (1965). The following is an example of indirect selection giving a better response than direct selection for a character measurable in only one sex.

Example 19.2 (*Data from Nagai et al.*, 1978). Mice were selected for two characters, nursing ability of females measured as the 12-day weight of the litter, and 6-week weight of individuals. Nursing ability will be designated as N and 6-week weight as W. The selection was done in two populations, P and Q, with different origins, and was continued for 12 generations. In each population one line was selected upwards for N, giving the direct response of N and the correlated response of W; a second line was selected for W giving the direct response of W and the correlated response of N. The responses are given in the table, correlated responses being shown in italics; all are in units of grams per generation.

Population		P		Q	
Character selected	N		W	N		W
Response* of N	0.080		*0.134*	0.054		*0.125*
Response* of W	*0.197*		0.680	*0.198*		0.868
Observed CR_N/R_N		1.675			2.315	
Realized h^2	0.16		0.40	0.11		0.43
Realized r_A		0.70			0.73	
$h_W r_A/h_N$		1.11			1.44	
Expected[+] CR_N/R_N		2.2			2.9	

* Correlated responses in italics.
[+] Assuming $i_W/i_N = 2$.

The genetic correlation in the P population is calculated by equation [19.7] as follows:

$$r_A^2 = \frac{CR_N}{R_N} \frac{CR_W}{R_W} = \frac{0.134}{0.080} \times \frac{0.197}{0.680} = 0.485; \; r_A = 0.70$$

Similarly in the Q population, $r_A = 0.73$. There was thus very good agreement between the two populations in the estimation of r_A.

As can be seen from the ratio CR/R in the table, indirect selection was substantially better than direct selection for improving nursing ability. The reasons for this are that W has a higher heritability than N and can be selected in both sexes. The heritabilities, estimated as realized heritabilities from the direct responses, are given in the table. Again the two populations show good agreement for both characters. The intensities of selection actually applied are not given, but the expected intensities can be deduced from the proportions selected. The same proportions were selected for both characters, but females only were selected for nursing ability while both sexes were selected for weight. The net intensity of selection was therefore expected to be twice as great for weight as for nursing ability.

The expected ratio of correlated to direct responses of nursing ability can now be calculated by equation [19.9] from the observed heritabilities and genetic correlation and the presumed intensities of selection. For population P it is

$$\frac{CR_N}{R_N} = r_A \frac{i_W h_W}{i_N h_N} = 0.70 \times 2 \times \sqrt{\frac{0.40}{0.16}} = 2.2$$

The ratio for population Q is 2.9. In both cases the ratio realized was somewhat less than the expectation, presumably because the intensities of selection realized were not as much as twice as great for weight as for nursing ability.

Though indirect selection has been presented above as an alternative to direct selection, the most effective method in theory is neither one nor the other but a combination of the two. The most effective use that can be made of a correlated character is in combination with the desired character, as an additional source of information about the breeding values of individuals. This, however, is a special case of a more general problem which will be dealt with in the final section of this chapter. First we shall show how the idea of indirect selection can be extended to cover selection in different environments.

Genotype–environment interaction

The concept of genetic correlation can be applied to the solution of some problems connected with the interaction of genotype with environment. The meaning of interaction between genotype and environment was explained in Chapter 8, where it was discussed as a source of variation of phenotypic values, which in most analyses is inseparable from the environmental variance. The chief problem which it raises, and which we are now in a position to discuss, concerns adaptation to local conditions. The existence of genotype–environment interaction may mean that the best genotype in one environment is not the best in another environment. It is obvious, for example, that the breed of cattle with the highest milk-yield in temperate climates is unlikely also to have the highest yield in tropical climates. But it is not so obvious whether smaller differences of environmental conditions also require locally adapted breeds; nor is it intuitively obvious how much of the improvement made in one environment will be carried over if the breed is then transferred to another environment. These matters have an important bearing on breeding policy. If selection is made under good conditions of feeding and management on the best farms and experimental stations, will the improvement achieved be carried over when the later generations are transferred to poorer conditions? Or would the selection be better done in the poorer conditions under which the majority of animals are required to live? The idea of genetic correlation provides the basis for a solution of these problems in the following way.

A character measured in two different environments is to be regarded not as one character but as two. The physiological mechanisms are to some extent different, and consequently the genes required for high performance are to some extent also different. For example, growth rate on a low plane of nutrition may be principally a matter of efficiency of food-utilization, whereas on a high plane of nutrition it may be principally a matter of appetite. By regarding performance in different environments as different characters with genetic correlation between them, we can in principle solve the problems outlined above from a knowledge of the heritabilities of the different characters and the genetic correlations between them. If the genetic correlation is high, then performance in two different environments represents very nearly the same character, determined by very nearly the same set of genes. If it is low, then the characters are to a great extent different, and high performance requires a different set of genes. Here we shall consider only two environments, but the idea can be extended to an indefinite number of different environments (Robertson, 1959b; Dickerson, 1962; Yamada, 1962).

Let us consider the problem of the 'carry-over' of the improvement from one environment to another. Suppose that we select for character X – say growth rate on a high plane of nutrition – and we look for improvement in character Y – say growth rate on a low plane of nutrition. The improvement of character Y is simply a correlated response, and the expected rate of improvement was given in equation [19.6] as

$$CR_Y = ih_X h_Y r_A \sigma_{PY}$$

The improvement of performance in an environment different from the one in which selection was carried out can therefore be predicted from a knowledge of the heritability of performance in each environment and the genetic correlation between the two performances. We can also compare the improvement expected by this means with that expected if we had selected directly for character Y, i.e., for performance in the environment for which improvement is wanted. This is simply a comparison of indirect with direct selection, which was explained in the previous section. The comparison is made from the ratio of the two expected responses given in equation [19.9], i.e.,

$$\frac{CR_Y}{R_Y} = r_A \frac{i_X h_X}{i_Y h_Y}$$

This shows how much we may expect to gain or lose by carrying out the selection in some environment other than the one in which the improved population is required to live. If we assume that the intensity of selection is not affected by the environment in which the selection is carried out, then the indirect method will be better if $r_A h_X$ is greater than h_Y, where h_X is the square-root of the heritability in the environment in which selection is made, and h_Y is the square-root of the heritability in the environment in which the population is required subsequently to live. If the genetic correlation is high, then the two characters can be regarded as being substantially the same; and if there are no special circumstances affecting the heritability or the intensity of selection, it will make little difference in which environment the selection is carried out. But if the genetic correlation is low, then it will be advantageous to carry out the selection in the environment in which the population is destined to live, unless the heritability or the intensity of selection in the other environment is very considerably higher.

This is the theoretical basis for dealing with selection in different environments. There have been several experiments testing the theory. In general they confirm the theory in finding correlated responses to be smaller than direct responses, i.e., selection is most effective if carried out in the environment for which the improvement is sought. These experiments, however, are not free of the inconsistencies mentioned earlier, particularly inconsistencies between the responses to upward and to downward selection. An experiment with mice is described briefly in the following example; for other examples, see two experiments with *Tribolium* in which the characters were larval growth (Yamada and Bell, 1969) and rate of egg-laying (Orozco, 1976).

Example 19.3 (*Data from Falconer*, 1960b). Mice were selected for growth from 3 to 6 weeks on two diets: 'good' which was the normal diet, and 'bad' which was the normal diet diluted with 50 per cent indigestible fibre. The bad diet reduced growth by about 20 per cent at the beginning of the experiment. The direct and correlated responses were measured in each generation, the direct responses from first-litter progeny grown on the diet of selection, and the correlated responses from second-litter progeny grown on the other diet. Selection was carried out in both directions. There were inconsistencies between selection in opposite directions and between the earlier and later generations. For the purpose of illustrating the theory, these inconsistencies are avoided by taking the results over the first four generations only, and expressing the responses as the divergence between upward and downward selection. Table (i) gives the information needed to calculate the genetic correlation from each pair of lines separately by equation [19.9] and from both pairs of lines together by equation [19.7]. The responses are grams per generation. As expected, both correlated responses were less than the direct responses, the ratios between the two indicating a genetic correlation of 0.66.

Table (i) Divergence (g) per generation to generation 4.

	Character	
	Growth on good diet	Growth on bad diet
Intensity of selection, i	1.66	1.40
Realized heritability, h^2	0.41	0.36
Direct response, R	0.90	1.20
Correlated response, CR	0.48	0.98
Genetic correlation, by eqn [19.9], r_A	0.67	0.65
Genetic correlation, by eqn [19.7], r_A	0.66	

Table (ii) Total response (g) to generation 7 as deviations from controls.

Selection		Response			Sensitivity	
Direction	Diet	Growth on good diet	Growth on bad diet	Mean of both diets	Effect of diet	(Control)
Up	good	2.3	0.6	1.45	5.4	(3.8)
Up	bad	1.6	3.1	2.35	3.5	(4.2)
Down	good	−2.8	−2.9	−2.85	3.6	(3.8)
Down	bad	−1.2	−3.2	−2.20	6.8	(4.2)

Table (ii) shows the responses of the upward- and downward-selected lines separately up to generation 7, when an unselected control was measured. Taking deviations from the control makes comparison of the upward and downward responses more meaningful. The purpose of the table is to illustrate a point to be explained in the next section, about the diet of selection that is expected to give the best average performance in both environments. As expected, the highest average growth was obtained by upward selection on the bad diet, and the lowest average growth by downward selection on the good diet. The last column in table (ii) illustrates another point to be explained in the next section, about the expected changes of environmental sensitivity. The figures under 'Effect of diet' are the difference in growth between mice of each selection line when reared on the good and on the bad diet, this difference being the sensitivity of the line to the environmental difference. The figures in parentheses refer to unselected controls. (The growths were adjusted to a common 3-week weight and are taken from the original data, not published.) As expected, the environmental sensitivity is greatest in the lines selected up in the good environment and down in the bad environment.

Environmental sensitivity.   The way in which genotype–environment interaction arises from differences in sensitivity to the environment was explained in Chapter 8. The genetic correlation provides a means of quantifying the interaction for the purpose of predicting responses to selection. Understanding responses to selection in different environments may, however, be helped by thinking about environmental sensitivity. A high genetic correlation means that all genotypes react in nearly the same way to environmental differences; a plot like that of Fig. 8.2 would have regression lines that were all nearly parallel. A low genetic correlation means that genotypes react differently and have regression lines with different slopes, i.e., individuals have different environmental sensitivities. How does selection act on these differences of sensitivity? It is convenient to refer to environments as 'good' or 'bad' according to whether they increase or decrease the character; in practice an increase is generally sought, so an environment that increases the character is 'good'. The effect of selection on sensitivity can be seen from Fig. 8.2. Upward selection in a good environment tends to pick individuals with high sensitivity, and downward selection in a bad environment does the same. In contrast, upward selection in a bad environment and downward selection in a good environment tend to pick individuals with low sensitivity. In other words, high sensitivity will be selected for when the selection and the environment act on the character in the same direction, and low sensitivity will be selected for when selection and environment act in opposite directions. These expected changes in environmental sensitivity were very clearly shown to take place when the fungus *Schizophyllum commune* was selected for growth rate at different temperatures (Jinks and Connolly, 1973, 1975). The mouse experiment described above also showed the expected changes of sensitivity.

What is wanted in practice is often not performance in a specific environment but performance in a range of environments, both good and bad, i.e., good average performance in different environments. Individuals cannot usually be measured in more than one environment, so selection for average performance has to be family selection with the families divided between the environments. Details of how the two phenotypes of a family should be combined in an index to give the maximum improvement of average performance are given by James

(1961). If measurements can only be made in one environment, consideration of environmental sensitivity suggests that the best average performance will be achieved by selecting in an environment that acts in the direction opposite to the selection, i.e., by upward selection in a bad environment or downward selection in a good environment (Jinks and Connolly, 1973). This expectation was borne out by the mouse experiment described in Example 19.3 and, partially at least, by the *Tribolium* experiment of Yamada and Bell (1969).

Index selection

When selection is applied to the improvement of the economic value of animals or plants, it is generally applied to several characters simultaneously and not just to one, because economic value depends on more than one character. This is usually referred to as *multiple trait selection*. For example, the profit made from a herd of pigs depends on their fertility, mothering ability, growth rate, efficiency of food-utilization, and carcass qualities. How, then, should selection be applied to the component characters in order to achieve the maximum improvement of economic value? There are several possible procedures. One might select in turn for each character singly in successive generations (*tandem selection*); or one might select for all the characters at the same time but independently, rejecting all individuals that fail to come up to a certain standard for each character regardless of their values for any other of the characters (*independent culling levels*). The method that is expected to give the most rapid improvement of economic value, however, is to apply the selection simultaneously to all the component characters together, appropriate weight being given to each character according to its relative economic importance, its heritability, and the genetic and phenotypic correlations between the different characters. The practice of selection for economic value is thus a matter of some complexity. The component characters have to be combined together into a score, or *index*, in such a way that selection applied to the index, as if the index were a single character, will yield the most rapid possible improvement of economic value.

The principles of index selection were introduced in Chapter 13 and will not be repeated in full here. The main difference in the index required here is that the breeding value to be predicted is not that of a single character but that of a composite of several characters evaluated in economic terms. The index is consequently more complex than the one developed in Chapter 13. We shall, however, start by considering the simpler problem of improving a single character by the use of an index, and then extend this to improving economic value.

Construction of the index

The objective of the selection, whatever it may be, will be referred to as *merit*, and the breeding value for merit will be symbolized by H. The index to be constructed for the improvement of merit is, as before,

$$I = b_1 P_1 + b_2 P_2 + \ldots + b_m P_m \qquad \ldots [19.10]$$

where P_1 to P_m are phenotypic measurements of m characters on which selection is to be based, and b_1 to b_m are the corresponding weighting factors to be determined. The b's are partial regression coefficients of H on I. Information from relatives can be included in the index, so the P's can be measurements of relatives in the manner explained in Chapter 13.

Single trait. First consider selection aimed at improving just one character. The purpose of applying index selection is then to use secondary characters as aids to improvement of the one desired character. The index equations, whose solution gives the values of the b's in the index, are exactly the same as equations [13.10], with character 1 as the character to be improved.

$$\left.\begin{aligned} b_1 P_{11} + b_2 P_{12} + \ldots + b_m P_{1m} &= A_{11} \\ b_1 P_{21} + b_2 P_{22} + \ldots + b_m P_{2m} &= A_{21} \\ &\vdots \\ b_1 P_{m1} + b_2 P_{m2} + \ldots + b_m P_{mm} &= A_{m1} \end{aligned}\right\} \quad \ldots [19.11]$$

The notation here is abbreviated as in Chapter 13. For example, P_{11} is the phenotypic variance of character 1, and P_{12} is the phenotypic covariance of characters 1 and 2; A_{11} and A_{12} are similarly the additive genetic variance and covariance. The variances and covariances can be expressed in terms of the heritabilities and correlations as follows, where the subscripts i and j refer to any two different characters and σ^2 is the phenotypic variance:

$$\left.\begin{aligned} P_{ii} &= \sigma_i^2 & ; \; A_{ii} &= h_i^2 \sigma_i^2 \\ P_{ij} &= r_P \sigma_i \sigma_j & ; \; A_{ij} &= r_A h_i h_j \sigma_i \sigma_j \end{aligned}\right\} \quad \ldots [19.12]$$

When the values of the variances and covariances have been entered, the solution of equations [19.11] provides the values of the weighting factors, b, to be used in the index in equation [19.10]. The construction of an index is illustrated later, in Example 19.4. The expected response to selection will be dealt with after the different forms of the index equations have been explained.

Economic value. Next consider the improvement of economic value. The economic value is the profit made from the sale of the individual. In practical breeding operations it is often possible to assign economic values to individuals. This is then the phenotypic value of merit, which is the character to be improved, and the index is constructed for the improvement of this single character. But the index equations, whose solution gives the values of the b's in the index, differ in one respect from what was described above. The economic values of individuals cannot be known at the time they are being considered for selection, and therefore cannot be included as a character in the index. The index equations are then as follows. In order to facilitate comparison, character 1 is still taken to be the character to be improved, in this case merit.

$$\left.\begin{aligned} b_2 P_{22} + b_3 P_{23} + \ldots + b_m P_{2m} &= A_{21} \\ b_2 P_{32} + b_3 P_{33} + \ldots + b_m P_{3m} &= A_{31} \\ &\vdots \\ b_2 P_{m2} + b_3 P_{m3} + \ldots + b_m P_{mm} &= A_{m1} \end{aligned}\right\} \quad \ldots [19.13]$$

The variances and covariances have to be estimated from past records of the economic values and the values of the characters in the index.

Multiple traits.   Finally, consider simultaneous selection for several characters. The objective is to improve the *aggregate breeding value*, or *net merit*, which is a particular combination of all the characters to be improved. Merit is now defined as

$$H = a_1 A_1 + a_2 A_2 + \ldots + a_n A_n \qquad\qquad \ldots [19.14]$$

Here the A's are breeding values for the n characters to be improved, and the a's are weighting factors which express the relative importance attached by the breeder to each character. The weighting factors can be economic values; that is to say, each a is the value in money units of 1 unit of the character. This is how an index is constructed if the aim is to improve economic value when there are no records of individuals' economic values, so that the index described above cannot be used. Assigning money values to the characters is, however, not necessarily the best method of improvement. Other criteria for weighting are discussed by Fowler, Bichard, and Pease (1976) in connection with the improvement of pigs. If the weighting factors are not in money units, they must express in some other way the relative importance to the breeder of 1 unit increase of each character. Yamada, Yokouchi, and Nishida (1975) describe indices constructed in this way.

The number of characters in the definition of merit (equation [19.14]) and in the index (equation [19.10]) may differ: there may be characters that are not in H but which may help to improve H through their correlations if included in I; and, conversely, there may be characters in H which cannot be measured and so are not in I. It is important to note, however, that if the aim is to improve economic value, then all the characters that influence economic value must be included in the definition of H.

The index equations, whose solution gives the b's to be used in the index, are obtained in the same way as was described in Chapter 13, by maximizing r_{HI}, the correlation between merit and the index. They are as follows:

$$\left.\begin{array}{l} b_1 P_{11} + b_2 P_{12} + \ldots + b_m P_{1m} = a_1 A_{11} + a_2 A_{12} + \ldots + a_n A_{1n} \\ b_1 P_{21} + b_2 P_{22} + \ldots + b_m P_{2m} = a_1 A_{21} + a_2 A_{22} + \ldots + a_n A_{2n} \\ \vdots \\ b_1 P_{m1} + b_2 P_{m2} + \ldots + b_m P_{mm} = a_1 A_{m1} + a_2 A_{m2} + \ldots + a_n A_{mn} \end{array}\right\} \ldots [19.15]$$

The variances and covariances can again be expressed in terms of the heritabilities and correlations by equations [19.12]. Example 19.5 below illustrates the construction of an index from these equations, simplified by considering only two characters.

Response
The index equations ([19.11], [19.13], [19.15]) are scaled in such a way that the regression of merit on index values is unity, e.g., $b_{HI} = 1$. The values of the b's in the index (equation [19.10]) are thus adjusted so that the metric values of the index I correspond numerically with the units in which merit H is expressed, whatever they are, when both are deviations from the mean. In

this way the index becomes a prediction of breeding value for merit. With $b_{HI} = 1$, it follows that $r_{HI} = \sigma_I/\sigma_H$, as shown in Chapter 13. The expected response to selection is the same as in Chapter 13, the predicted change in merit being given by

$$R_H = i r_{IH} \sigma_H \qquad \qquad \ldots [19.16]$$

or, if the index has not been rescaled,

$$R_H = i \sigma_I \qquad \qquad \ldots [19.17]$$

The variance of the index, from which σ_I can be evaluated, is as in equation [13.12] when there is only one character in merit, i.e., for single-trait selection. Extended to include multiple-trait selection it is as follows. To simplify the notation, let $\sum_1 aA$ be the sum of the terms on the right-hand side of the first equation of [19.15], $\sum_2 aA$ that of the second equation, etc. Then the variance is

$$\sigma_I^2 = b_1 \sum\nolimits_1 aA + b_2 \sum\nolimits_2 aA + \ldots + b_m \sum\nolimits_m aA \qquad \ldots [19.18]$$

The standard deviation of the index, σ_I, provides a simple way of comparing the relative efficiencies of different indices for improving merit because, as can be seen from equation [19.17], the response of merit is simply proportional to σ_I.

The response of any one of the component characters of the index or of merit can be predicted as follows. Suppose we want to predict the response of character 1. This is a correlated response of character 1 to selection for the index, and it is predicted by adaptation of equation [19.5a], putting character 1 in place of Y and the index I in place of X. The response of the index is the same as that of merit given in equation [19.17]. The correlated response of character 1 then becomes

$$CR_1 = b_{(A)11} i \sigma_I = \frac{\text{cov}_{(A)1I}}{\sigma_I^2} i \sigma_I = \frac{i}{\sigma_I} \text{cov}_{(A)1I} \qquad \ldots [19.19]$$

Here $\text{cov}_{(A)1I}$ is the additive genetic covariance of character 1 with the index, and it is obtained as follows. Multiplying equation [19.10] by A_1 gives the sum of products of A_1 with I as $b_1 P_1 A_1 + b_2 P_2 A_1 + \ldots$. If each P is now written as $(A + E)$, the products AE drop out because breeding values and environments are uncorrelated. The required covariance can now be seen to be as follows, where the variances and covariances are written in the notation of the index equations:

$$\text{cov}_{(A)1I} = b_1 A_{11} + b_2 A_{12} + \ldots + b_m A_{1m} \qquad \ldots [19.20]$$

These variances and covariances must be known for construction of the index, so substitution into equation [19.19] gives the predicted response.

The following two examples will make clearer what is involved in constructing an index and will bring in one or two points of interest that have not been explained above.

Example 19.4 The use of an index for the improvement of a single character will be illustrated from an experiment to be described in Example 19.5. Suppose the character to be improved is body weight in mice, and we consider using the correlated character tail length in an index. Let body weight be character 1 and tail length character 2. The parameters needed to construct the index are given in the table, with

Character	h^2	h	σ_P^2	σ_P	r_A	r_P
1 = Weight (g)	0.36	0.60	6.37	2.52		
					0.29	0.45
2 = Tail length (cm)	0.44	0.67	0.28	0.53		

$$P_{11} = 6.37 \qquad P_{22} = 0.28 \qquad P_{12} = P_{21} = 0.6010$$
$$A_{11} = 2.2932 \qquad A_{22} = 0.1232 \qquad A_{12} = A_{21} = 0.1557$$

values taken from Rutledge, Eisen, and Legates (1973). The index equations for solution, from equations [19.11], are

$$b_1 P_{11} + b_2 P_{12} = A_{11}$$
$$b_1 P_{21} + b_2 P_{22} = A_{21}$$

The values of the variances and covariances, calculated by equations [19.12], are given in the table. Substituting these in the above equations gives

$$6.37 b_1 + 0.6010\, b_2 = 2.2932$$
$$0.6010\, b_1 + 0.28\, b_2 = 0.1557$$

and the solution is

$$b_1 = 0.386 \; ; b_2 = -0.272$$

The index for selection is therefore

$$I = 0.386\ W - 0.272\ T$$

where W and T are the weight (g) and tail length (cm) respectively. The index can be rescaled for convenience by dividing all through by b_1, to give

$$I' = W - 0.705\ T$$

The index values are altered by the rescaling, but not the order of merit of the individuals.

Note that tail length is given a negative weighting in an index for increasing body weight. In other words, tail length is an indicator of environment, rather than breeding value, for weight. The reason for this is that the environmental correlation, which is 0.56, is much higher than the genetic correlation of 0.29. This illustrates a point not made previously, that a character may be useful in an index as an indicator of environmental deviations rather than of breeding values.

The usefulness of a secondary character can be judged from its weighting coefficient b_2. It can be shown that $b_2 = 0$ if the genetic and phenotypic regressions of character 2 on character 1 are the same; or, in terms of correlations, if $r_A/r_P = h_1/h_2$. Under these conditions a secondary character will give no benefit; in fact, errors of estimation will make it worse than useless (Sales and Hill, 1976).

To predict the response to selection, we have to calculate the variance of the index. For this purpose the unscaled index must be used. The variance, by equation [19.18], is

$$\sigma_I^2 = (0.386 \times 2.2932) + (-0.272 \times 0.1557) = 0.8428$$

and $\sigma_I = 0.918$. The response could then be predicted by equation [19.17] if the intensity of selection were known. In order to see how useful the secondary character would be, we can compare the expected responses to index selection and to simple selection, assuming the intensity of selection is the same. The response to simple selection, by equation [11.3], is $ih^2\sigma_P$. The ratio of the responses, index selection/

simple selection, is therefore $0.918/0.907 = 1.012$. The index would be only 1 per cent better than selection for body weight alone.

Finally, what would be the expected change of tail length resulting from selection for the index? The correlated response of character 1 is given by equations [19.19] and [19.20], but in this case we want the response of character 2. First, by equation [19.20],

$$\text{cov}_{(A)2I} = b_2 A_{22} + b_1 A_{21} = -0.0335 + 0.0601 = +0.0266$$

Then, by equation [19.19], the expected response of tail length is

$$CR_2 = +i(0.0266/0.9185) = +0.03i \text{ cm per generation.}$$

A very small increase of tail length is expected. It might seem at first that the negative weighting of tail length in the index should result in a decrease. But the correlated response depends on the genetic correlation, which is positive.

Example 19.5 The experiment with mice, from which the data for Example 19.4 were taken, applied index selection with the object of changing both body weight and tail length, and compared the observed with the expected responses (Rutledge, Eisen, and Legates, 1973). The objective was to change the body conformation by increasing one character and decreasing the other. Four lines were selected for seven generations, two selected for increased body weight and decreased tail length, and two selected in the opposite direction. The construction of the index for increasing body weight and decreasing tail length will be explained. The parameters needed are given in the table of Example 19.4. In addition we need the 'economic' weighting of the two characters. An equal change in standard-deviation units was desired for each character. The 'economic' weights, a, assigned were therefore the reciprocals of the phenotypic standard deviations, and these were $a_1 = 0.40$ and $a_2 = -1.89$.

The index equations, from equation [19.15], are

$$b_1 P_{11} + b_2 P_{12} = a_1 A_{11} + a_2 A_{12}$$
$$b_1 P_{21} + b_2 P_{22} = a_1 A_{21} + a_2 A_{22}$$

Substituting the variances and covariances, and the weights, leads to

$$6.37 b_1 + 0.601 b_2 = 0.9173 - 0.2943 = 0.6230$$
$$0.601 b_1 + 0.28 b_2 = 0.0623 - 0.2328 = -0.1706$$

and the solution is

$$b_1 = 0.195 \; ; b_2 = -1.027$$

The index for selection is thus

$$I = 0.195 \, W - 1.027 \, T$$

where W and T are an individual's body weight and tail length respectively. For selection in the opposite direction, to decrease W and increase T, the signs in the index are simply reversed.

The variance of the index, by equation [19.18], is

$$\sigma_I^2 = (0.195 \times 0.6230) + (-1.027 \times -0.1706)$$
$$= 0.2967$$
$$\sigma_I = 0.5447$$

The intensity of selection realized, averaged over the four lines, was $i = 1.01$. The expected response based on the selection actually applied was, by equation [19.17],

$$R_H = 1.01 \times 0.5447 = 0.55 \text{ index units per generation.}$$

The observed responses were 0.26 and 0.30 in the lines selected for increased W with decreased T, and 0.42 and 0.45 in the lines selected in the opposite direction. The reason for the observed responses being somewhat less than expected is probably that the parameters for construction of the index were not accurately estimated.

(Some of the quantities calculated in this example differ a little from those given in the original paper. The reason for this is that the parameters used in the paper were derived from the base population before selection, but the published parameters used here are those of unselected lines maintained concurrently with the selection.)

Effect of selection on genetic correlations

There is one important consequence of multiple-trait selection to be discussed before we leave the subject. Just as the heritabilities are expected to change after selection has been applied for some time, so also are the genetic correlations. If selection has been applied to two characters simultaneously, the genetic correlation between them is expected eventually to become negative, for the following reason. Those pleiotropic genes that affect both characters in the desired direction will be strongly acted on by selection and brought rapidly toward fixation. They will then contribute little to the variances or to the covariance of the two characters. The pleiotropic genes that affect one character favourably and the other adversely will, however, be much less strongly influenced by selection and will remain for longer at intermediate frequencies. Most of the remaining covariance of the two characters will therefore be due to these genes, and the resulting genetic correlation will be negative. The consequence of a negative genetic correlation, whether produced by selection in this way or present from the beginning, is that the two characters may each show a heritability that is far from zero, and yet when selection is applied to them simultaneously neither responds. We have already discussed, in Chapter 12, what is essentially the same situation resulting from the combined effects of artificial and natural selection: a selection limit is reached even though the character to which artificial selection is applied still shows a substantial amount of additive genetic variance.

The theoretical expectation that selection should change the genetic correlation has been tested in several experiments but, as so often with genetic correlations, the evidence is conflicting. For a review and discussion of this question see Sheridan and Barker (1974).

20 METRIC CHARACTERS UNDER NATURAL SELECTION

Throughout the discussion of the genetic properties of metric characters, which has occupied the major part of the book, very little attention has been given to the effects of natural selection, and something must now be done to remedy this omission. The absence of differential viability and fertility was specified as a condition in the theoretical development of the subject: that is to say, natural selection was assumed to be absent. Though for many purposes this assumption may lead to no serious error, a complete understanding of metric characters will not be reached until the effects of natural selection can be brought into the picture. The operation of natural selection on metric characters has, however, a much wider interest than just as a complication that may disturb the simple theoretical picture and the predictions based on it. It is to natural selection that we must look for an explanation of the genetic properties of metric characters which hitherto we have accepted with little comment. The genetic properties of a population are the product of natural selection in the past, together with mutation and random drift. It is by these processes that we must account for the existence of genetic variability; and it is chiefly by natural selection that we must account for the fact that characters differ in their genetic properties, some having proportionately more additive variance than others, some showing inbreeding depression while others do not. These, however, are very wide problems which are still far from solution, and in this concluding chapter we can do little more than indicate their nature. Before considering the ways in which natural selection affects metric characters, we shall give a brief account of natural selection itself and what it means.

Natural selection

Fitness and its components

The 'character' that natural selection selects for is *fitness*. The fitness of an individual is the contribution of genes that it makes to the next generation, or the number of its progeny represented in the next generation. *Relative fitness* is the fitness of an individual relative to the population mean, i.e., $W/\bar{W}$, if W is the individual's fitness. If a population is neither expanding nor contracting in numbers, the mean fitness of its individuals is 1 and then absolute fitness and relative fitness are the same. There are difficulties in defining fitness precisely. One such difficulty lies in separating the fitness of an individual from that of its parents. In a mammal, for example, the survival of the juvenile progeny depends partly on their viability, which is an aspect of their own fitness, and partly on the parental care that they received, which is an aspect of the parents' fitness. This overlap of fitness from one generation to the next means that there is no precise time in an individual's life at which we can say that its attributes reflect its own fitness rather than that of its parents. The point of separation between the generations must therefore be a more or less arbitrary choice.

The mean fitness of a population is a concept that has to be used with great care. It was said above that the individuals of a population have a mean fitness of 1 if the population is neither increasing nor decreasing in numbers. This seems simple enough. But whether a population increases or decreases or remains constant in numbers depends to a large extent on the environmental resources available to it. Natural selection between individuals within a population may change the genetic constitution of the population, but the mean fitness will not change if the population is already at the limit of the carrying capacity of its environment. When the mean fitness is referred to in what follows, it will be assumed that the population is not limited by environmental resources.

The fitness of an individual is the final outcome of all its developmental and physiological processes. The differences between individuals in these processes are seen in variation of the measurable attributes which can be studied as metric characters. Thus the variation of each metric character reflects to a greater or lesser degree the variation of fitness; and the variation of fitness can theoretically be broken down into variation of metric characters. Consider, for example, a mammal such as the mouse. Figure 20.1 illustrates the hierarchy of characters contributing to the fitness of females. Nearly all of the characters shown have been studied genetically as metric characters. Fitness itself can be broken down

1	2	3	4
		Viability	Disease resistance / Predator avoidance
	Total number of offspring born (Fertility)	Mating success	
		Litter size	Ovulation rate / Embryo survival
Fitness		Frequency of litters / Number of litters	
	Quality of offspring weaned (Maternal performance)	Milk-yield	Mammary-gland size
		Maternal behaviour	

Fig. 20.1. Some of the components of fitness of a mammal such as the mouse, to show the heirarchy of causes of variation. Variation of each of these metric characters is associated, to a greater or lesser degree, with variation of fitness.

into two major components, the total number of offspring produced and the quality of these offspring, which might be measured as their weaning weights. The variation of the major components, if properly measured, would account for all the variation in fitness. The variation of the major components can in turn be attributed to other characters, some of which are shown in column 3 of the diagram. These again are influenced by others, a few of which are in column 4. The characters in column 4 are themselves influenced by many others. Among these, for example, are physiological functions such as the output of the various gonadotrophic hormones which influence ovulation rate, embryo survival, and milk-yield. There are, in addition, characters whose influences on fitness are less direct and less obvious, but which are correlated with some of the components of fitness. Body size, for example, is correlated with several, perhaps all, of the characters in column 4. The problem we have to examine is how all these metric characters are affected by selection for fitness, and how the action of natural selection is related to the character's position in the hierarchy.

Measurement of fitness. It is very difficult to measure fitness directly, particularly the fitness of individuals. It is less difficult to measure the major components of fitness separately. The overall fitness can then be estimated by combining the values of the components. A measure related to the mean fitness of a strain can be obtained by rearing it in competition with a tester strain which is genetically marked so that the offspring produced by crossing can be identified. The relative numbers of progeny produced by the two strains then provide a 'competitive index' of the strain under test. For details of the application of this method to *Drosophila*, see Latter and Robertson (1962). In industrialized societies of man, by far the most important component of fitness is the number of children reared, since mortality between childhood and the end of reproduction is very low. The size of completed families therefore provides a fairly good measure of the fitness of the two parents jointly though, of course, it includes infant and childhood survival in the parents' fitness. An improved measure can be obtained by taking account of the rate of reproduction, i.e., the average age of the parents at the birth of their children (see Waller, 1971).

Relationships between metric characters and fitness

We are now in a position to discuss the ways in which natural selection affects characters of different sorts, as described above. We have to consider what sort of selection is being applied to the character, what this will do to the frequencies of the genes concerned, and how the genetic properties of the character are thereby influenced. We shall be concerned mainly with populations that are approximately in equilibrium, and so the properties of equilibrium populations will be described first.

Equilibrium populations

A population in equilibrium is one in which gene frequencies are not changing at any loci. Consequently the mean values of all metric characters remain constant and, despite continued natural selection, fitness does not increase. Or, to put it in another way, the population has reached a selection limit for fitness. Probably few real populations are strictly in equilibrium at selection limits, because the attainment of the limit would require the environment to have remained unchanged over a long time. Nevertheless, most populations studied are probably near enough to equilibrium that we can infer from them what the genetic properties of an equilibrium population are.

Abundant evidence proves that virtually all metric characters are genetically variable in populations that are more or less in equilibrium, including characters that affect fitness. There must therefore be genetic variance of fitness. But, since selection for fitness produces no response, there can be no additive genetic variance of fitness; so all the genetic variance of fitness must be non-additive, i.e., variance due to dominance and epistatic interactions. The array of gene frequencies in an equilibrium population is the best, in the circumstances, for maximizing fitness.

Now, if selection is applied to any metric character that is not fitness itself, the gene frequencies at loci affecting the character must change if there is a response. Fitness must therefore be reduced as a correlated response, unless the character selected is controlled entirely by genes with no effects on fitness. This expectation is amply born out by experience: experimental selection for metric characters almost always results in a reduction of one or more of the major components of fitness. To give just one example: the mean fitness of *Drosophila* was estimated as a competitive index after five generations of selection for abdominal bristle number (Latter and Robertson, 1962). There were two lines selected upwards and two downwards. The mean fitness, relative to an unselected control, was 79 per cent in the upward selected lines and 65 per cent in the downward selected lines.

If artificial selection is carried out and is then suspended before much of the variation has been lost by fixation, natural selection must tend to bring the gene frequencies back toward their equilibrium values, and the mean of the character artificially selected is expected to revert toward its original value. This tendency for natural selection to resist changes of gene frequency is known as *genetic homeostasis* (Lerner, 1954). Its effect can often be seen in experimental selection when the weighted selection differential is less than the unweighted (see Example 11.5).

If the environment to which an equilibrium population is adapted changes, the array of gene frequencies is no longer optimal. The changed environmental circumstances alter the relative weighting of the components of fitness, so that fitness now has some additive variance and can respond to natural selection. The application of artificial selection can be thought of as changed environmental circumstances in this way, altering the weighting of the components in the combined natural and artificial fitness. The components of fitness in human populations have changed drastically in the recent past as a result of medicine and contraception, psychological and behavioural factors having largely re-placed physiological factors as determinants of fitness.

'Fitness profiles'

The question about how a metric character is related to fitness can be put more precisely as follows: if two individuals differ in their genotypic values of the character, how do they differ in fitness? Suppose we could measure the fitness and

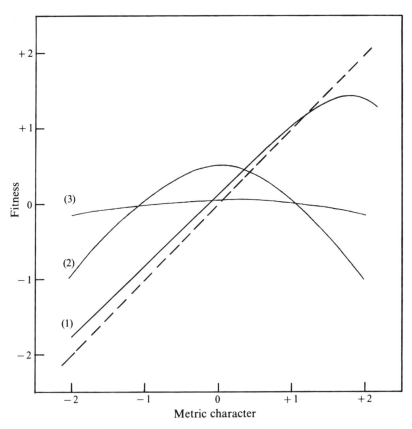

Fig. 20.2. 'Fitness profiles' as explained in the text. The scales on both axes are standard deviations from the means.

the genotypic value of each individual. If we now plot fitness against the metric character we should get what may be called a 'fitness profile'. Figure 20.2, which is based on the ideas of A. Robertson (1955), illustrates three such profiles, representing characters having different relationships with fitness. These will first be described briefly and then considered in turn more fully.

The broken diagonal line is the profile that would be obtained if the metric character measured was fitness itself. Curve (1) is the profile of a major component of fitness, such as the total number of offspring born. As this increases from its lowest values, fitness increases almost linearly and at a rate nearly equal to fitness itself. But at the upper end of the range there is a point above which a further increase of the character leads to a reduction of fitness. This bending down of the profile at the upper end results from interactions with other components. To take number born as an example: at the low end of the range each additional young born results in nearly one additional offspring reaching adulthood. But as the number born increases, their 'quality' is progressively reduced by limitations of maternal performance until, above a certain number, the reduced quality outweighs the extra numbers, and fewer offspring survive to adulthood. In profile (2) the fittest individuals are those with values of the character at or near the mean. Body size of many organisms is probably of this sort. The reasons for characters having an intermediate optimum will be discussed later. Many characters must be expected to have profiles falling between curves (1) and (2), with optima at some distance from the mean. Finally, profile (3) represents a character that is neutral, or nearly so, with respect to fitness. There are almost no differences of fitness among individuals with different values of the character. The number of abdominal bristles in *Drosophila* is a character of this sort. The evidence for characters being neutral will be given later.

In earlier chapters characters have been referred to as being 'closely connected' with fitness, but the precise meaning of the 'close connection' was not explained. It can now be seen that characters closely connected to fitness are the major components, with fitness profiles like curve (1). The closeness of the connection falls off as the profile approaches that of curve (2). The generalization was made in earlier chapters that characters closely connected with fitness tend to have low heritabilities and to be severely affected by inbreeding depression. The reasons for these genetic properties can be understood from the effect of natural selection on these characters, as follows.

Major components

The essential feature of characters with fitness profiles like curve (1) is that the population mean of the character is below its optimal value for fitness: individuals above the mean for the character are of above-average fitness. Natural selection favours these individuals but, despite the correlated selection differential on the character, the mean of the character is unchanged. One might say that natural selection is trying to increase the character but cannot do so. This situation could arise from two genetic causes: either from genes at more or less intermediate frequencies that are overdominant with respect to fitness, or from deleterious recessives maintained at low frequencies by mutation balancing the selection. In either case, the variance of the character would be mainly non-additive, and dominance would be directional; the character would consequently have a low heritability and be subject to inbreeding depression. A comparison of twelve characters in *Drosophila* showed that they all conformed to this expectation (Kearsey and Kojima, 1967). All the characters that were measures of a major component of fitness showed epistatic interaction and strong directional dominance; all the others – measures of body size and bristle numbers – showed little or no dominance or interaction.

It is not known whether the non-additive genetic variance of the major components comes mainly from overdominant loci or from rare recessives. Genes at intermediate frequencies cause more variation than genes at low frequencies. It is possible therefore that most of the variance comes from overdominant loci and most of the inbreeding depression from rare recessives (Crow, 1952).

Fitness may be thought of as an index by which natural selection selects simultaneously for all the major components. We should then expect additive genetic correlations between characters that are major components of fitness to be negative, for the reasons given at the end of the previous chapter. But deleterious genes would be likely to have deleterious effects on more than one component of fitness, and so to contribute positively to the genetic correlation.

Therefore if the genetic variance of the major components is mainly due to rare recessives, the net genetic correlation between the major components might be positive.

Characters with intermediate optimum

Characters with a fitness profile like curve (2) in Fig. 20.2 are said to have an intermediate optimum because individuals with values of the character near the population mean have the highest fitness, and selection appears to favour intermediates. Selection that appears to favour intermediate values is known as *stabilizing selection*. There is, however, an ambiguity in the way the terms 'intermediate optimum' and 'stabilizing selection' are used. They may be used purely as a description of the fitness profile in which intermediates have the highest fitness. Or they may be used in an operative or functional sense, meaning that the character is a criterion of selection, intermediates being favoured because they have that value of the character. When artificial stabilizing selection is applied, the selection is functional because the character is a criterion of selection. But with natural selection, the shape of the fitness profile is not enough to tell us that functional stabilizing selection is operating. The reasons for this will be clearer when we consider examples of characters with intermediate optima.

The evidence from which stabilizing natural selection is usually inferred rests on showing that the phenotypic variance is reduced by the selection. This is done by comparing the variances before and after selection, or among the survivors and the dead, or under different intensities of natural selection. An example of the last comparison is shown in Fig. 20.3(*a*). Sometimes more direct evidence can be obtained by showing that intermediates have higher values of one of the major components of fitness, such as survival. To prove that intermediates have the highest fitness, however, is not enough to tell us how natural selection acts on the character and on the genes that affect the character. For this it is necessary to know why intermediates are fittest, and the answer to this question is not always obvious. The effect of the selection depends on whether the value of the character itself is a direct cause of fitness, or whether the connection with fitness is through pleiotropic effects of the genes. The problem will be best explained by considering some examples, the first being a direct causal effect on fitness and the last having no causal effect.

1. A straightforward example of a character having a direct causal effect on fitness would be any measure of the thermal insulation of a mammalian coat. The conflicting needs for conserving heat during inactivity and dissipating heat during activity are balanced at an intermediate coat density. Intermediates are favoured because that value of the character is best. The selection is functional or true stabilizing selection. A change of the mean in either direction would reduce the mean fitness, and the only way by which the mean fitness could be increased is by reducing the variance. (This example, it must be said, is conjectural: it might be found that over the range of variation in any real population the differences in fitness associated with the character were fairly small.)

2. Clutch size in birds is a well-known example of an intermediate optimum The number of eggs laid (clutch size) is a direct cause of fitness, and if no other factor were involved the birds laying the largest clutches would be fittest. But the

number of offspring that can be reared is limited by the available food supply, and an intermediate clutch size results in the largest number reared. The mean clutch size has been proved to be at or near the optimum in several species (Lack, 1966). The intermediate optimum might be said to be imposed by the environment or, alternatively, be said to result from a strong interaction between the two major components of fitness, number and quality of the young. The selection is again true stabilizing selection, but it differs from the previous example in one respect: the mean fitness would be improved by an increase of clutch size if at the same time the birds' skill in obtaining food could be increased.

3. Next consider a character that itself has very little effect on fitness but which appears to have an intermediate optimum as a result of its correlations with other characters that do affect fitness. Body size of mice provides a plausible though not fully proved example. Body size in females is positively correlated with the size of litter that they bear (Falconer, 1965b), and so with the major component of fitness, number born. If this were the only factor, the optimum body size would be above the mean. But body size is negatively correlated with reactivity, or wildness. Large mice are placid and unreactive to disturbance, whereas small mice are alert and react vigorously to disturbance (MacArthur, 1949; Falconer, 1953). Under natural conditions one must suppose that larger mice would be less good at escaping predators than small mice. The intermediate optimum body size thus results from opposing correlations with different components of fitness, a positive correlation with number born and a negative correlation with length of life. The stabilizing selection in this case is spurious because the criterion of selection is not body size but the characters correlated with it.

4. Finally, an example that illustrates well the difficulties in interpreting an intermediate optimum is provided by sternopleural bristle number in *Drosophila melanogaster* (Kearsey and Barnes, 1970). The character has the clear appearance of being subject to stabilizing selection, but the apparent selection has nothing whatever to do with the character itself. The sternopleural bristles are small bristles on the sides of the thorax of the adult flies. The population studied was derived from a cross of two strains selected in opposite directions; it had a mean of 18.5 bristles in females, with a range of approximately 10 to 45. The evidence for stabilizing selection is given in Fig. 20.3(a), which shows the distributions of bristle number in flies grown in uncrowded conditions (solid line) and in crowded conditions (broken line). In crowded conditions the natural selection for larval survival is much stronger and its effect in eliminating flies with the more extreme bristle numbers is clearly seen. A fitness profile can be constructed from the reduction in frequency of flies with different bristle numbers. This profile, however, is not of fitness itself but of one major component, larval survival. Figure 20.3(b) shows the profile, which has a sharp peak of fitness at the mean bristle number. The important point about this fitness profile, however, is that the selection which gives rise to the intermediate optimum takes place in the larval stages, before the flies have developed any bristles. The superior fitness of intermediates is therefore in no way caused by the character itself, but must result from pleiotropic effects of the genes on the larval characters that contribute to survival. The stabilizing selection is spurious, as in the previous example, but in

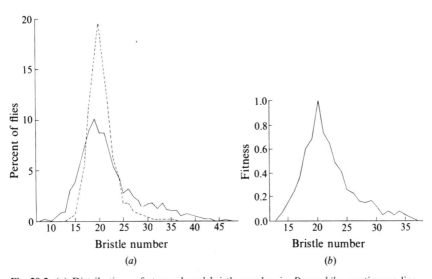

Fig. 20.3. (*a*) Distributions of sternopleural bristle number in *Drosophila*: continuous line, under weak natural selection; broken line, under strong natural selection.
(*b*) Fitness profile derived from the frequency distributions in (*a*). The fitness scale is relative to the fittest bristle class. The fitness measured is only the component, survival from egg to adult, and not the whole of fitness. (*Adapted from Kearsey and Barnes, 1970.*)

this case we do not know what the real criteria of selection are, nor why adults with intermediate bristle numbers are fittest in the larval stages.

Effects of stabilizing selection. The question to which the foregoing discussion has been leading is: How does stabilizing selection affect the genes that affect the character subject to it? The answer to this question is too complex to give in detail here, but the main conclusions reached by theoretical study are the following (for details, see Robertson, 1956; Lewontin, 1964; Curnow, 1964; Bulmer, 1971, 1976). First, it depends on whether the stabilizing selection is real or spurious. If it is spurious there are no means of knowing its effects without knowing what are the components of fitness being selected. One possibility is the following (Robertson, 1956). The genes affecting the major components of fitness may have pleiotropic effects on the character in question, and these effects on the character may be more or less additive. Individuals with intermediate values of the character must then be heterozygous at more loci than extreme individuals. If the loci have overdominant effects on fitness, then natural selection favours heterozygotes and in consequence appears to favour intermediates for the character. The consequence of selection acting in this way would be the maintenance of the genetic variation of the character. In the case of the sternopleural bristles described above, however, it was shown from the shape of the fitness profile that the apparent stabilizing selection could not be accounted for in this way (Kearsey and Barnes, 1970).

Real stabilizing selection has two main effects. First, it favours genotypes with the least variability (Curnow, 1964). Therefore, unless the least variable genotypes are heterozygotes, it tends to fix genes that confer the greatest developmental stability, irrespective of whether they affect the mean of the character or not. Stabilizing selection is thus expected to increase canalization of development (Waddington, 1957). Second, it tends to reduce the genetic variance of the character and this it does in two ways (Bulmer, 1971, 1976). The more immediate, and at first the largest, effect is through the creation of gametic phase disequilibrium. The selection causes allelic effects at different loci to be negatively correlated in individual genotypes. The covariance term in equation [8.10] is therefore negative and the genetic variance is reduced. In so far as the genes are linked, the selection tends to build up 'balanced' combinations in which linked genes are in predominantly repulsion linkage, so that the effect of the chromosome as a whole is minimized (see Mather, 1941; Lewontin, 1964). The selection, however, has to be very strong, or the linkage very close, to keep the genes in combinations that are appreciably different from a random arrangement (see Wright, 1969, p. 92). The second way by which the genetic variance is reduced is through changes of gene frequencies at loci that affect the mean of the character. Provided the genotypes do not differ much in variability, and provided the loci do not affect fitness in any other way than through the character, then stabilizing selection tends to change the gene frequencies toward fixation, and so to reduce the genetic variance (Robertson, 1956). However, the gene frequencies change only slowly, unless selection is intense or the genes have large effects. The consequences of real stabilizing selection are thus to reduce both environmental and genetic variance. Both these effects have been observed experimentally; for example, with sternopleural bristle number in *Drosophila* (Gibson and Bradley, 1974), and with pupa weight in *Tribolium* (Kaufman, Enfield, and Comstock, 1977).

Disruptive selection is the opposite of stabilizing selection, intermediates being selected against. Its expected effects are the opposite of stabilizing selection, an increase of both genetic and environmental variance, and these effects have again been observed experimentally. For example, when applied to larval development-time in *Drosophila* (Prout, 1962), the phenotypic variance was decreased by stabilizing selection and increased by disruptive selection. In this case, however, the components of variance did not change in quite the expected way: stabilizing selection reduced the additive genetic variance but not the environmental variance, while disruptive selection increased the environmental but not the genetic variance. For a review of disruptive selection, see Thoday (1972).

Neutral characters

Evidence that a character is neutral with respect to fitness, having a profile like curve (3) in Fig. 20.2, can be obtained from a 'perturbation' experiment. A population is subjected to a few generations of directional selection for the character; then, when the mean has been changed some way from its original value, selection is suspended and the population is allowed to breed at random, subject only to natural selection. If the mean does not revert to its original value, or does so only very slowly, it can be concluded that the character is neutral, or nearly so. Strictly speaking, reversed selection should also be applied, to prove that the mean can be brought back. This is the evidence for abdominal bristle number in *Drosophila* being cited earlier as an example of a neutral character (Latter and Robertson, 1962). If a character is proved to be neutral in this way, it does not mean that the genes affecting the character have no pleiotropic effects on fitness. It means only that, over the range covered by the perturbation, the character itself is not subject to natural selection. On the principle of genetic homeostasis, the gene frequencies must tend to revert if the perturbation is large, and the mean of the character may then move some way back toward its original value as a correlated response. To say that a character is neutral with respect to fitness is not to say that what is measured has no function. The abdominal bristles of *Drosophila* doubtless have a function, perhaps an important one, in the life of the adult fly. All that their neutrality means is that the precise number is not important for the fitness of the fly.

Response of fitness to selection

The foregoing section dealt with populations that are in equilibrium, or nearly so. Let us now consider populations that are not in equilibrium, in which fitness is not maximal, and in which there is some additive genetic variance of fitness. How fast will fitness increase under natural selection? This problem has an elegant solution, known as *Fisher's fundamental theorem*, which states that the increase of fitness in one generation is equal to the additive genetic variance of fitness. This theorem has given rise to a great deal of discussion about its validity and its generality. A proof of the theorem and an explanation of why it has caused so much difficulty is given by Price (1972). The conclusion can be very simply demonstrated for populations with non-overlapping generations, by consideration of the weighted selection differential, as follows.

The response to selection for a metric character is predicted from the selection differential by equation [11.2] $(R = h^2S)$. The selection differential S is the weighted mean superiority of the selected parents, the weights being the relative contribution of progeny from which the response is evaluated. The weighted selection differential is thus given by

$$S = \frac{\sum k(X - \bar{X})}{N}$$

where $\bar{X}$ is the population mean, X is the value of an individual parent, k the number of its progeny, and $N = \sum k$ is the total number of progeny. The number of offspring, k, of any parent is that parent's absolute fitness, W. So the weighted selection differential on fitness becomes

$$S_W = \frac{\sum W(W - \bar{W})}{N}$$

$$= \frac{\sum W^2}{N} - \frac{\sum W}{N}\bar{W}$$

$$= \overline{W^2} - (\bar{W})^2$$

$$= \sigma_W^2 = V_{P(W)} \qquad \qquad \dots [20.1]$$

(The last step in the derivation was explained in connection with equation [3.4]). Thus the selection differential on fitness is equal to the phenotypic variance of fitness, and we may note incidentally that the intensity of selection is equal to the phenotypic standard deviation:

$$i_W = S_W/\sigma_W = \sigma_W \qquad \qquad \ldots [20.2]$$

Substituting equation [20.1] into equation [11.2] shows that the response of fitness to natural selection is equal to the additive genetic variance of fitness:

$$R_W = h_W^2 V_{P(W)} = V_{A(W)} \qquad \qquad \ldots [20.3]$$

Correlated responses. A question of more interest, perhaps, than the change of fitness itself is the change of a metric character resulting from natural selection. For example, it used to be thought that natural selection was tending to reduce human intelligence because children from larger families have lower IQs than those from smaller families; in other words, there seemed to be a correlated selection differential for reduced IQ. In this case, however, the supposition about the selection differential is false for two reasons: first, because parents with zero fitness have no children and so cannot appear in the data; and second, because it is based on the correlation between parents' fitness and children's IQ. When the parents' IQ is compared with their own subsequent number of children as a measurement of fitness, the correlated selection differential is found to be slightly positive (Waller, 1971).

If a correlated selection differential is observed, can anything be deduced about the change in the character from natural selection? The correlated selection differential, S_Y', on a character Y is the weighted mean of Y, the weight being the individual's fitness. Thus

$$S_Y' = \frac{\sum W(Y - \bar{Y})}{N}$$

and by a derivation similar to that of equation [20.1] this leads to

$$S_Y' = \mathrm{cov}_{P(YW)} \qquad \qquad \ldots [20.4]$$

i.e., the correlated selection differential is equal to the phenotypic covariance of the character with fitness. The correlated response can now be obtained from equation [19.8c], i.e., $CR_Y/S_Y' = \mathrm{cov}_{(A)}/\mathrm{cov}_{(P)}$. From this and equation [20.4], the correlated response is found to be

$$CR_Y = \mathrm{cov}_{A(YW)} \qquad \qquad \ldots [20.5a]$$

i.e., the additive covariance of the character Y with fitness. Alternatively, the covariance can be written in terms of the genetic correlation and heritabilities. This gives

$$CR_Y = r_A h_Y h_W \sigma_Y \sigma_W \qquad \qquad \ldots [20.5b]$$

where σ denotes the phenotypic standard deviation. Thus, to predict a correlated response to natural selection it would be necessary to know the genetic correlation between the character and fitness, and the heritability of the character and of fitness. (Equation [20.5a] was derived by Robertson (1966) in connection with an analogous problem in dairy cattle, namely the correlated responses expected from the overall selection applied by farmers.)

Origin of variation by mutation

Both random drift and stabilizing selection tend to reduce genetic variation. This raises the question of whether mutation produces new variation at a rate sufficient to counteract these tendencies. The rate of origin of new variation can be estimated from inbred lines, by seeing how much variation accumulates when an inbred line is random-mated, or is selected for a quantitative character without further inbreeding. The new variation of abdominal and sternopleural bristle number in *Drosophila* has been estimated in this way (Clayton and Robertson, 1955; Paxman, 1957). In both characters, the average amount of variation generated by mutation over one generation was about one-thousandth part of the genetic variation in the base population. In other words, it would take about 1,000 generations for mutation to restore the genetic variance to its original level. Roughly similar estimates (summarized by Lande, 1976) have been obtained for characters in mice and maize. Though mutation produces new variation exceedingly slowly it is not negligible as a source for restoring the variation lost by random drift or stabilizing selection.

Consider first the balance against loss of variance by random drift. If all the genetic variance is additive, the proportion of the original variance remaining when the inbreeding coefficient is F, is given by $1 - F$ (Table 15.1). In one generation the variance is reduced by the proportion ΔF, which is the inbreeding increment and, by equation [3.7], $\Delta F = 1/2N_e$, where N_e is the effective population size. Therefore the loss of variance by random drift will be balanced by its gain from mutation, at the rate given above, when the population size is $N_e = 500$. Thus mutation seems to be enough to maintain the variance of metric characters in populations of moderate size. Now consider the loss from stabilizing selection. The rate of loss depends on the shape of the fitness profile and the intensity of the selection, and on the number of loci contributing to the variance, or the magnitude of the individual effects of the genes. By making reasonable assumptions about these parameters, and taking the rate of origin of variation by mutation given above, it has been shown that mutation can maintain the amount of variation observed in characters subject to stabilizing selection (Lande, 1976). Heterozygote advantage is, of course, another way by which variance is conserved. But it seems from the conclusions outlined above, that it is not necessary to invoke heterozygote advantage for maintaining the variance of characters subject to stabilizing selection.

The mutation of genes affecting a major component of fitness has been studied in a different way. The component was the larval survival from egg to adult of *Drosophila* (Mukai et al., 1972). By the use of special marker stocks, second chromosomes were kept intact through 40 generations and mutations allowed to accumulate. The homozygous effects of these chromosomes on larval survival were tested at 10-generation intervals, from which the rate of origin of mutants on chromosome-2 was deduced. There were two classes of mutated genes: those that were lethal in homozygotes, and those that were mildly detrimental, with viabilities down to about 0.6 of normal. After 40 generations, 24 per cent of chromosomes carried one or more lethals, and the rate of origin was 0.6 per cent per chromosome per generation. The detrimentals were about 10 to 20 times as frequent, with a rate of origin of about 10 per cent per chromosome per generation. Scaling up by 5/2 to represent the whole genome, this means that after one generation of mutation, about 25 per cent of gametes carry a new mildly detrimental gene. This is clearly enough to maintain the genetic variance in the face of selection against the detrimental genes. It may be noted that, whereas lethals were found to be almost completely recessive, mildly detrimental genes were not. This inverse relationship between dominance and severity of effect is a general feature of deleterious genes in *Drosophila*; see the review by Simmons and Crow (1977).

The genes causing quantitative variation

Finally, some comment must be made on the nature of the genes that cause variation of metric characters. Virtually everything that can be measured as a 'character', in almost any organism, has been found to be genetically variable. Estimates of the number of loci causing the variation of any one character, though these estimates are very rough and not very meaningful, indicate numbers of the order of 10 to 100 at least. The variation of all characters together clearly requires allelic differences at many hundreds or thousands of loci, even though many genes affect more than one character. What do these genes do in the development and life of the organism? It seems most unlikely that they are there solely to 'control' the character through which we recognize them. Their effects on metric characters are much more likely to be secondary to some other function. Are these effects on metric characters characteristic of any particular category of genes? There is at present no convincing evidence that they are (Thoday, 1977). It seems most likely, therefore, that loci of all sorts influence metric characters, no matter what the function of their gene product is. Let us, however, look at this question a little more closely.

The genes whose variation we know about fall, broadly speaking, into two categories. First there are deleterious genes maintained at low frequencies by the balance of selection against mutation. Some of these are the mutants of Mendelian genetics, which have easily recognizable phenotypic effects. All the genes may have pleiotropic effects on metric characters. Though they are mainly recessive or nearly recessive in their effects on fitness, they may be additive in their effects on metric characters. These genes may be responsible for a large part of the variance of the major components of fitness, but it does not seem possible that they can be the major cause of variation of characters not closely connected with fitness, because the experimental evidence does not point to this variation being due to genes at low frequencies.

The second category of genes whose variation we know about are the genes involved in the polymorphisms of enzymes, structural proteins, and antigens. These genes are at intermediate frequencies, and in this respect are more likely causes of quantitative variation. The reasons for their being at intermediate frequencies are disputed, as was explained in Chapter 2. Many people believe that they are maintained by some form of balancing selection. If this view is right, these genes must affect fitness, and must therefore affect the characters that influence fitness. There is, however, very little evidence that the protein polymorphisms do affect metric characters. There are two ways in which effects can be looked for.

First, by comparing the mean of the character in different genotypes of the protein locus, and second, by looking for changes in their gene frequencies resulting from artificial selection for the character and too great to be accounted for by random drift. For example, Niemann-Sørensen and Robertson (1961) looked for effects of ten blood-group loci in cows by comparing body weights, milk-yields, and fat percentages in the milk of individuals with different alleles. Two significant effects were found, of which the clearest was an increase of fat percentage associated with one particular allele of the B blood-group system. The magnitude of the effect was 0.23 phenotypic standard deviations, and the segregation of this allele accounted for about 2.8 per cent of the genetic variance.

As an example of the second method, Garnett and Falconer (1975) looked for differences of gene frequency at five polymorphic enzyme loci in mice, resulting from selection for large and small body size. No differences greater than were expected from random drift were found. There is evidence of a different kind that the mildly detrimental genes affecting larval survival in *Drosophila* are not structural genes for enzymes (Mukai and Cockerham, 1977). This conclusion is based on estimates of the mutation rates of five enzyme loci on the second chromosome. The total number of structural genes on the chromosome is assumed to be the same as the number of salivary chromosome bands. If they all mutate at the estimated rate, they would account for only 10 to 20 per cent of the observed rate of mutation of detrimental genes. This is not to say that no enzyme variants affect larval survival, but it shows that the majority of the genes affecting larval survival cannot be enzyme variants.

If the deleterious recessives and the known protein polymorphisms are not the main sources of variation of metric characters, where else could we look? There seem to be three possibilities. The first, and most obvious, is that they are variants of regulatory genes, but this gets us little further toward understanding their function. (See McDonald and Ayala, 1978, for a discussion of regulatory genes in this connection.) Second, many of the known protein polymorphisms are enzymes whose function is in the day-to-day living of the adult organism. The variation of many metric characters, however, depends on developmental processes in morphogenesis, and we know very little yet about the genes that control development. Third, it is possible that quantitative variation is in part caused by genes that do not code for proteins, or by some of the various classes of DNA whose function is not yet known. It is known that abdominal bristle number in *Drosophila melanogaster* is affected by the *bobbed* locus which codes for ribosomal RNA (Frankham, Briscoe, and Nurthen, 1978). The new variants are believed to have arisen by unequal crossing over giving an altered number of tandem repeats of the reiterated sequences. These authors suggest that many metric characters may be affected by variation arising in this way at other loci with multiple copies tandemly repeated.

To answer the question about what kinds of loci cause variation of metric characters, the segregation of individual loci must be made recognizable, so that their individual effects can be studied. For example, loci affecting sternopleural bristles in *Drosophila* have been studied in this way (Spickett and Thoday, 1966). Five loci were identified, each with characteristic effects on the bristles.

APPENDIX TABLES

Appendix Table A Truncated normal distribution – large sample. p = proportion of population with values exceeding the truncation point T. x = deviation of T from the mean, in standard-deviation units. i = mean deviation of individuals with values exceeding T, in standard-deviation units from the population mean. For values of p greater than 50 per cent: take x and i tabulated for $(1 - p)$; give x a negative sign; multiply i by $(1 - p)/p$, retaining the positive sign. Errors from linear interpolation of p are positive, the largest in both x and i being approximately $+0.001$ when $p > 0.10$ per cent. (Abridged from Falconer, 1965a.)

$p\%$	x	i	$p\%$	x	i	$p\%$	x	i
0.01	3.719	3.960	0.75	2.432	2.761	10	1.282	1.755
0.02	3.540	3.790	0.80	2.409	2.740	11	1.227	1.709
0.03	3.432	3.687	0.85	2.387	2.720	12	1.175	1.667
0.04	3.353	3.613	0.90	2.366	2.701	13	1.126	1.627
0.05	3.291	3.554	0.95	2.346	2.683	14	1.080	1.590
0.06	3.239	3.507	1.00	2.326	2.665	15	1.036	1.554
0.07	3.195	3.464				16	0.994	1.521
0.08	3.156	3.429	1.0	2.326	2.665	17	0.954	1.489
0.09	3.121	3.397	1.2	2.257	2.603	18	0.915	1.458
0.10	3.090	3.367	1.4	2.197	2.549	19	0.878	1.428
			1.6	2.144	2.502	20	0.842	1.400
			1.8	2.097	2.459	21	0.806	1.372
0.10	3.090	3.367	2.0	2.054	2.421	22	0.772	1.346
0.12	3.036	3.317	2.2	2.014	2.386	23	0.739	1.320
0.14	2.989	3.273	2.4	1.977	2.353	24	0.706	1.295
0.16	2.948	3.234	2.6	1.943	2.323	25	0.674	1.271
0.18	2.911	3.201	2.8	1.911	2.295	26	0.643	1.248
0.20	2.878	3.170	3.0	1.881	2.268	27	0.613	1.225
0.22	2.848	3.142	3.2	1.852	2.243	28	0.583	1.202
0.24	2.820	3.117	3.4	1.825	2.219	29	0.553	1.180
0.26	2.794	3.093	3.6	1.799	2.197	30	0.524	1.159
0.28	2.770	3.070	3.8	1.774	2.175	31	0.496	1.138
0.30	2.748	3.050	4.0	1.751	2.154	32	0.468	1.118
0.32	2.727	3.030	4.2	1.728	2.135	33	0.440	1.097
0.34	2.706	3.012	4.4	1.706	2.116	34	0.412	1.078
0.36	2.687	2.994	4.6	1.685	2.097	35	0.385	1.058
0.38	2.669	2.978	4.8	1.665	2.080	36	0.358	1.039
0.40	2.652	2.962	5.0	1.645	2.063	37	0.332	1.020
0.42	2.636	2.947				38	0.305	1.002
0.44	2.620	2.932				39	0.279	0.984
0.46	2.605	2.918	5.0	1.645	2.063	40	0.253	0.966
0.48	2.590	2.905	5.5	1.598	2.023	41	0.228	0.948
0.50	2.576	2.892	6.0	1.555	1.985	42	0.202	0.931
			6.5	1.514	1.951	43	0.176	0.913
			7.0	1.476	1.918	44	0.151	0.896
0.50	2.576	2.892	7.5	1.440	1.887	45	0.126	0.880
0.55	2.543	2.862	8.0	1.405	1.858	46	0.100	0.863
0.60	2.512	2.834	8.5	1.372	1.831	47	0.075	0.846
0.65	2.484	2.808	9.0	1.341	1.804	48	0.050	0.830
0.70	2.457	2.784	9.5	1.311	1.779	49	0.025	0.814
0.75	2.432	2.761	10.0	1.282	1.755	50	0.000	0.798

Appendix Table B Truncated normal distribution – small sample. The tabulated values are the intensity of selection, i, when n individuals are selected from a total of N. Errors from linear interpolation of N are negative, the largest being approximately -0.0075; interpolation of n gives positive errors, maximum about $+0.006$. (Abridged from Becker, 1975, where much more extensive tables may be found.)

n	N 2	3	4	5	6	7	8	10	12	n
1	0.564	0.846	1.029	1.163	1.267	1.352	1.424	1.539	1.629	1
2	—	0.423	0.663	0.829	0.954	1.055	1.138	1.270	1.372	2
3	—	—	0.343	0.553	0.704	0.821	0.916	1.065	1.179	3
4	—	—	—	0.291	0.477	0.616	0.725	0.893	1.019	4
5	—	—	—	—	0.253	0.422	0.550	0.739	0.877	5
6	—	—	—	—	—	0.225	0.379	0.595	0.748	6
7	—	—	—	—	—	—	0.203	0.457	0.627	7
8	—	—	—	—	—	—	—	0.318	0.509	8
9	—	—	—	—	—	—	—	0.171	0.393	9
10	—	—	—	—	—	—	—	—	0.274	10

n	N 14	16	18	20	25	30	40	50	60	n
1	1.703	1.766	1.820	1.867	1.965	2.043	2.161	2.249	2.319	1
2	1.456	1.525	1.585	1.638	1.745	1.829	1.957	2.052	2.127	2
3	1.271	1.347	1.412	1.469	1.584	1.674	1.810	1.911	1.990	3
4	1.119	1.201	1.271	1.332	1.455	1.550	1.694	1.799	1.882	4
5	0.986	1.075	1.150	1.214	1.345	1.446	1.596	1.705	1.792	5
6	0.866	0.962	1.042	1.110	1.248	1.354	1.510	1.624	1.713	6
7	0.755	0.858	0.943	1.016	1.161	1.271	1.434	1.552	1.644	7
8	0.650	0.760	0.851	0.928	1.081	1.196	1.365	1.487	1.582	8
10	0.447	0.577	0.681	0.767	0.936	1.061	1.242	1.372	1.472	10
15	—	0.118	0.282	0.405	0.624	0.777	0.991	1.139	1.252	15
20	—	—	—	—	0.336	0.530	0.782	0.951	1.076	20

n	N 70	80	100	150	200	250	300	350	400	n
1	2.377	2.427	2.508	2.649	2.746	2.819	2.878	2.927	2.968	1
2	2.189	2.242	2.328	2.478	2.580	2.657	2.718	2.769	2.813	2
3	2.055	2.111	2.201	2.357	2.463	2.543	2.607	2.660	2.705	3
4	1.950	2.008	2.101	2.263	2.372	2.455	2.520	2.574	2.621	4
5	1.862	1.922	2.018	2.185	2.297	2.382	2.449	2.504	2.552	5
6	1.786	1.848	1.947	2.118	2.233	2.320	2.388	2.445	2.493	6
8	1.659	1.724	1.828	2.007	2.127	2.217	2.288	2.346	2.396	8
10	1.553	1.621	1.730	1.916	2.040	2.132	2.206	2.266	2.317	10
15	1.342	1.417	1.536	1.738	1.871	1.970	2.048	2.112	2.166	15
20	1.175	1.257	1.386	1.601	1.742	1.846	1.928	1.995	2.051	20
25	1.032	1.121	1.259	1.488	1.636	1.745	1.830	1.900	1.958	25

GLOSSARY OF SYMBOLS

Numbers in square brackets refer to Chapters where the meaning applies.
Some meanings with restricted use are not listed.

A_1, A_2 Alleles at a locus under consideration.
A Breeding value.
a Genotypic value of homozygote A_1A_1, as deviation from the mid-homozygote value.

B or b As subscript, indicates between families or groups.
b Regression coefficient; e.g., b_{OP} = regression of offspring on parent.

CR Correlated response to selection.

D Dominance deviation.
d Genotypic value of heterozygote as deviation from the mid-homozygote value.

E Environmental deviation.
e^2 $= 1 - h^2$.
E_c Common environment; i.e., environmental deviation of family mean from population mean.
E_g Environment due to permanent, or general, effects.
E_s Environment due to temporary, or special, effects.
E_w Within-family environment; i.e., environmental deviation of individual from family-mean.

F Coefficient of inbreeding.
F_1 First generation of cross between lines or populations.
F_2 Second generation of cross, by random mating among F_1.
FS Full sibs.
f Coancestry = coefficient of kinship.
f Subscript referring to females.
f [13] Subscript meaning between families.

G Genotypic value.
GCA General combining ability.

H Frequency of heterozygous genotype, A_1A_2.
H [14] Amount of heterosis; i.e., deviation of cross mean from mid-parent value.
HS Half sibs.
h^2 Heritability ('narrow sense').

I	Interaction deviation, due to epistasis.
I	[13, 19] Index for selection.
i	Intensity of selection; i.e., selection differential in units of phenotypic standard deviation.
k	Numbers in various contexts. In [4, 10, 20] family size, i.e., number in family.
L	[2] Load.
L	[4, 11] Generation length.
M	Population mean.
m	[18] Population mean.
m	Subscript referring to males.
m	[10] Correlation between breeding values of mates.
N	Population size; i.e., number of breeding individuals in a population or line.
N	[10, 13] Number of families.
N_e	Effective population size.
n	Numbers in various contexts. In [10, 13] specifically number of offspring per family.
O	Offspring.
P	Parent.
P	Mid-parent.
P	Frequency of homozygous genotype $A_1 A_1$.
P	Panmictic index, $= 1 - F$.
P	Phenotypic value.
p	Gene frequency of A_1, the allele that increases the character.
p	[11, 18] Proportion selected, or exceeding point of truncation of a normal distribution.
pg	The pygmy gene of mice, used in several examples.
Q	Frequency of homozygous genotype $A_2 A_2$.
q	Gene frequency of A_2, the allele that reduces the character.
R	Response to selection – specifically to individual selection.
R	[12] Total range; i.e., difference in mean between two populations at opposite selection limits.
r	[8] Repeatability; i.e., correlation between repeated measurements of the same individual.
r	Coefficient of relationship; i.e., correlation of breeding values between related individuals.
r	[10] Phenotypic correlation between mates.
r	[19] Correlations between two characters: r_A = correlation of breeding values, r_E = environmental correlation, r_P = phenotypic correlation.
S	Selection differential in actual units of measurements.
SCA	Specific combining ability.
s	Coefficient of selection against a specified genotype.
s	[13] Subscript referring to sib-selection.
T	Total in various contexts.
t	Time in number of generations. As subscript it means 'at generation t'.
t	Phenotypic correlation (intraclass) between members of families.
u	Mutation rate (from A_1 to A_2).

u [9, 15] Coefficient of the dominance variance in the covariance of relatives.

V Variance (causal component) of the value of deviation indicated by a subscript: V_P = phenotypic, V_G = genotypic, V_A = additive genetic, V_D = dominance, V_I = Interaction (epistatic), V_{NA} = non-additive genetic, V_E = environmental.

v Mutation rate (from A_2 to A_1).

W or w As subscript, indicates within families or groups.

W [20] Fitness under natural selection.

X Subscript denoting any particular individual, e.g., [5] F_X = inbreeding coefficient of individual X.

X One of two correlated characters.

x [11, 18] The normal deviate; i.e., deviation, in standard-deviation units, of point of truncation from population mean.

Y The other of two correlated characters.

y Difference in gene frequency between two lines.

z Height of the ordinate of a normal distribution, in standard-deviation units.

α Average effect of a gene substitution.

α_1, α_2 Average effects of alleles A_1 and A_2 respectively.

Δ Change of, as Δq = change of gene frequency, ΔF = rate of inbreeding.

Σ Summation of the quantity following the sign.

σ Standard deviation (σ^2 = variance) of the quantity indicated by subscript.

Equivalence of symbols used by Mather and Jinks as defined in Mather and Jinks (1977, p. 219)

Mather and Jinks	This book
d	a
$[d]$	Σa
D	$\Sigma a^2 = 2V_A$ when all $p = q = \frac{1}{2}$ (Equation [8.7]).
D_R	$2V_A$ in random-breeding population.
E_w	V_{Ew}
E_b	V_{Ec}
E_1	V_{Ew}
E_2	$V_{Ec} + \frac{1}{n}V_{Ew}$
h	d
$[h]$	Σd
H	$\Sigma d^2 = 4V_D$ when all $p = q = \frac{1}{2}$ (equation [8.7]).
H_R	$4V_D$ in random-breeding population.
u	p
v	q

REFERENCES

Numbers in square bracket refer to the chapters in which the work is cited

Allard, R.W., Jain, S.K., and **Workman P.L.** (1968) 'The genetics of inbreeding populations'. *Adv. Genet.*, **14**, 55–131. [5]

Allard, R.W., Kahler, A.L., and **Weir, B.S.** (1972) 'The effect of selection on esterase allozymes in a barley population'. *Genetics*, **72**, 489–503. [5]

Allen, J.A. (1975) 'Further evidence for apostatic selection by wild passerine birds: 9:1 experiments'. *Heredity*, **36**, 173–80. [2]

Allison, A.C. (1954) 'Notes on sickle-cell polymorphism'. *Ann. Hum. Genet.*, **19**, 39–57. [2]

—— (1956) 'The sickle-cell and haemoglobin *C* genes in some African populations'. *Ann. Hum. Genet.*, **21**, 67–89. [2]

Ayala, F.J. and **Campbell, C.A.** (1974) 'Frequency-dependent selection'. *Ann. Rev. Ecology and Systematics*, **5**, 115–38. [2]

Bailey, D.W. (1959) 'Rates of subline divergence in highly inbred strains of mice'. *J. Hered.*, **50**, 26–30. [15]

Baptist, R. and **Robertson, A.** (1976) 'Asymmetrical responses to automatic selection for body size in *Drosophila melanogaster*'. *Theor. Appl. Genet.*, **47**, 209–13. [12]

Barker, J.S.F. and **Robertson, A.** (1966) 'Genetic and phenotypic parameters for the first three lactations in Friesian cows'. *Anim. Prod.*, **8**, 221–40. [8, 10, 19]

Barlett, M.S., and **Haldane, J.B.S.** (1935) 'The theory of inbreeding with forced heterozygosis'. *J. Genet.* **31**, 327–40. [5]

Becker, W.A. (1975) *Manual of Quantitative Genetics* (3rd edn). Students Book Corporation, Pullman, Washington, USA. [Introd., App. B]

Bereskin, B. *et al.* (1968) 'Inbreeding and swine productivity traits'. *J. Anim. Sci.*, **27**, 339–50. [14]

Berger, E. (1976) 'Heterosis and the maintenance of enzyme polymorphism'. *Am. Nat.*, **110**, 823–39. [2]

Bohren, B.B. (1974) 'Designing artificial selection experiments for specific objectives'. *Genetics*, **80**, 205–20. [12]

Bohren, B.B., Hill, W.G., and **Robertson, A.** (1966) 'Some observations on asymmetrical correlated responses'. *Genet. Res.*, **7**, 44–57. [19]

Bohren, B.B., McKean, H.E., and **Yamada, Y.** (1961) 'Relative efficiencies of heritability estimates based on regression of offspring on parent'. *Biometrics*, **17**, 481–91. [10]

Bowman, J.C. and **Falconer, D.S.** (1960) 'Inbreeding depression and heterosis of litter size in mice'. *Genet. Res.*, **1**, 262–74. [14]

Brumby, P.J. (1958) 'Monozygotic twins and dairy cattle improvement'. *Anim. Breed. Abstr.*, **26**, 1–12.   [10]

Bulmer, M.G. (1971) 'The effect of selection on genetic variability'. *Am. Nat.*, **105**, 201–11.   [11, 20]

(1976) 'The effects of selection on genetic variability: a simulation study'. *Genet. Res.*, **28**, 101–17.   [11, 20]

Buri, P. (1956) 'Gene frequency in small populations of mutant *Drosophila*'. *Evolution*, **10**, 367–402.   [3, 4]

Calhoon, R.E. and **Bohren, B.B.** (1974) 'Genetic gains from reciprocal recurrent and within-line selection for egg production in the fowl'. *Theor. Appl. Genet.*, **44**, 364–72.   [16]

Chai, C.K. (1957) 'Developmental homeostasis of body growth in mice'. *Am. Nat.*, **91**, 49–55.   [15]

Chi, R.K., Eberhart, S.A., and **Penny, L.H.** (1969) 'Covariances among relatives in a maize variety (*Zea may* L.)'. *Genetics*, **63**, 511–20.   [9]

Christian, J.C., Kang, K.W., and **Norton, J.A.** (1974) 'Choice of an estimate of genetic variance from twin data'. *Am. J. Hum. Genet.*, **26**, 154–61.   [10]

Clarke, B. (1969) 'The evidence for apostatic selection'. *Heredity*, **24**, 347–52.   [2]

Clarke, B.C. (1979) 'The evolution of genetic diversity'. *Proc. R. Soc. Lond. B.*, **205**, 453–74.   [2]

Clayton, G.A., Morris, J.A., and **Robertson, A.** (1957) 'An experimental check on quantitative genetical theory. I. Short-term responses to selection'. *J. Genet.*, **55**, 131–51.   [8, 10, 11, 12, 13]

Clayton, G.A. and **Robertson, A.** (1955) 'Mutation and quantitative variation'. *Am. Nat.*, **89**, 151–8.   [8, 20]

(1957) 'An experimental check on quantitative genetical theory. II. The long-term effects of selection'. *J. Genet.*, **55**, 152–70.   [12]

Cloninger, C.R., Rice, J., and **Reich, T.** (1979) 'Multifactorial inheritance with cultural transmission and assortative mating. II. A general model of combined polygenic and cultural inheritance'. *Am. J. Hum. Genet.*, **31**, 176–98.   [10]

Cockerham, C.C. (1954) 'An extension of the concept of partitioning hereditary variance for analysis of covariances among relatives when epistasis is present'. *Genetics*, **39**, 859–82.   [8]

(1956) 'Effects of linkage on the covariances between relatives'. *Genetics*, **41**, 138–41.   [9]

(1963) 'Estimation of genetic variances'. pp. 53–94 in Hanson, W.D. and Robinson, H.F. (eds) *Statistical Genetics and Plant Breeding*. Nat. Acad. Sci. Nat. Res. Council Publ. No. 982, Washington, D.C., USA.   [8]

Comstock, R.E., Robinson, H.F., and **Harvey, P.H.** (1949) 'A breeding procedure designed to make maximum use of both general and specific combining ability'. *J. Amer. Soc. Agron.*, **41**, 360–7.   [16]

Cornelius, P.L. and **Dudley, J.W.** (1974) 'Effects of inbreeding by selfing and full-sib mating in a maize population'. *Crop Sci.*, **14**, 815–19.   [14]

Crow, J.F. (1948) 'Alternative hypotheses of hybrid vigor'. *Genetics*, **33**, 477–87. [16]

(1952) 'Dominance and overdominance'. pp. 282–297 in Gowen, J.W. (ed.), *Heterosis*, Iowa State College, Ames, Iowa, USA.   [16, 20]

(1957) 'Possible consequences of an increased mutation rate'. *Eugen. Quart.*, **4**, 67–80.   [2]

(1972) 'The dilemma of nearly neutral mutations: how important are they for evolution and human welfare?' *J. Hered.*, **63**, 306–16.   [2]

Crow, J.F. and **Kimura, M.** (1970) *An Introduction to Population Genetics Theory*, Harper and Row, New York.   [Introd., 1, 4, 5, 8, 9, 10, 14]

Curnow, R.N. (1964) 'The effect of continued selection of phenotypic intermediates on gene frequency'. *Genet. Res.,* **5**, 341–53. [20]

Curnow, R.N. and **Smith, C.** (1975) 'Multifactorial models for familial diseases in man'. *J. Roy. Statist. Soc. A,* **138**, 131–69. [18]

Dempster, E.R. and **Lerner, I.M.** (1950) 'Heritability of threshold characters'. *Genetics,* **35**, 212–36. [18]

Dickerson, G.E. (1952) 'Inbred lines for heterosis tests?' pp. 330–351 in Gowen, J.W. (ed.), *Heterosis,* Iowa State College, Ames, Iowa, USA. [16]

— (1962) 'Implications of genetic–environmental interaction in animal breeding'. *Anim. Prod.,* **4**, 47–63. [19]

— (1969) 'Experimental approaches in utilising breed resources'. *Anim. Breed. Abstr.,* **37**, 191–202. [16]

Dobzhansky, Th. and **Pavlovsky, O.** (1955) 'An extreme case of heterosis in a Central American population of *Drosophila tropicalis*'. *Proc. Natnl. Acad. Sci. Wash.,* **41**, 289–95. [2]

Dobzhansky, Th. and **Spassky, B.** (1969) 'Artificial and natural selection for two behavioural traits in *Drosophila pseudoobscura*'. *Proc. Natnl Acad. Sci.,* **62**, 75–80. [12]

Dudley, J.W. (1977) '76 generations of selection for oil and protein percentage in maize'. pp. 459–73 in Pollak, E., Kempthorne, O., and Bailey, T.B. (eds), *Proc. Int. Conf. Quantitative Genetics,* Iowa State Univ., Ames, Iowa, USA. [12]

Eaves, L. (1976) 'The effect of cultural transmission on continuous variation'. *Heredity,* **37**, 41–57. [10]

Eaves, L.J., Last, K.A., Young, P.A., and **Martin, N.G.** (1978) 'Model-fitting approaches to the analysis of human behaviour'. *Heredity,* **41**, 249–320. [10]

Eberhart, S.A. (1977) 'Quantitative genetics and practical corn breeding'. pp. 491–502 in Pollak, E., Kempthorne, O., and Bailey, T.B. (eds), *Proc. Int. Conf. Quantitative Genetics,* Iowa State Univ., Ames, Iowa, USA. [16]

Eklund, J. and **Bradford, G.E.** (1977) 'Genetic analysis of a strain of mice plateaued for litter size'. *Genetics,* **85**, 529–42. [12, 14]

Elandt-Johnson, R.C. (1971) *Probability Models and Statistical Methods in Genetics.* Wiley, New York. [1]

Emery, A.E.H. (1976) *Methodology in Medical Genetics.* Churchill Livingstone, Edinburgh. [18]

Emigh, T.H. and **Pollak, E.** (1979) 'Fixation probabilities and effective population numbers in diploid populations with overlapping generations'. *Theor. Pop. Biol.,* **15**, 86–107. [4]

Emsley, A., Dickerson, G.E., and **Kashyap, T.S.** (1977) 'Genetic parameters in progeny-test selection for field performance of strain-cross layers'. *Poult. Sci.,* **56**, 121–46. [10, 19]

Enfield, F.D. (1977) 'Selection experiments in tribolium designed to look at gene action issues'. pp. 177–90 in Pollak, E., Kempthorne, O., and Bailey, T.B. (eds), *Proc. Int. Conf. Quantitative Genetics,* Iowa State Univ., Ames, Iowa, USA. [12]

England, F. (1977) 'Response to family selection based on replicated trials'. *J. Agric. Sci.,* **88**, 127–34. [13]

Falconer, D.S. (1953) 'Selection for large and small size in mice'. *J. Genet.,* **51**, 470–501. [11, 20]

— (1954) 'Asymmetrical responses in selection experiments'. pp. 16–41 in *Symposium on Genetics of Population Structure.* Internat. Union Biol. Sci., Naples. Series B, No. 15. [2, 11, 12, 17]

— (1955) 'Patterns of response in selection experiments with mice'. *Cold Spring Harbor Symp. Quant. Biol.,* **20**, 178–96. [11, 12]

(1960a) 'The genetics of litter size in mice'. *J. Cell. Comp. Physiol.*, **56** (*Suppl.* 1), 153–67. [14]

(1960b) 'Selection of mice for growth on high and low planes of nutrition'. *Genet. Res.*, **1**, 91–113. [19]

(1963) 'Quantitative inheritance'. pp. 193–216 in Burdette, W.J. (ed.), *Methodology in Mammalian Genetics*, Holden-Day, San Francisco. [10]

(1965a) 'The inheritance of liability to certain diseases, estimated from the incidence among relatives'. *Ann. Hum. Genet.*, **29**, 51–76. [18, App. A]

(1965b) 'Maternal effects and selection response'. pp. 763–74 in Geerts, S.J. (ed.), *Genetics Today*, Proc. XI Internat. Congr. Genetics, Vol. 3. Pergamon, Oxford. [10, 12, 20]

(1967) 'The inheritance of liability to diseases with variable age of onset, with particular reference to diabetes mellitus'. *Ann. Hum. Genet.*, **31**, 1–20. [18]

(1971) 'Improvement of litter size in a strain of mice at a selection limit'. *Genet. Res.*, **17**, 215–35. [5, 12, 14]

(1973) 'Replicated selection for body weight in mice'. *Genet. Res.*, **22**, 291–321. [4, 10, 12]

(1977) 'Some results of the Edinburgh selection experiments with mice'. pp. 101–15 in Pollak, E., Kempthorne, O., and Bailey, T.B. (eds), *Proc. Int. Conf. Quantitative Genetics*, Iowa State Univ., Ames, Iowa, USA. [12]

Felsenstein, J. (1976) 'The theoretical population genetics of variable selection and migration'. *Ann. Rev. Genet.*, **10**, 253–80. [2]

Festing, M. (1973) 'A multivariate analysis of subline divergence in the shape of the mandible in C57BL/Gr mice'. *Genet. Res.*, **21**, 121–32. [15]

Fisher, R.A. (1918) 'The correlation between relatives on the supposition of Mendelian inheritance'. *Trans. Roy. Soc. Edinburgh*, **52**, 399–433. [Introd., 7]

(1941) 'Average excess and average effect of a gene substitution'. *Ann. Eugen. (Lond.)*, **11**, 53–63. [7]

Fisher, R.A. and **Yates, F.** (1963) *Statistical Tables* (6th edn). Longman, London and New York. [11]

Fowler, V.R., Bichard, M., and **Pease, A.** (1976) 'Objectives in pig breeding'. *Anim. Prod.*, **23**, 365–87. [19]

Frankham, R., Briscoe, D.A., and **Nurthen, R.K.** (1978) 'Unequal crossing over at the rRNA locus as a source of quantitative genetic variation'. *Nature, Lond.*, **272**, 80–1. [20]

Frankham, R., Jones, L.P., and **Barker, J.S.F.** (1968) 'The effects of population size and selection intensity in selection for a quantitative character in *Drosophila*. III. Analyses of the lines'. *Genet. Res.*, **12**, 267–83. [12]

Fredeen, H.T. and **Jonsson, P.** (1957) 'Genic variance and covariance in Danish Landrace swine as evaluated under a system of individual feeding of progeny test groups'. *Zeits. Tierzücht. u. Zücht. Biol.*, **70**, 348–63. [10]

Gama, E.E.G. and **Hallauer, A.R.** (1977) 'Relation between inbred and hybrid traits in maize'. *Crop. Sci.*, **17**, 703–6. [16]

Gardner, C.O. (1977) 'Quantitative genetic studies and population improvement in maize and sorghum'. pp. 475–89 in Pollak, E., Kempthorne, O., and Bailey, T.B. (eds), *Proc. Int. Conf. Quantitative Genetics*, Iowa State Univ., Ames, Iowa, USA. [16]

Garnett, I. and **Falconer, D.S.** (1975) 'Protein variation in strains of mice differing in body size'. *Genet. Res.*, **25**, 45–57. [4, 20]

Gibson, J.B. and **Bradley, B.P.** (1974) 'Stabilizing selection in constant and fluctuating environments'. *Heredity*, **33**, 293–302. [20]

Gowe, R.S., Robertson, A., and **Latter, B.D.H.** (1959) 'Environment and poultry breeding problems. 5. The design of poultry control strains'. *Poult. Sci.*, **38**, 462–71. [4]

Greaves, J.H., Redfern, R., Ayres, P.B., and **Gill, J.E.** (1977) 'Warfarin resistance: a balanced polymorphism in the Norway rat'. *Genet. Res.*, **30**, 257–63. [2]

Green, E.L. (1962) 'Quantitative genetics of skeletal variations in the mouse. II. Crosses between four inbred strains'. *Genetics*, **47**, 1085–96. [18]

Green, E.L. (ed.) (1968) *Handbook of Genetically Standardized JAX Mice*. The Jackson Laboratory, Bar Harbor, Maine. [14]

Grewal, M.J. (1962) 'The rate of genetic divergence of sublines in the C57BL strain of mice'. *Genet. Res.*, **3**, 226–37. [15]

Griffing, B. (1956*a*) 'A generalised treatment of the use of diallel crosses in quantitative inheritance'. *Heredity*, **10**, 31–50. [15]

 (1956*b*) 'Concept of general and specific combining ability in relation to diallel crossing systems'. *Aust. J. Biol. Sci.*, **9**, 463–93. [15]

Grundbacher, F.J. (1974) 'Heritability estimates and genetic and environmental correlations for the human immunoglobulins G, M, and A'. *Am. J. Hum. Genet.*, **26**, 1–12. [10, 19]

Haldane, J.B.S. (1932) *The Causes of Evolution*. Longmans, Green, London. [Introd.]

 (1936) 'The amount of heterozygosis to be expected in an approximately pure line'. *J. Genet.*, **32**, 375–91. [5]

 (1949) 'The rate of mutation of human genes'. Proc. 8th Internat. Congr. Genetics, *Hereditas, Suppl. Vol.* 1949, 267–73. [2]

 (1955) 'The complete matrices for brother–sister and alternate parent–offspring mating involving one locus'. *J. Genet.*, **53**, 315–24. [5]

Hallauer, A.R. and **Sears, J.H.** (1973) 'Changes in quantitative traits associated with inbreeding in a synthetic variety of maize'. *Crop. Sci.*, **13**, 327–30. [14]

Hancock, J. (1954) 'Monozygotic twins in cattle'. *Adv. Genet.*, **6**, 141–81. [10]

Hardy, G.H. (1908) 'Mendelian proportions in a mixed population'. *Science*, **28**, 49–50. [1]

Hayman, B.I. and **Mather, K.** (1953) 'The progress of inbreeding when homozygotes are at a disadvantage'. *Heredity*, **7**, 165–83. [5]

Henderson, C.R. (1963) 'Selection index and expected genetic advance'. pp. 141–63 in Hanson, W.D. and Robinson, H.F. (eds) *Statistical Genetics and Plant Breeding*, Nat. Acad. Sci. Nat. Res. Council Publ., No. 982, Washington, D.C. [13]

Hill, W.G. (1971) 'Design and efficiency of selection experiments for estimating genetic parameters'. *Biometrics*, **27**, 293–311. [12]

 (1972*a*) 'Estimation of genetic change. I. General theory and design of control populations'. *Anim. Breed. Abstr.*, **40**, 1–15. [11]

 (1972*b*) 'Estimation of genetic change. II. Experimental evaluation of control populations'. *Anim. Breed. Abstr.*, **40**, 193–213. [11]

 (1972*c*) 'Estimation of realized heritabilities from selection experiments. I. Divergent selection'. *Biometrics*, **28**, 747–65. [12]

 (1972*d*) 'Estimation of realized heritabilities from selection experiments. II. Selection in one direction'. *Biometrics*, **28**, 767–80. [12]

 (1977) 'Variation in response to selection'. pp. 343–65 in Pollak, E., Kempthorne, O., and Bailey, T.B. (eds), *Proc. Int. Conf. Quantitative Genetics*, Iowa State Univ., Ames, Iowa, USA. [12]

 (1979) 'A note on effective population size with overlapping generations'. *Genetics*, **92**, 317–22. [4]

Hill, W.G. and **Nicholas, F.W.** (1974) 'Estimation of heritability by both regression of

offspring on parent and intra-class correlation of sibs in one experiment'.
  Biometrics, **30**, 447–68.   [10]
Hill, W.G. and **Robertson, A.** (1966) 'The effect of linkage on limits to artificial selection'.
  Genet. Res., **8**, 269–94.   [12]
  (1968) 'The effects of inbreeding at loci with heterozygote advantages' *Genetics*, **60**,
  615–28.   [14]
Hill, W.G. and **Thompson, R.** (1977) 'Design of experiments to estimate offspring–
  parent regression using selected parents'. *Anim. Prod.*, **24**, 163–8.   [10]
Hollingsworth, M.J. and **Maynard Smith, J.** (1955) 'The effects of inbreeding on rate of
  development and on fertility in *Drosophila subobscura*'. *J. Genet.*, **53**, 295–314.
                                                                                                    [14]
Hunt, H.R., Hoppert, C.A., and **Erwin, W.G.** (1944) 'Inheritance of susceptibility to
  caries in albino rats (*Mus norvegicus*)'. *J. Dental Res.*, **23**, 385–401.   [17]
Huntley, R.M.C. (1966) 'Heritability of intelligence'. pp. 201–18 in Meade, J.E. and
  Parkes, A.S. (eds), *Genetic and Environmental Factors in Human Ability*, Oliver
  and Boyd, Edinburgh.   [10]
Hyde, J.S. (1973) 'Genetic homeostasis and behaviour: analysis, data, and theory'.
  Behaviour Genetics, **3**, 233–45.   [15]
Jacquard, A. (1974) *The Genetic Structure of Populations* (translated by D. and
  B. Charlesworth). Springer-Verlag, Berlin.   [Introd.]
James, J.W. (1961) 'Selection in two environments'. *Heredity*, **16**, 145–52.   [19]
Jinks, J.L. and **Connoly, V.** (1973) 'Selection for specific and general response to
  environmental differences'. *Heredity*, **30**, 33–40.   [19]
  (1975) 'Determination of the environmental sensitivity of selection lines by the
  selection environment'. *Heredity*, **34**, 401–6.   [19]
Jinks, J.L. and **Pooni, H.S.** (1976) 'Predicting the properties of recombinant inbred lines
  derived by single seed descent'. *Heredity*, **36**, 253–66.   [16]
Jódar, B. and **López-Fanjul, C.** (1977) 'Optimum proportions selected with unequal sex
  numbers'. *Theor. Appl. Genet.*, **50**, 57–61.   [12]
Jones, L.P., Frankham, R., and **Barker, J.S.F.** (1968) 'The effects of population size and
  selection intensity in selection for a quantitative character in *Drosophila*.
  II. Long-term response to selection'. *Genet. Res.*, **12**, 249–66.   [12]
Kaufman, P.K., Enfield, F.D., and **Comstock, R.E.** (1977) 'Stabilizing selection for pupa
  weight in *Tribolium castaneum*'. *Genetics*, **87**, 327–41.   [20]
Kearsey, M.J. and **Barnes, B.W.** (1970) 'Variation for metrical characters in *Drosophila*
  populations. II. Natural selection'. *Heredity*, **25**, 11–21.   [20]
Kearsey, M.J. and **Kojima, K.** (1967) 'The genetic architecture of body weight and egg
  hatchability in *Drosophila melanogaster*'. *Genetics*, **56**, 23–37.   [8, 20]
Kempthorne, O. (1954) 'The correlation between relatives in a random mating population'. *Proc. Roy. Soc. London, B.*, **143**, 103–13.   [8]
  (1955a) 'The theoretical values of correlations between relatives in random mating
  populations'. *Genetics*, **40**, 153–67.   [8, 9]
  (1955b) 'The correlations between relatives in random mating populations'. *Cold
  Spring Harbor Symp. Quant. Biol.*, **20**, 60–75.   [8, 9]
  (1957) *An Introduction to Genetic Statistics*. Wiley, New York.   [Introd.]
Kempthorne, O. and **Tandon, O.B.** (1953) 'The estimation of heritability by regression of
  offspring on parent'. *Biometrics*, **9**, 90–100.   [10]
Kerr, W.E. and **Wright, S.** (1954a) 'Experimental studies of the distribution of gene
  frequencies in very small populations of *Drosophila melanogaster*: I. Forked'.
  Evolution, **8**, 172–177.   [4]
  (1954b) 'Experimental studies of the distribution of gene frequencies in very small

populations of *Drosophila melanogaster*. III. Aristapedia and spineless'. *Evolution*, **8**, 293–302.   [4]

Kidwell, J.F. and **Kidwell, M.M.** (1966) 'The effects of inbreeding on body weight and abdominal chaeta number in *Drosophila melanogaster*'. *Can. J. Genet. Cytol.*, **8**, 207–15.   [14]

Kimura, M. (1955) 'Solution of a process of random genetic drift with a continuous model'. *Proc. Natnl Acad. Sci. Wash.*, **41**, 144–50.   [3]

(1956) 'Rules for testing stability of a selective polymorphism'. *Proc. Natnl Acad. Sci., Wash.*, **42**, 336–40.   [2]

(1979) 'Model for effectively neutral mutations in which selective constraint is incorporated'. *Proc. Natnl Acad. Sci., Wash.*, **76**, 3440–4.   [4]

Kimura, M. and **Crow, J.M.** (1978) 'Effect of overall phenotypic selection on genetic change at individual loci'. *Proc. Natnl Acad. Sci., Wash.*, **75**, 6168–71.   [11]

Kimura, M. and **Ohta, T.** (1971) *Theoretical Aspects of Population Genetics*. Princeton Univ., Princeton.   [4]

King, J.W.B. (1950) 'Pygmy, a dwarfing gene in the house mouse'. *J. Hered.*, **41**, 249–52.   [7]

(1955) 'Observations on the mutant "pygmy" in the house mouse'. *J. Genet.*, **53**, 487–97.   [7, 17]

Krosigk, C.M. von, Havenstein, G.B., Flock, D.K., and **McClary, C.F.** (1973) 'Estimates of response to selection in populations of white leghorns under reciprocal recurrent selection'. pp. 256–71 in *4th Europ. Poult. Conf., London, 1972*, British Poultry Science Ltd.   [16]

Lack, D. (1966) *Population Studies of Birds*. Clarendon Press, Oxford.   [20]

Lalouel, J.M., Morton, N.E., MacLean, C.J., and **Jackson, J.** (1977) 'Recurrence risks in complex inheritance with special regard to pyloric stenosis'. *J. Med. Genet.*, **14**, 408–14.   [18]

Lande, R. (1976) 'The maintenance of genetic variability by mutation in a polygenic character with linked loci'. *Genet. Res.*, **26**, 221–35.   [20]

Latter, B.D.H. (1965) 'The response to artificial selection due to autosomal genes of large effect. I. Changes in gene frequency at an additive locus'. *Aust. J. Biol. Sci.*, **18**, 585–98.   [11]

Latter, B.D.H. and **Robertson, A.** (1960) 'Experimental design in the estimation of heritability by regression methods'. *Biometrics*, **16**, 348–53.   [10]

(1962) 'The effects of inbreeding and artificial selection on reproductive fitness'. *Genet. Res.*, **3**, 110–38.   [20]

Lerner, I.M. (1954) *Genetic Homeostasis*. Oliver and Boyd, Edinburgh.   [15, 20]

Lewontin, R.C. (1964) 'The interaction of selection and linkage. II. Optimum models'. *Genetics*, **50**, 757–82.   [20]

(1974) *The Genetic Basis of Evolutionary Change*. Columbia Univ., New York.   [Introd., 2]

Li, C.C. (1976) *First Course in Population Genetics*. Boxwood, Pacific Grove, California.   [Introd., 5]

Lin, C.Y. (1978) 'Index selection for genetic improvement of quantitative characters'. *Theor. Appl. Genet.*, **52**, 49–56.   [13]

Livesay, E.A. (1930) 'An experimental study of hybrid vigor or heterosis in rats'. *Genetics*, **15**, 17–54.   [15]

Lush, J.L. (1947) 'Family merit and individual merit as bases for selection'. *Am. Nat.*, **81**, 241–61; 362–79.   [13]

MacArthur, J.W. (1949) 'Selection for small and large body size in the house mouse'. *Genetics*, **34**, 194–209.   [12, 17, 20]

McBride, G. and **Robertson, A.** (1963) 'Selection using assortative mating in *Drosophila melanogaster*'. *Genet. Res.*, **4**, 356–69.   [10]

McCarthy, J.C. (1965) 'Effects of concurrent lactation on litter size and prenatal mortality in an inbred strain of mice'. *J. Reprod. Fertil.*, **9**, 29–39.   [14]

McDonald, J.F. and **Ayala, F.J.** (1978) 'Gene regulation in adaptive evolution'. *Can. J. Genet. Cytol.*, **20**, 159–175.   [20]

McKusick, V.A. (1978) *Medical Genetic Studies of the Amish: Selected Papers Assembled, with Commentary.* Johns Hopkins Univ. Baltimore, USA.   [4]

McLaren, A. and **Michie, D.** (1954) 'Factors affecting vertebral variation in mice. I. Variation within an inbred strain'. *J. Embryol. Exp. Morph.*, **2**, 149–60.
                                                                                                          [15]

    (1956) 'Variability of response in experimental animals'. *J. Genet.*, **54**, 440–55.
                                                                                                          [15]

McNew, R.W. and **Bell, A.E.** (1976) 'Comparisons of crossbred and purebred selection for a heterotic trait in highly selected populations of *Tribolium*'. *J. Hered.*, **67**, 275–83.   [16]

Mangelsdorf, P.C. (1951) 'Hybrid corn: its genetic basis and its significance in human affairs'. pp. 555–71 in Dunn, L.C. (ed.), *Genetics in the 20th century.* Macmillan Co., New York.   [6]

Mather, K. (1941) 'Variation and selection of polygenic characters'. *J. Genet.*, **41**, 159–93.
                                                                                                          [20]

    (1953) 'Genetical control of stability in development'. *Heredity*, **7**, 297–336. [15]

Mather, K. and **Jinks, J.L.** (1971) *Biometrical Genetics.* 2nd edn. Chapman and Hall, London.   [Introd., 8, 16, 17]

    (1977) *Introduction to Biometrical Genetics.* Chapman and Hall, London.
                                                                                                [Introd., 8, 16]

Maynard Smith, J. (1975) *The Theory of Evolution* (3rd edn). Penguin Books, Harmondsworth, Middlesex.   [2]

Metcalfe, J.A. and **Turner, J.R.G.** (1971) 'Gene frequencies in the domestic cats of York: evidence of selection'. *Heredity*, **26**, 259–68.   [1]

Mi, M.P. and **Rashad, M.N.** (1975) 'Genetic parameters of dermal patterns and ridge counts'. *Hum. Hered.*, **25**, 249–57.   [10]

Mikami, H. and **Fredeen, H.T.** (1979) 'A genetic study of cryptorchidism and scrotal hernia in pigs'. *Can. J. Genet. Cytol.*, **21**, 9–19.   [18]

Minvielle, F. (1979) 'Comparing the means of inbred lines with the base population: a model with overdominant loci'. *Genet. Res.*, **33**, 89–92.   [14]

Moll, R.H., Lonnquist, J.H., Fortuno, J.V., and **Johnson, E.C.** (1965) 'The relationship of heterosis and genetic divergence in maize'. *Genetics*, **52**, 139–44.   [14]

Morley, F.H.W. (1951) 'Selection for economic characters in Australian Merino sheep'. *N.S.W. Dept. Agric. Bull.*, No. 73.   [8]

    (1954) 'Selection for economic characters in Australian Merino sheep. IV. The effect of inbreeding'. *Austr. J. Agric. Res.*, **5**, 305–16.   [14]

Morton, N.E. (1955) 'The inheritance of human birth weight'. *Ann. Hum. Genet.*, **20**, 125–34.   [8]

    (1974) 'Analysis of family resemblance. I. Introduction'. *Am. J. Hum. Genet.*, **26**, 318–30.   [10]

    (1978) 'Effect of inbreeding on IQ and mental retardation'. *Proc. Natnl Acad. Sci., Wash.*, **75**, 3906–8.   [14]

Morton, N.E. and **MacLean, C.J.** (1974) 'Analysis of family resemblance. III. Complex segregation of quantitative traits'. *Am. J. Hum. Genet.*, **26**, 489–503.   [18]

Mourant, A.E. (1954) *The Distribution of the Human Blood Groups.* Blackwell, Oxford.
                                                                                                          [1]

Mukai, T., Chigusa, S.I., Mettler, L.E., and **Crow, J.F.** (1972) 'Mutation rate and dominance of genes affecting viability in *Drosophila melanogaster*'. Genetics, **72**, 335–55. [20]

Mukai, T. and **Cockerham, C.C.** (1977) 'Spontaneous mutation rates at enzyme loci in *Drosophila melanogaster*'. *Proc. Natnl Acad. Sci., Wash.*, **74**, 2514–17. [20]

Muller, H.J. and **Oster, I.I.** (1957) 'Principles of back mutation as observed in *Drosophila* and other organisms'. pp. 407–15 in *Advances in Radiobiology. Proc. 5th int. Conf. Radiobiol. (Stockh.)*, 1956. Oliver and Boyd, Edinburgh. [2]

Nagai, J., Eisen, E.J., Emsley, J.A.B., and **McAllister, A.J.** (1978) 'Selection for nursing ability and adult weight in mice'. *Genetics*, **88**, 761–80. [19]

Nagylaki, T. (1978) 'The correlation between relatives with assortative mating'. *Ann. Hum. Genet.*, **42**, 131–7. [10]

Nance, W.E. (1976) 'Note on the analysis of twin data'. *Am. J. Hum. Genet.*, **28**, 297–8. [10]

Nei, M. (1975) *Molecular Population Genetics and Evolution*. North-Holland, Amsterdam. [4]

Nevo, E. (1978) 'Genetic variation in natural populations: patterns and theory'. *Theor. Pop. Biol.*, **13**, 121–77. [2]

Niemann-Sørensen, A. and **Robertson, A.** (1961) 'The association between blood groups and several production characteristics in three Danish cattle breeds'. *Acta Agric. Scand.*, **11**, 163–96. [20]

Nordskog, A.W. (1978) 'Some statistical properties of an index of multiple traits'. *Theor. Appl. Genet.*, **52**, 91–4. [13]

Ohta, T. (1974) 'Mutational pressure as the main cause of molecular evolution and polymorphism'. *Nature, Lond.*, **252**, 351–4. [2]

Ollivier, L. (1974) 'Optimum replacement rates in animal breeding'. *Anim. Prod.*, **19**, 257–271. [11]

Orozco, F. (1976) 'A dynamic study of genotype–environment interaction with egg laying of *Tribolium castaneum*'. *Heredity*, **37**, 157–71. [19]

Osborne, R. (1957) 'The use of sire and dam family averages in increasing the efficiency of selective breeding under a hierarchical mating system'. *Heredity*, **11**, 93–116. [13]

Otsuka, Y., Eberhart, S.A., and **Russell, W.A.** (1972) 'Comparisons of prediction formulas for maize hybrids'. *Crop Sci.*, **12**, 325–31. [16]

Paxman, G.J. (1957) 'A study of spontaneous mutation in *Drosophila melanogaster*'. *Genetica*, **29**, 39–57. [20]

Penrose, L.S. (1954) 'Some recent trends in human genetics'. *Proc. 9th Internat. Congr. Genet., Caryologia*, **6** (Suppl.), 521–530 [8]

Perkins, J.M. and **Jinks, J.L.** (1968) 'Environmental and genotype–environmental components of variability. III. Multiple lines and crosses'. *Heredity*, **23**, 339–56. [8]

Petit, C. and **Nouaud, D.** (1976) 'Ecological competition and the advantage of the rare type in *Drosophila melanogaster*. *Evolution*, **29**, 763–76. [2]

Plum, M. (1954) 'Computation of inbreeding and relationship coefficients'. *J. Hered.*, **45**, 92–4. [5]

Ponzoni, R.W. and **James, J.W.** (1978) 'Possible biases in heritability estimates from intraclass correlation'. *Theor. Appl. Genet.*, **53**, 25–7. [10]

Pooni, H.S. and **Jinks, J.L.** (1979) 'Sources and biases of the predictors of the properties of recombinant inbreds produced by single seed descent'. *Heredity*, **42**, 41–8. [16]

Pooni, H.S., Jinks, J.L., and **Jayasekara, N.E.M.** (1978) 'An investigation of gene action and genotype × environment interaction in two crosses of *Nicotiana* *rustica* by triple test cross and inbred line analysis'. *Heredity*, **41**, 83–92. [9]

Powers, L. (1950) 'Determining scales and the use of transformations in studies on weight per locule of tomato fruit'. *Biometrics*, **6**, 145–63. [17]

(1952) 'Gene recombination and heterosis'. pp. 298–319 in Gowen, J.W. (ed.), *Heterosis*. Iowa State Coll., Ames, Iowa, USA. [14]

Preston, T.R. and **Willis, M.B.** (1970) *Intensive Beef Production*. Pergamon, Oxford. [10]

Price, G.R. (1972) 'Fisher's "fundamental theorem" made clear'. *Ann. Hum. Genet.*, **36**, 129–40. [20]

Prout, T. (1962) 'The effects of stabilizing selection on the time of development in *Drosophila melanogaster*'. *Genet. Res.*, **3**, 364–82. [20]

(1965) 'The estimation of fitnesses from genotypic frequencies'. *Evolution*, **19**, 546–51. [1]

Race, R.R. and **Sanger, R.** (1954) *Blood Groups in Man* (2nd edn). Blackwell, Oxford. [1]

Raine, D.N., Cooke, J.R., Andrews, W.A., and **Mahon, D.F.** (1972) 'Screening for inherited metabolic disease by plasma chromatography (Scriver) in a large city'. *Brit. Med. J.*, 1972, **3**, 7–13. [1]

Rao, D.C., Morton, N.E., and **Yee, S.** (1974) 'Analysis of family resemblance. II. A linear model for familial correlation'. *Am. J. Hum. Genet.*, **26**, 331–59. [10]

(1976) 'Resolution of cultural and biological inheritance by path analysis'. *Am. J. Hum. Genet.*, **28**, 228–42. [10]

Rasmuson, M. (1952) 'Variation in bristle number of *Drosophila melanogaster*'. *Acta Zoolog., Stockh.*, **33**, 277–307. [15]

Reeve, E.C.R. (1955a) 'Inbreeding with homozygotes at a disadvantage'. *Ann. Hum. Genet.*, **19**, 332–46. [5]

(1955b) 'The variance of the genetic correlation coefficient'. *Biometrics*, **11**, 357–74. [19]

(1961) 'A note on non-random mating in progeny tests'. *Genet. Res.*, **2**, 195–203. [10]

Reeve, E.C.R. and **Robertson, F.W.** (1953) 'Studies in quantitative inheritance. II. Analysis of a strain of *Drosophila melanogaster* selected for long wings'. *J. Genet.*, **51**, 276–316. [10, 19]

(1954) 'Studies in quantitative inheritance. VI. Sternite chaeta number in *Drosophila*: a metameric quantitative character'. *Z. indukt. Abstam.- u.Vererblehre*, **86**, 269–88. [8]

Reich, T., James, J.W., and **Morris, C.A.** (1972) 'The use of multiple thresholds in determining the mode of transmission of semi-continuous traits'. *Ann. Hum. Genet.*, **36**, 163–84. [18]

Rendel, J.M. and **Robertson, A.** (1950) 'Estimation of genetic gain in milk yield by selection in a closed herd of dairy cattle'. *J. Genet.*, **50**, 1–8. [11]

Rendel, J.M., Robertson, A., Asker, A.A., Khishin, S.S., and **Ragab, M.T.** (1957) 'The inheritance of milk production characteristics'. *J. Agric. Sci.*, **48**, 426–32. [8]

Rich, S.S., Bell, A.E., and **Wilson, S.P.** (1979) 'Genetic drift in small populations of *Tribolium*'. *Evolution*, **33**, 579–84. [3]

Roberts, D.F., Billewicz, W.Z., and **McGregor, I.A.** (1978) 'Heritability of stature in a West African population'. *Ann. Hum. Genet.*, **42**, 15–24. [10]

Roberts, R.C. (1960) 'The effects on litter size of crossing lines of mice inbred without selection'. *Genet. Res.*, **1**, 239–52. [14]

(1966a) 'The limits to artificial selection for body weight in the mouse. I. The limits attained in earlier experiments'. *Genet. Res.*, **8**, 347–60. [12]

(1966b) 'The limits to artificial selection for body weight in the mouse. II. The genetic nature of the limits'. *Genet. Res.*, **8**, 361–75. [12]

(1967a) 'The limits to artificial selection for body weight in the mouse. III. Selection from crosses between previously selected lines'. *Genet. Res.*, **9**, 73–85. [12]

(1967b) 'The limits to artificial selection for body weight in the mouse. IV. Sources of new genetic variance – irradiation and outcrossing'. *Genet. Res.*, **9**, 87–98. [12]

Robertson, A. (1952) 'The effect of inbreeding on the variation due to recessive genes'. *Genetics*, **37**, 189–207. [15]

(1954) 'Inbreeding and performance in British Friesian cattle'. *Proc. Brit. Soc. Anim. Prod.*, 1954, 87–92. [14]

(1955) 'Selection in animals: synthesis'. *Cold Spring Harbor Symp. Quant. Biol.*, **20**, 225–9. [20]

(1956) 'The effect of selection against extreme deviants based on deviation or on homozygosis'. *J. Genet.*, **54**, 236–48. [20]

(1959a) 'Experimental design in the evaluation of genetic parameters'. *Biometrics*, **15**, 219–26. [10]

(1959b) 'The sampling variance of the genetic correlation coefficient'. *Biometrics*, **15**, 469–85. [19]

(1960) 'A theory of limits in artificial selection'. *Proc. Roy. Soc. London, B.*, **153**, 234–49. [12]

(1962) 'Selection for heterozygotes in small populations'. *Genetics*, **47**, 1291–1300. [4]

(1965) 'The interpretation of genotypic ratios in domestic animal populations'. *Anim. Prod.*, **7**, 319–24. [3]

(1966) 'A mathematical model of the culling process in dairy cattle'. *Anim. Prod.*, **8**, 95–108. [20]

(1970) 'A theory of limits in artificial selection with many linked loci'. pp. 246–88 in Kojima, K. (ed.) *Mathematical Topics in Population Genetics*, Vol. 1, Springer-verlag, Berlin. [12]

(1977a) 'The effect of selection on the estimation of genetic parameters'. *Z. Tierzüchtg. Züchtgsbiol.*, **94**, 131–5. [10, 11]

(1977b) 'The non-linearity of offspring–parent regression'. pp. 297–304 in Pollak, E., Kempthorne, O., and Bailey, T.B. (eds), *Proc. Int. Conf., Quantitative Genetics*, Iowa State Univ., Ames, Iowa, USA. [12]

Robertson, F.W. (1955) 'Selection response and the properties of genetic variation'. *Cold Spring Harbor Symp. Quant. Biol.*, **20**, 166–77. [12]

(1957a) 'Studies in quantitative inheritance. X. Genetic variation of ovary size in *Drosophila*'. *J. Genet.*, **55**, 410–27. [8, 10]

(1957b) 'Studies in quantitative inheritance. XI. Genetic and environmental correlation between body size and egg production in *Drosophila melanogaster*'. *J. Genet.*, **55**, 428–43. [8, 10]

Robertson, F.W. and **Reeve, E.C.R.** (1952) 'Heterozygosity, environmental variation and heterosis'. *Nature Lond.*, **170**, 296. [15]

Robinson, P. and **Bray, D.F.** (1965) 'Expected effects on the inbreeding coefficient and rate of gene loss of four methods of reproducing finite diploid populations'. *Biometrics*, **21**, 447–58. [4]

Robson, E.B. (1955) 'Birth weight in cousins'. *Ann. Hum. Genet.*, **19**, 262–8. [8]

Russell, E.S. (1949) 'A quantitative histological study of the pigment found in the coat-color mutants of the house mouse. IV. The nature of the effects of genic substitution in five major allelic series'. *Genetics*, **34**, 146–66. [7]

Rutledge, J.J., Eisen, E.J., and **Legates, J.E.** (1973) 'An experimental evaluation of genetic correlation'. *Genetics*, **75**, 709–26.   [10, 19]

Sales, J. and **Hill, W.G.** (1976) 'Effect of sampling errors on efficiency of selection indices. 2. Use of information on associated traits for improvement of a single important trait'. *Anim. Prod.*, **23**, 1–14.   [19]

Schäfer, W. (1937) Über die Zunahme der Isozygotie (Gleicherbarkeit) bie fortgesetzter Bruder-Schwester-Inzucht'. *Z. indukt. Abstamm.-u.Vererblehre*, **72**, 50–79.   [5]

Schlager, G. and **Dickie, M.M.** (1967) 'Spontaneous mutations and mutation rates in the house mouse'. *Genetics*, **57**, 319–30.   [2]

Schull, W.J. (1962) 'Inbreeding and maternal effects in the Japanese'. *Eugen. Quart.*, **9**, 14–22.   [14]

Searle, A.G. (1949) 'Gene frequencies in London's cats'. *J. Genet.*, **49**, 214–20.   [1]

Searle, S.R. (1965) 'The value of indirect selection: I. Mass selection'. *Biometrics*, **21**, 682–707.   [19]

   (1971) *Linear Models*. Wiley, New York.   [10]

Shank, D.B. and **Adams, M.W.** (1960) 'Environmental variability within inbred lines and single crosses of maize'. *J. Genet.*, **57**, 119–26.   [15]

Sheridan, A.K. and **Barker, J.S.F.** (1974) 'Two-trait selection and the genetic correlation. II. Changes in the genetic correlation during two-trait selection'. *Aust. J. Biol. Sci.*, **27**, 89–101.   [19]

Sheridan, A.K., Frankham, R., Jones, L.P., Rathie, K.A., and **Barker, J.S.F.** (1968) 'Partitioning of variance and estimation of genetic parameters for various bristle number characters of *Drosophila melanogaster*'. *Theor. Appl. Genet.*, **38**, 179–87.   [19]

Sidwell, G.M., Everson, D.O., and **Terrill, C.E.** (1962) 'Fertility, prolificacy and lamb livability of some pure breeds and their crosses'. *J. Anim. Sci.*, **21**, 875–9.)
                                                    [16]

   (1964) Lamb weights in some pure breeds and crosses'. *J. Anim. Sci.*, **23**, 105–10.
                                                    [16]

Simmonds, N.W. (1979) *Principles of Crop Improvement*. Longman, London. [15, 16]

Simmons, M.J. and **Crow, J.F.** (1977) 'Mutations affecting fitness in Drosophila populations'. *Ann. Rev. Genet.*, **11**, 49–78.   [20]

Slizynski, B.M. (1955) 'Chiasmata in the male mouse'. *J. Genet.*, **53**, 597–605.   [5]

Smith, C. (1962) 'Estimation of genetic change in farm livestock using field records'. *Anim. Prod.*, **4**, 239–51.   [11]

Smith, C., King, J.W.B., and **Gilbert, N.** (1962) 'Genetic parameters of British Large White pigs'. *Anim. Prod.*, **4**, 128–43.   [10, 19]

Smith, C.A.B. (1970) 'A note on testing the Hardy–Weinberg Law'. *Ann. Hum. Genet.*, **33**, 377–83.   [1]

Smith, H.H. (1952) 'Fixing transgressive vigor in. *Nicotiana rustica*'. pp. 161–74 in Gowen, J.W. (ed.), *Heterosis*, Iowa State Coll., Ames, Iowa, USA.   [14]

Snedecor, G.W. and **Cochran, W.G.** (1967) *Statistical Methods*. (6th edn). Iowa State Univ., Ames, Iowa, USA.   [10]

Spickett, S.G. and **Thoday, J.M.** (1966) 'Regular responses to selection. III. Interaction between located polygenes'. *Genet. Res.*, **7**, 96–121.   [12, 20]

Sprague, G.F. and **Tatum, L.A.** (1942) 'General *vs.* specific combining ability in single crosses of corn'. *J. Amer. Soc. Agron.*, **34**, 923–32.   [15]

Stern, C. (1973) *Principles of Human Genetics* (3rd edn). Freeman, San Francisco.
                                                    [2, 10]

Stockard, C.R. (1941) *The Genetic and Endocrine Basis for Differences in Form and Behaviour*. Wistar Institute, Philadelphia.   [12]

Strang, G.S. and **Smith, C.** (1979) 'A note on the heritability of litter traits in pigs'. *Anim. Prod.*, **28**, 403–6. [10]

Tantawy, A.O. and **Reeve, E.C.R.** (1956) 'Studies in quantitative inheritance. IX. The effects of inbreeding at different rates in *Drosophila melanogaster*'. *Z. indukt. Abstamm.-u. Vererblehre*, **87**, 648–67. [14, 15]

Thoday, J.M. (1972) 'Disruptive selection'. *Proc. Roy. Soc., London, B*, **182**, 109–43. [20]

—— (1977) 'Effects of specific genes'. pp. 141–59 in Pollak, E., Kempthorne, O., and Bailey, T.B. (eds), *Proc. Int. Conf. Quantitative Genetics*, Iowa State Univ., Ames, Iowa, USA. [20]

Thoday, J.M., Gibson, J.B., and **Spickett, S.G.** (1964) 'Regular responses to selection. 2. Recombination and accelerated response'. *Genet. Res.*, **5**, 1–19. [12]

Thompson, R. (1976) 'The estimation of maternal genetic variances'. *Biometrics*, **32**, 903–17. [9]

Turner, H.N. and **Young, S.Y.** (1969) *Quantitative Genetics in Sheep Breeding*. Macmillan, Melbourne, Australia. [10]

Waddington, C.H. (1942) 'Canalisation of development and the inheritance of acquired characters'. *Nature, Lond.*, **150**, 563. [18]

—— (1953) 'Genetic assimilation of an acquired character'. *Evolution*, **7**, 118–26. [18]

—— (1957) *The Strategy of the Genes*. Allen and Unwin, London. [20]

Wallace, B. (1958) 'The comparison of observed and calculated zygotic distributions'. *Evolution*, **12**, 113–15. [1]

—— (1963) 'The elimination of an autosomal lethal from an experimental population of *Drosophila melanogaster*'. *Am. Nat.*, **97**, 65–6. [2]

—— (1968) *Topics in Population Genetics*. Norton, New York. [1, 2]

Wallace, B. and **Vetukhiv, M.** (1955) 'Adaptive organization of the gene pools of *Drosophila* populations'. *Cold Spring Harbor Symp. Quant. Biol.*, **20**, 303–9. [14]

Waller, J.H. (1971) 'Differential reproduction: its relation to IQ test score, education, and occupation'. *Social Biol.*, **18**, 122–36. [20]

Warwick, E.J. and **Lewis, W.L.** (1954) 'Increase in frequency of a deleterious recessive gene in mice'. *J. Hered.*, **45**, 143–5. [7]

Weatherspoon, J.H. (1970) 'Comparative yields of single, three-way, and double crosses of maize'. *Crop Sci.*, **10**, 157–9. [16]

Weinberg, W. (1908) 'Über den Nachweis der Vererbung beim Menschen'. *Jh. Ver. vaterl Naturk. Württemb.*, **64**, 369–82. [1]

Weir, B.S. and **Cockerham, C.C.** (1977) 'Two-locus theory in quantitative genetics'. pp. 247–269 in Pollak, E., Kempthorne, O., and Bailey, T.B. (eds), *Proc. Int. Conf. Quantitative Genetics*, Iowa State Univ., Ames, Iowa, USA. [15]

—— (1979) 'Estimation of linkage disequilibrium in randomly mating populations'. *Heredity*, **42**, 105–11. [1]

White, J.M. (1972) 'Inbreeding effects upon growth and material ability in laboratory mice'. *Genetics*, **70**, 307–17. [14]

Willham, R.L. (1963) 'The covariance between relatives for characters composed of components contributed by related individuals'. *Biometrics*, **19**, 18–27. [9]

Wills, C. (1978) 'Rank-order selection is capable of maintaining all genetic polymorphisms'. *Genetics*, **89**, 403–17. [2]

Wright, S. (1921) 'Systems of mating'. *Genetics*, **6**, 111–78. [Introd., 10]

—— (1933) 'Inbreeding and homozygosis'. *Proc. Natnl Acad. Sci. Wash.*, **19**, 411–20. [5]

—— (1942) 'Statistical genetics and evolution'. *Bull. Amer. Math. Soc.*, **48**, 223–46. [4]

(1951) 'The genetical structure of populations'. *Ann. Eugen. Lond.*, **15**, 323–54.
                                                                                    [4]
(1952) 'The theoretical variance within and among subdivisions of a population
        that is in a steady state'. *Genetics*, **37**, 312–21.   [3]
(1968) *Evolution and the Genetics of Populations.* Vol. 1, *Genetic and Biometric
        Foundations.* Univ. of Chicago, Chicago.   [Introd., 17]
(1969) *Evolution and the Genetics of Populations.* Vol. 2, *The Theory of Gene
        Frequencies.* Univ. of Chicago, Chicago.   [Introd., 4, 5, 15, 20]
(1977) *Evolution and the Genetics of Populations.* Vol. 3, *Experimental Results and
        Evolutionary Deductions.* Univ. of Chicago, Chicago.   [Introd., 15]
(1978) *Evolution and the Genetics of Populations.* Vol. 4, *Variability within and among
        Natural Populations.* Univ. of Chicago, Chicago.   [Introd.]

Wright, S. and **Kerr, W.E.** (1954) 'Experimental studies of the distribution of gene
        frequencies in very small populations of *Drosophila melanogaster.* II. Bar'.
        Evolution, **8**, 225–40.   [4]
Wright, S. and **McPhee, H.C.** (1925) 'An approximate method of calculating coefficients
        of inbreeding and relationship from livestock pedigrees'. *J. Agric. Res.,* **31**, 377–
        83.   [5]
Yamada, Y. (1962) 'Genotype by environment interaction and genetic correlation of the
        same trait under different environments'. *Jap. J. Genet.,* **37**, 498–509.   [19]
Yamada, Y. and **Bell, A.E.** (1969) 'Selection for larval growth in *Tribolium* under two
        levels of nutrition'. *Genet. Res.,* **13**, 175–95.   [19]
Yamada, Y., Yokouchi, K., and **Nishida, A.** (1975) 'Selection index when genetic gains of
        individual traits are of primary concern'. *Jap. J. Genet.,* **50**, 33–41.   [19]
Yasuda, N. and **Kimura, M.** (1968) 'A gene-counting method of maximum likelihood for
        estimating gene frequencies in ABO and ABO-like systems'. *Ann. Hum. Genet.,*
        31, 409–20.   [1]
Yoon, C.H. (1955) 'Homeostasis associated with heterozygosity in the genetics of time of
        vaginal opening in the house mouse'. *Genetics,* **40**, 297–309.   [15]
Zeleny, C. (1922) 'The effect of selection for eye facet number in the white bar-eye race of
        Drosophila melanogaster'. Genetics, **7**, 1–115.   [6]
Zuberi, M.I. and **Gale, J.S.** (1976) 'Variation in wild populations of *papaver dubium'.*
        Heredity, **36**, 359–68.   [8]

New references

Frankham, R. (1980). 'Origin of genetic variation in selection lines'. pp. 56–68 in
        Robertson, A. (ed.) *Selection Experiments in Laboratory and Domestic Animals.*
        Commonwealth Agricultural Bureaux, Slough, UK.
Hill, W.G. (1982a) 'Rates of change in quantitative traits from fixation of new mutations'.
        Proc. Natnl. Acad. Sci., Wash., **79**, 142–5.
Hill, W.G. (1982b) 'Predictions of response to artificial selection from new mutations'.
        Genet. Res., **40**, 255–78.

INDEX

(Characters are indexed under the relevant organism)